Dietary Antioxidants and Prevention of Non-Communicable Diseases

Dietary Antioxidants and Prevention of Non-Communicable Diseases

Special Issue Editor

Giuseppe Grosso

MDPI • Basel • Beijing • Wuhan • Barcelona • Belgrade

MDPI

Special Issue Editor
Giuseppe Grosso
NNEdPro Global Centre for Nutrition and Health, St John's Innovation Centre
UK

Editorial Office
MDPI
St. Alban-Anlage 66
Basel, Switzerland

This is a reprint of articles from the Special Issue published online in the open access journal *Antioxidants* (ISSN 2076-3921) from 2017 to 2018 (available at: http://www.mdpi.com/journal/antioxidants/special_issues/noncommunicable_diseases#published)

For citation purposes, cite each article independently as indicated on the article page online and as indicated below:

LastName, A.A.; LastName, B.B.; LastName, C.C. Article Title. *Journal Name* **Year**, *Article Number*, Page Range.

ISBN 978-3-03897-226-6 (Pbk)
ISBN 978-3-03897-227-3 (PDF)

Cover image courtesy of Giuseppe Grosso.

Contents

About the Special Issue Editor

Giuseppe Grosso's research focuses on evidence-based nutrition, a recently emerged field in the context of health technology assessments applied to food and nutrition. His main interests include the impact of dietary and lifestyle habits on common non-communicable diseases. In particular, he has produced over 100 papers on the effects of dietary patterns (i.e., Mediterranean diet) and specific antioxidant-rich foods (i.e., coffee, tea), as well as individual antioxidants (i.e., polyphenols, n-3 PUFA) on cardiovascular and metabolic diseases, cancer, and depression. Dr. Grosso has conducted his research on cohorts of individuals in both Mediterranean and non-Mediterranean countries, collaborating with several research institutions. He is interested in evidence synthesis aimed at generating policy-oriented research in the area of public health nutrition. He is currently working as research fellow at the Integrated Cancer Registry of Catania-Messina-Siracusa-Enna, southern Italy. He graduated cum-laude as an MD and PhD.

Preface to "Dietary Antioxidants and Prevention of Non-Communicable Diseases"

Over the last years, exogenous antioxidants have received great attention because of their potential beneficial effects on human health. Antioxidants are contained in foods commonly consumed by all populations worldwide, and accumulating epidemiological evidence has demonstrated the association between both antioxidants and the intake of antioxidant-rich foods and human health. For instance, recent evidence suggests that a high consumption of antioxidants and antioxidant-rich foods is associated with a decreased risk of overall and CVD-related mortality, certain cancers, CVD, and mood disorders.

This book includes 12 peer-reviewed papers, including five original research papers and seven literature reviews. They present the most recent information regarding the dietary intake of antioxidants and antioxidant-rich foods, as well as their effects in terms of the prevention and treatment of non-communicable diseases. Importantly, numerous reviews summarize and highlight the preventive role of antioxidants contained in grape seeds, cocoa, Moringa oleifera leaves, coffee, mulberry fruit, and brown rice. Moreover, the molecular mechanisms and signalling pathways through which exogenous antioxidants exert beneficial effects are underlined.

As the Guest Editor, I would like to acknowledge the authors for their valuable contributions and the reviewers for their constructive remarks. Special thanks to the publishing team of the journal *Antioxidants* for their professional help in the timely completion of this Special Issue.

Giuseppe Grosso
Special Issue Editor

antioxidants

MDPI

Editorial

Dietary Antioxidants and Prevention of Non-Communicable Diseases

Giuseppe Grosso

NNEdPro Global Centre for Nutrition and Health, St John's Innovation Centre, Cambridge CB4 0WS, UK; g.grosso@nnedpro.org.uk; Tel.: +39-095-378-2182

Received: 18 July 2018; Accepted: 18 July 2018; Published: 19 July 2018

According to the recent report of the World Health Organization (WHO), the global burden of non-communicable diseases (NCDs) has been rising over the last century, with the leading causes of disability being depression, diabetes, cardiovascular diseases (CVDs), and certain cancers [1]. Besides genetic, environmental, and socioeconomic factors, exploring dietary factors influencing these conditions became of primary importance in order to better define effective strategies for reducing the burden of disease [2]. In fact, higher adherence to healthy and equilibrated dietary patterns has been shown to be implicated in prevention of NCDs [3].

Numerous epidemiological studies have demonstrated the association between oxidative stress and NCDs. Oxidative stress is commonly known as an imbalance in the production of reactive oxygen species (ROS) and the biological antioxidant defense system. Over the last years, exogenous antioxidants have received great attention because of their potential beneficial effects toward human health. Contained in foods commonly consumed in all populations worldwide, antioxidants represent an attractive explanation of their beneficial effects. However, antioxidants are contained not only of fruits and vegetables, which are characteristic components of healthy dietary patterns, but also in other plant-derived foods, such as tea, coffee, and cocoa. Therefore, the evaluation of dietary habits of a population, and in particular the intake of antioxidants and adherence to healthy dietary patterns is crucial. Indeed, several studies explored the dietary intake of antioxidants and antioxidant-rich foods in a Mediterranean area, as well as their association with fluid and beverage intake [4–6], demonstrating that relatively healthy dietary habits are common in southern Italy.

Accumulating epidemiological evidence have demonstrated the association between both antioxidants and antioxidant-rich foods intake and human health. For instance, recent meta-analysis and cohort studies showed that high consumption of antioxidants and antioxidant-rich foods is associated with decreased risk of overall and CVD-related mortality [7], certain cancers [8], CVD [9], and mood disorders [10]. Importantly, numerous reviews summarized and highlighted the preventive role of antioxidants contained in grape seeds [11], cocoa [12], *Moringa oleifera* leaves [13], coffee [14], mulberry fruit [15], and brown rice [16].

Nevertheless, nowadays studies exploring the possible effects of antioxidants toward human health, should take into consideration the differences in dietary intake of polyphenols in various populations, differences in food processing (loss of phenolic content), absorption, bioavailability, and metabolism of polyphenols. The study by Khairallah et al. focused on the antioxidant effect of phenolic extracts from polyphenol-rich potato, demonstrating that the colonic microbial digestion of potato-based polyphenols could lead to improved colonic health, as this process generates phenolic metabolites with significant antioxidant potential [17].

Several molecular mechanisms may account for the beneficial effects of polyphenols. The antioxidant effects of dietary polyphenols can be attributed to the regulation of redox enzymes through reducing reactive oxygen species (ROS) production and modulation of the II-phase enzymes responsible for the cellular oxidative response. Indeed, the study by Marmouzi et al. showed that antioxidant compounds from hybrid oat lines prevent against hyperglycemia-induced oxidative stress via the modulation of expression of key II-phase enzymes [18]. Additionally, antioxidants may exert chemo-preventive effects through a variety of mechanisms, including the elimination of carcinogenic agents, the modulation of pathways responsible for cancer cell signaling and cell cycle progression, and by the promotion of apoptosis. A review by Losada-Echeberría et al. summarized the evidence of the antitumor effects of plant polyphenols on breast cancer, with special attention to their activity on estrogen receptors (ERs) and human epidermal growth factor receptor 2 (HER2) targets and also covering different aspects, such as redox balance, uncontrolled proliferation, and chronic inflammation [19].

Further evidence from both epidemiological and experimental studies is needed in order to better characterize antioxidants that may exert beneficial effects toward the prevention of chronic diseases associated with oxidative stress and inflammation.

Conflicts of Interest: The author declares no conflict of interest.

References

1. Disease, G.B.D.; Injury, I.; Prevalence, C. Global, regional, and national incidence, prevalence, and years lived with disability for 328 diseases and injuries for 195 countries, 1990–2016: A systematic analysis for the global burden of disease study 2016. *Lancet* **2017**, *390*, 1211–1259. [CrossRef]
2. Collaborators, G.B.D.R.F. Global, regional, and national comparative risk assessment of 84 behavioural, environmental and occupational, and metabolic risks or clusters of risks, 1990-2016: A systematic analysis for the global burden of disease study 2016. *Lancet* **2017**, *390*, 1345–1422. [CrossRef]
3. Grosso, G.; Bella, F.; Godos, J.; Sciacca, S.; Del Rio, D.; Ray, S.; Galvano, F.; Giovannucci, E.L. Possible role of diet in cancer: Systematic review and multiple meta-analyses of dietary patterns, lifestyle factors, and cancer risk. *Nutr. Rev.* **2017**, *75*, 405–419. [CrossRef] [PubMed]
4. Platania, A.; Castiglione, D.; Sinatra, D.; Urso, M.; Marranzano, M. Fluid intake and beverage consumption description and their association with dietary vitamins and antioxidant compounds in italian adults from the mediterranean healthy eating, aging and lifestyles (MEAL) study. *Antioxidants* **2018**, *7*, 56. [CrossRef] [PubMed]
5. Mule, S.; Falla, M.; Conti, A.; Castiglione, D.; Blanco, I.; Platania, A.; D'Urso, M.; Marranzano, M. Macronutrient and major food group intake in a cohort of southern italian adults. *Antioxidants* **2018**, *7*, 58. [CrossRef] [PubMed]
6. Castiglione, D.; Platania, A.; Conti, A.; Falla, M.; D'Urso, M.; Marranzano, M. Dietary micronutrient and mineral intake in the mediterranean healthy eating, ageing, and lifestyle (MEAL) study. *Antioxidants* **2018**, *7*, 79. [CrossRef] [PubMed]
7. Grosso, G.; Micek, A.; Godos, J.; Pajak, A.; Sciacca, S.; Galvano, F.; Giovannucci, E.L. Dietary flavonoid and lignan intake and mortality in prospective cohort studies: Systematic review and dose-response meta-analysis. *Am. J. Epidemiol.* **2017**, *185*, 1304–1316. [CrossRef] [PubMed]
8. Grosso, G.; Godos, J.; Lamuela-Raventos, R.; Ray, S.; Micek, A.; Pajak, A.; Sciacca, S.; D'Orazio, N.; Del Rio, D.; Galvano, F. A comprehensive meta-analysis on dietary flavonoid and lignan intake and cancer risk: Level of evidence and limitations. *Mol. Nutr. Food Res.* **2017**, *61*. [CrossRef] [PubMed]
9. Hooper, L.; Kroon, P.A.; Rimm, E.B.; Cohn, J.S.; Harvey, I.; Le Cornu, K.A.; Ryder, J.J.; Hall, W.L.; Cassidy, A. Flavonoids, flavonoid-rich foods, and cardiovascular risk: A meta-analysis of randomized controlled trials. *Am. J. Clin. Nutr.* **2008**, *88*, 38–50. [CrossRef] [PubMed]
10. Godos, J.; Castellano, S.; Ray, S.; Grosso, G.; Galvano, F. Dietary polyphenol intake and depression: Results from the mediterranean healthy eating, lifestyle and aging (MEAL) study. *Molecules* **2018**, *23*, 999. [CrossRef] [PubMed]
11. Ma, Z.F.; Zhang, H. Phytochemical constituents, health benefits, and industrial applications of grape seeds: A mini-review. *Antioxidants* **2017**, *6*, 71. [CrossRef] [PubMed]
12. Ramos, S.; Martin, M.A.; Goya, L. Effects of cocoa antioxidants in type 2 diabetes mellitus. *Antioxidants* **2017**, *6*, 84. [CrossRef] [PubMed]
13. Vergara-Jimenez, M.; Almatrafi, M.M.; Fernandez, M.L. Bioactive components in moringa oleifera leaves protect against chronic disease. *Antioxidants* **2017**, *6*, 91. [CrossRef] [PubMed]

14. Yamagata, K. Do coffee polyphenols have a preventive action on metabolic syndrome associated endothelial dysfunctions? An assessment of the current evidence. *Antioxidants* **2018**, *7*, 26. [CrossRef] [PubMed]
15. Zhang, H.; Ma, Z.F.; Luo, X.; Li, X. Effects of mulberry fruit (*Morus alba* L.) consumption on health outcomes: A mini-review. *Antioxidants* **2018**, *7*, 69. [CrossRef] [PubMed]
16. Ravichanthiran, K.; Ma, Z.F.; Zhang, H.; Cao, Y.; Wang, C.W.; Muhammad, S.; Aglago, E.K.; Zhang, Y.; Jin, Y.; Pan, B. Phytochemical profile of brown rice and its nutrigenomic implications. *Antioxidants* **2018**, *7*, 71. [CrossRef] [PubMed]
17. Khairallah, J.; Sadeghi Ekbatan, S.; Sabally, K.; Iskandar, M.M.; Hussain, R.; Nassar, A.; Sleno, L.; Rodes, L.; Prakash, S.; Donnelly, D.J.; et al. Microbial biotransformation of a polyphenol-rich potato extract affects antioxidant capacity in a simulated gastrointestinal model. *Antioxidants* **2018**, *7*, 43. [CrossRef] [PubMed]
18. Marmouzi, I.; Karym, E.M.; Saidi, N.; Meddah, B.; Kharbach, M.; Masrar, A.; Bouabdellah, M.; Chabraoui, L.; El Allali, K.; Cherrah, Y.; et al. In vitro and in vivo antioxidant and anti-hyperglycemic activities of moroccan oat cultivars. *Antioxidants* **2017**, *6*, 102. [CrossRef] [PubMed]
19. Losada-Echeberria, M.; Herranz-Lopez, M.; Micol, V.; Barrajon-Catalan, E. Polyphenols as promising drugs against main breast cancer signatures. *Antioxidants* **2017**, *6*, 88. [CrossRef] [PubMed]

antioxidants

MDPI

Article

Dietary Micronutrient and Mineral Intake in the Mediterranean Healthy Eating, Ageing, and Lifestyle (MEAL) Study

Dora Castiglione [1], Armando Platania [1], Alessandra Conti [1], Mariagiovanna Falla [1], Maurizio D'Urso [2] and Marina Marranzano [1,*]

[1]   Department of Medical and Surgical Sciences and Advanced Technologies "G.F. Ingrassia", University of Catania, 95123 Catania, Italy; doracastiglione29@gmail.com (D.C.); armplt@hotmail.it (A.P.); alessandra_conti@ymail.com (A.C.); mariagiovannafalla@yahoo.com (M.F.)
[2]   Provincial Health Authority of Catania, 95127 Catania, Italy; ma.durso76@gmail.com
*   Correspondence: marranz@unict.it; Tel.: +39-095-378-2180

Received: 14 May 2018; Accepted: 21 June 2018; Published: 23 June 2018

Abstract: Background: Dietary vitamins and minerals are essential compounds for the proper functioning of metabolic enzymes, regulation of gene transcription, and powering the body's defense against oxidative stress. The aim of the present study was to investigate micronutrient consumption separately by age and sex, major dietary sources, and percentage of individuals meeting the recommended requirements according to Italian (Livelli di Assunzione di Riferimento di Nutrienti (LARN)) and European (European Food Safety Agency (EFSA)) agencies. Methods: Data were obtained from the Mediterranean Healthy Eating, Ageing, and Lifestyle (MEAL) study, which included a sample of 1838 individuals randomly collected in the city of Catania, southern Italy. A validated food frequency questionnaire was used to collect information on diet. Results: Intake of vitamin A, vitamin C, and vitamin B group (except vitamin B9) was in line with other reports and was adequate according to the guidelines, while the percentage of individuals meeting the guidelines for vitamin D, vitamin E, and vitamin B9 was about 3%, 10%, and 40%, respectively. Among minerals, intake of iron, magnesium, and selenium was adequate for most of the sample, while the percentage of individuals meeting the recommendations for calcium, sodium, and potassium intake was about 20%, 8%, and 35%, respectively. Conclusions: An important percentage of the population would benefit from campaigns raising awareness of micronutrient deficiency or excessive consumption potentially affecting their health.

Keywords: micronutrients; vitamins; minerals; Italy; population; dietary guidelines

1. Introduction

Adopting a healthy diet has been shown to decrease the risk of certain noncommunicable diseases, such as cardiovascular disease (CVD) and cancer [1,2]. Among the most important components of the diet, micronutrients and minerals comprise organic vitamins and inorganic trace elements necessary for homoeostasis of the body and cannot be synthetized endogenously [3]. Vitamins may serve as co-factors for many important metabolic enzymes, regulate gene transcription, and power the body's defense against oxidative stress [4,5]. Additionally, minerals may act as co-factors for enzymatic processes and the correct functioning of body cells [6]. Altogether, vitamins and minerals are considered crucial for a healthy diet and are included in international dietary guidelines of relevant interest for patients, health-care providers, and public health policy-makers.

Among the most important vitamins, vitamin A (retinol and carotene) is important for the maintenance of epithelial cell integrity, growth, and development and participates in immune functions and normal vision [7]. Vitamin E plays a protective role against lipoproteins and polyunsaturated fatty acids (PUFA), protects cellular and intracellular membranes from damage, influences the activity of some enzymes, inhibits platelet aggregation, and is involved in erythrocyte maintenance [8]. Vitamin C is involved in the antioxidant defense system and has important implications for the immune system [9]. It is involved in the metabolism of cholesterol and in many biochemical reactions, including the synthesis of catecholamines, carnitine, and collagen [10]. Vitamin D, together with calcium and magnesium, has a relevant role in bone maintenance and development [11]. Vitamin D is a fat-soluble vitamin important in regulating serum calcium and phosphorus, iron, phosphate, magnesium, and zinc homoeostasis. Additionally, vitamin D plays an important role in cell differentiation and proliferation, and exerts beneficial effects on the immune and nervous systems. Vitamin D can be obtained either from dietary sources or by direct exposure to sunlight; under normal conditions, levels of vitamin D remain within a target range of 20–60 ng/mL. A person is in a state of vitamin D deficiency if the blood level is $\leq$20 ng/mL. The world is in a state of D hypovitaminosis, primarily due to less exposure to sunlight [12]. Other factors that influence vitamin D level are age, gender, ethnicity, skin color, season, clothing, and housing. Besides its effect on bone health, several observational studies suggested that vitamin D deficiency may be associated with many types of cancer, CVD, and metabolic disorders [13–15]. The B vitamins (thiamine (B1), riboflavin (B2), niacin (B3), pantothenic acid (B5), vitamin B6, folate (B9), and vitamin B12) are water-soluble, and they are essential for many body functions, including catabolic metabolism and anabolic metabolism. Importantly, the active forms of thiamine, riboflavin, and niacin are essential co-enzymes playing a role in mitochondrial aerobic respiration and cellular energy production, and therefore deficiency of any one B vitamin could dysregulate the aforementioned processes [16–20].

Among the most important minerals, dietary sodium is an essential compound responsible for maintenance of plasma volume and plasma osmolality. Nonetheless, high dietary sodium is an important risk factor for hypertension and cardiovascular and kidney disease, therefore dietary guidelines recommend that sodium intake does not exceed 2–2.4 g/day [21]. Potassium has important biological functions in neural transmission, vascular tone, and muscle contraction [22]. Zinc and selenium are essential nutrients for the antioxidant defense system. Zinc participates in many metabolic processes as a catalytic, and as a structural component and regulator of gene expression [23]. Selenium is a key component of several selenium-based proteins with essential enzymatic functions that comprise thyroid hormone metabolism, and plays a role in anti-inflammatory and anti-oxidant responses [24]. Iron is crucial for the delivery of oxygen to the cells, because it is part of hemoglobin [25]. It has important implications in antimicrobial activity exerted by phagocytes, neurotransmitter synthesis, and synthesis and function of DNA, collagen, and bile acids. Certain amounts of iron must be delivered in the diet in order to replace the iron that is lost from the body through blood loss and exfoliation of skin and gastrointestinal cells. Women, especially adolescents, who adhere to low-energy diets are at high risk for iron deficiency. Among the major food sources of iron are cereals, vegetables, nuts, eggs, fish, and meat. Iron is also added to fortified foods in many countries and is available as a supplement. Recommended iron intake for women of childbearing age is 18 mg/day, while for men and postmenopausal women it is 8 mg/day. The estimated average requirement of iron is 8.1 mg/day for fertile women, 6 mg/day for men, and 5 mg/day for postmenopausal women [26]. Magnesium is a major mineral that exists in the human body, 70% in the skeleton and the rest in the cells. Magnesium, as a constituent of chlorophyll, is contained in large quantities in green leafy vegetables. Magnesium plays an important role in muscle contraction, gland secretion, and nerve transmission, which is also required for normal cardiac electrophysiology. It also has protective effects on the cardiovascular system through enhancing endothelium-dependent vasodilation, improving lipid metabolism and profile, reducing systemic inflammation, and inhibiting platelet function [27,28]. In particular, magnesium administered with taurine lowers blood pressure, improves insulin resistance, delays atherogenesis, prevents against arrhythmias, and stabilizes platelets. In the general population, magnesium deficiency is rather common, as its intake has reduced over the years [28]. Magnesium can protect from CVD, and abnormally low circulating magnesium (<0.65 mmol/L) is a risk factor for cardiac arrest [29]. Calcium may act together with vitamin D in improving bone health, and it has been associated with a decreased risk for various types of cancers [30], despite high circulating levels possibly being a risk factor for CVD [31,32].

Better knowledge of micronutrient consumption in the Italian population is necessary to prevent and/or delay the adverse effects that result from an inadequate diet. Levels of vitamin and mineral consumption and comparisons among countries may help to characterize the level of quality of nutritional requirements of populations and identify potential gaps for healthy and proper dietary intake of such compounds. The aim of the present study was to investigate micronutrient consumption separately by age and sex, major dietary sources, and percentage of individuals meeting the recommended requirements according to Italian (Livelli di Assunzione di Riferimento di Nutrienti (LARN)) [33] and European (European Food Safety Agency (EFSA)) agencies [34].

2. Materials and Methods

2.1. Study Design and Population

The Mediterranean Healthy Eating, Ageing, and Lifestyle (MEAL) study is an observational study primarily focused on nutritional habits in a sample of individuals living in Sicily, southern Italy. The theoretical sample comprised a sample of 2044 men and women 18 years of age or older, randomly selected in the area of the city of Catania. The study enrollment was performed between 2014 and 2015 by selecting from lists of registered patients among a pool of general practitioners. Full details regarding the study protocol were published in detail previously [35]. The theoretical sample size was set at 1500 individuals to provide a specific relative precision of 5% (Type I error, 0.05; Type II error, 0.10), taking into account an anticipated 70% participation rate. Out of 2405 individuals invited to participate in the study, 2044 participants (response rate of 85%) made up the final sample. All participants were informed about the aims of the study and provided written informed consent. All the study procedures were carried out in accordance to the Declaration of Helsinki (1989) of the World Medical Association. The study protocol has been approved by the concerning ethical committee (protocol number: 802/23 December 2014).

2.2. Data Collection and Dietary Assessment

Electronic data collection was performed by face-to-face computer-assisted personal interviews. Demographics and health status were assessed according to standard procedures [36]. The dietary intake assessment was executed by the administration of 2 food frequency questionnaires (a long and a short version) that were previously validated for the Sicilian population [37,38]. For the purposes of this study, data retrieved from the long version was used. We used food composition tables of the Research Center for Foods and Nutrition in order to identify and calculate food intake, energy content, and micronutrient intake [39]. Intake of seasonal foods referred to consumption during the period in which the food was available and then adjusted by proportional intake during 1 year. Food frequency questionnaires (FFQs) with unreliable dietary intake (<1000 or >6000 kcal/day) were excluded (*n* = 107), leaving a total of 1838 individuals included in the final analysis.

2.3. Adherence to Dietary Recommendations

To investigate adherence to healthy dietary requirements for micronutrients, the European recommendations from EFSA [34] and those proposed by the Italian Society of Human Nutrition (LARN) [33] were considered for the present study.

2.4. Statistical Analysis

Frequencies are expressed as absolute numbers and percentages; continuous variables are expressed as means and standard errors, medians, and ranges. Differences between groups for continuous variables were calculated using Student's *t*-test and ANOVA for continuous variables distributed normally, and Mann–Whitney U-test and Kruskall–Wallis test for variables distributed not normally. All reported P values were based on 2-sided tests and compared to a significance level of 5%. Finally, SPSS 17 (SPSS Inc., Chicago, IL, USA) software was used for all statistical analysis.

3. Results

Tables 1 and 2 show the intake of vitamins and minerals in the study population, in total and separately by sex and age.

With regard to vitamins, men showed significant energy-adjusted higher intake of some vitamins from the B complex (B1, B6, B9, and B12), vitamin C, and vitamin E. When considering differences between sexes within age groups, there was significantly higher intake in men than women of vitamin B1 among younger individuals, while for vitamins B6 and B12, vitamin C, and vitamin E, there was higher intake among older individuals (Table 1). Consequently, when considering the whole sample, younger individuals (20–50 years old) had significantly higher intake of vitamin B1 and vitamin C, while among women there was significant intake of vitamin E in the same age group (Table 1). Regarding minerals, there was no significant difference in intake between sexes and age groups, with the exception of iron and potassium, which were more consumed by men than women (Table 2).

Figure 1 shows the major dietary sources of the micronutrients investigated in this study. Among the most interesting findings, grains were the highest contributors of selenium, but also sodium; fruits of potassium and vitamin C (especially citrus fruits); and vegetables of vitamins B9, E, and A (the latter especially from leafy vegetables). Meat also contributed sodium, while legumes and nuts contributed folate and most minerals. Fish was an important source of vitamins B12 and D, while dairy products were sources of calcium and phosphorus.

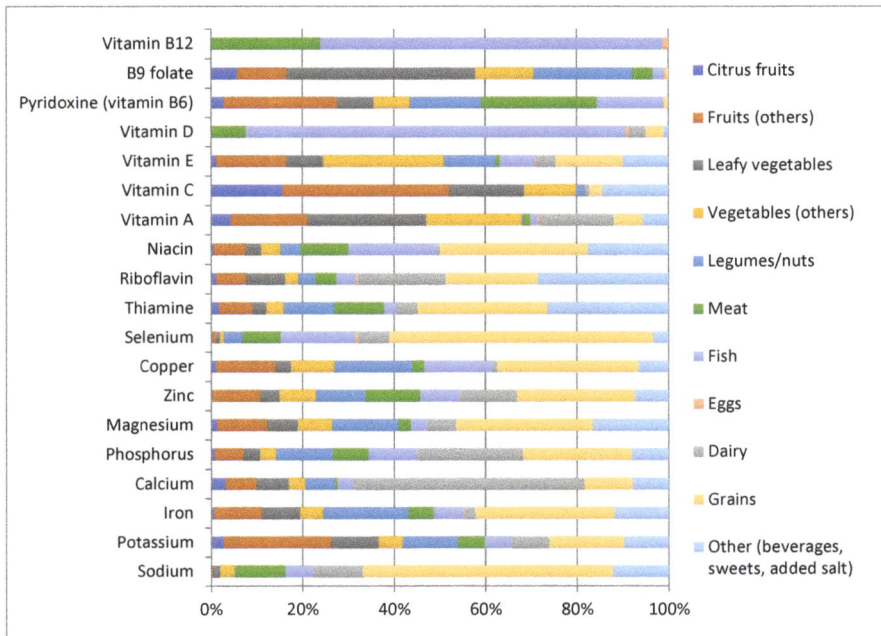

Figure 1. Major dietary sources of the micronutrients investigated in the MEAL study.

Table 1. Total, sex-, and age-specific consumption of dietary vitamins for the participants of the Mediterranean Healthy Eating, Ageing, and Lifestyle (MEAL) study (n = 1838). * Denotes significant difference between sex ($p < 0.05$); † denotes significant difference between age groups. SE, standard error.

	Total			<20 years			20–50			50–70			>70 years		
	n	Mean (SE)	Median (Range)	n	Mean (SE)	Median (Range)	n	Mean (SE)	Median (Range)	n	Mean (SE)	Median (Range)	n	Mean (SE)	Median (Range)
Vitamin A															
Total	1838	868.72 (10.00)	776.04 (152.21, 4949.53)	53	930.31 (65.40)	842.47 (256.18, 2743.98)	963	873.19 (14.84)	770.37 (152.21, 4949.53)	597	873.40 (16.15)	785.75 (192.61, 2832.82)	225	822.69 (23.56)	759.64 (194.54, 2886.16)
Men	772	887.26 (16.22)	791.16 (172.82, 4949.53)	30	959.30 (61.61)	873.31 (417.78, 1772.20)	384	885.03 (25.90)	757.77 (172.82, 4949.53)	265	879.78 (24.49)	796.64 (231.20, 2331.72)	93	894.57 (38.26)	859.58 (353.52, 2263.34)
Women	1066	855.30 (12.60)	768.46 (152.21, 3211.70)	23	892.51 (129.05)	714.87 (256.18, 2743.98)	579	865.34 (17.74)	778.92 (152.21, 3211.70)	332	868.31 (21.50)	780.27 (192.61, 2832.82)	132	772.05 (29.09) *	722.17 (194.54, 2886.16)
Vitamin B1															
Total	1838	1.77 (0.02)	1.61 (0.41, 8.79)	53	1.91 (0.10)	1.81 (0.93, 4.05)	963	1.80 (0.03)	1.63 (0.41, 8.79)	597	1.75 (0.03)	1.59 (0.42, 6.11)	225	1.63 (0.04) †	1.52 (0.41, 4.50)
Men	772	1.82 (0.03)	1.63 (0.53, 8.79)	30	2.09 (0.12)	2.12 (1.03, 3.70)	384	1.87 (0.05)	1.65 (0.64, 8.79)	265	1.76 (0.05)	1.58 (0.53, 5.75)	93	1.69 (0.07)	1.54 (0.62, 4.50)
Women	1066	1.73 (0.02) *	1.59 (0.41, 6.11)	23	1.67 (0.15) *	1.48 (0.93, 4.05)	579	1.75 (0.03) *	1.61 (0.41, 5.74)	332	1.75 (0.04)	1.60 (0.42, 6.11)	132	1.59 (0.05) *	1.49 (0.41, 4.42)
Vitamin B2															
Total	1838	2.24 (0.02)	2.06 (0.48, 10.44)	53	2.36 (0.12)	2.16 (0.97, 5.00)	963	2.27 (0.03)	2.08 (0.48, 10.44)	597	2.23 (0.04)	2.03 (0.48, 7.28)	225	2.14 (0.05)	2.00 (0.48, 5.12)
Men	772	2.29 (0.04)	2.05 (0.87, 10.44)	30	2.55 (0.14)	2.39 (1.29, 4.73)	384	2.33 (0.06)	2.08 (0.96, 10.44)	265	2.22 (0.06)	1.99 (0.87, 7.05)	93	2.19 (0.08)	2.10 (0.94, 5.12)
Women	1066	2.21 (0.03)	2.06 (0.48, 7.28)	23	2.11 (0.19)	1.99 (0.97, 5.00)	579	2.22 (0.04)	2.08 (0.48, 6.57)	332	2.24 (0.05)	2.06 (0.48, 7.28)	132	2.10 (0.06)	1.99 (0.48, 4.98)

Table 1. *Cont.*

	Total			<20 years			20–50			50–70			>70 years		
	n	Mean (SE)	Median (Range)	n	Mean (SE)	Median (Range)	n	Mean (SE)	Median (Range)	n	Mean (SE)	Median (Range)	n	Mean (SE)	Median (Range)
Vitamin B3															
Total	1838	21.96 (0.19)	20.78 (4.13, 78.42)	53	20.90 (0.80)	19.94 (4.13, 38.57)	963	22.00 (0.27)	20.68 (7.44, 71.83)	597	22.32 (0.34)	21.22 (6.99, 78.42)	225	21.11 (0.40)	20.40 (8.18, 45.74)
Men	772	22.28 (0.31)	20.81 (4.13, 78.42)	30	21.21 (1.05)	21.16 (4.13, 32.47)	384	22.32 (0.46)	20.61 (8.23, 71.83)	265	22.67 (0.58)	21.07 (6.99, 78.42)	93	21.38 (0.62)	20.84 (9.81, 45.74)
Women	1066	21.73 (0.23)	20.77 (7.20, 65.67)	23	20.49 (1.24)	19.09 (14.28, 38.57)	579	21.79 (0.33)	20.68 (7.44, 65.67)	332	22.04 (0.39)	21.42 (7.20, 50.48)	132	20.93 (0.53)	20.16 (8.18, 39.88)
Vitamin B6															
Total	1838	2.55 (0.02)	2.43 (0.43, 8.46)	53	2.63 (0.12)	2.52 (0.43, 5.36)	963	2.55 (0.03)	2.39 (1.04, 8.46)	597	2.57 (0.04)	2.46 (0.84, 7.19)	225	2.49 (0.05)	2.40 (1.07, 4.76)
Men	772	2.60 (0.03)	2.46 (0.43, 8.46)	30	2.75 (0.16)	2.57 (0.43, 4.95)	384	2.59 (0.05)	2.42 (1.08, 8.46)	265	2.61 (0.06)	2.48 (1.03, 7.19)	93	2.61 (0.08)	2.52 (1.29, 4.76)
Women	1066	2.51 (0.03) *	2.40 (0.84, 6.52)	23	2.48 (0.19)	2.29 (1.35, 5.36)	579	2.52 (0.04)	2.37 (1.04, 6.52)	332	2.54 (0.04)	2.46 (0.84, 5.10)	132	2.41 (0.06) *	2.36 (1.07, 4.41)
Vitamin B9															
Total	1838	391.50 (3.86)	366.83 (61.31, 2535.14)	53	390.56 (20.41)	364.10 (61.31, 808.99)	963	395.44 (5.88)	363.44 (83.83, 2535.14)	597	388.86 (6.13)	374.15 (88.26, 1471.10)	225	381.84 (8.59)	369.55 (100.34, 829.71)
Men	772	401.17 (6.48)	370.70 (61.31, 2535.14)	30	409.11 (25.12)	406.79 (61.31, 745.20)	384	402.70 (10.23)	370.06 (142.01, 2535.14)	265	398.02 (10.33)	363.36 (111.22, 1471.10)	93	401.23 (13.68)	378.32 (202.42, 810.86)
Women	1066	384.49 (4.71) *	364.69 (83.83, 1378.24)	23	366.35 (33.72)	319.99 (137.33, 808.99)	579	390.62 (7.05)	356.29 (83.83, 1378.24)	332	381.54 (7.31)	376.80 (88.26, 925.62)	132	368.17 (10.91)	357.82 (100.34, 829.71)
Vitamin B12															
Total	1838	6.18 (0.10)	5.50 (0.57, 120.52)	53	6.40 (0.37)	6.15 (2.33, 14.92)	963	6.23 (0.17)	5.36 (0.57, 120.52)	597	6.17 (0.14)	5.58 (1.59, 36.02)	225	5.92 (0.18)	5.53 (1.24, 19.68)
Men	772	6.45 (0.21)	5.56 (1.24, 120.52)	30	6.48 (0.53)	5.88 (2.33, 14.92)	384	6.47 (0.37)	5.29 (2.00, 120.52)	265	6.54 (0.26)	5.80 (1.62, 36.02)	93	6.15 (0.29)	5.69 (1.24, 19.68)
Women	1066	5.98 (0.10) *	5.40 (0.57, 31.75)	23	6.29 (0.53)	6.33 (2.92, 12.07)	579	6.08 (0.14)	5.47 (0.57, 31.75)	332	5.87 (0.15) *	5.30 (1.59, 16.52)	132	5.76 (0.22)	5.41 (1.50, 14.60)

Table 1. *Cont.*

	Total			<20 years			20–50			50–70			>70 years		
	n	Mean (SE)	Median (Range)	*n*	Mean (SE)	Median (Range)	*n*	Mean (SE)	Median (Range)	*n*	Mean (SE)	Median (Range)	*n*	Mean (SE)	Median (Range)
Vitamin C															
Total	1838	158.09 (2.29)	132.65 (5.29, 1097.80)	53	151.40 (9.94)	139.84 (20.96, 308.49)	963	162.11 (3.53)	132.63 (5.29, 1097.80)	597	158.22 (3.62)	134.94 (6.09, 569.98)	225	142.15 (4.67) †	125.81 (17.67, 378.71)
Men	772	164.10 (3.78)	135.84 (12.73, 1097.80)	30	167.32 (13.70)	166.42 (20.96, 308.49)	384	165.47 (6.12)	131.28 (24.88, 1097.80)	265	163.82 (5.76)	136.48 (12.73, 569.98)	93	158.16 (7.62)	141.57 (47.61, 366.63)
Women	1066	153.74 (2.83) *	129.83 (5.29, 700.31)	23	130.62 (13.46)	107.79 (44.12, 297.64)	579	159.88 (4.25)	134.02 (5.29, 700.31)	332	153.74 (4.59)	132.82 (6.09, 500.62)	132	130.87 (5.70) *	120.41 (17.67, 378.71)
Vitamin D															
Total	1838	5.16 (0.10)	3.93 (0.20, 56.42)	53	5.47 (0.43)	4.84 (0.60, 13.64)	963	5.02 (0.15)	3.78 (0.20, 56.42)	597	5.37 (0.19)	4.01 (0.56, 45.67)	225	5.09 (0.27)	4.19 (0.58, 30.24)
Men	772	5.19 (0.18)	3.93 (0.56, 56.42)	30	5.11 (0.60)	4.28 (0.60, 13.64)	384	4.89 (0.25)	3.72 (0.56, 56.42)	265	5.61 (0.32)	4.28 (0.56, 45.67)	93	5.30 (0.48)	4.40 (0.58, 30.24)
Women	1066	5.13 (0.13)	3.91 (0.20, 34.12)	23	5.94 (0.62)	5.62 (2.14, 12.96)	579	5.11 (0.18)	3.83 (0.20, 34.12)	332	5.19 (0.22)	3.79 (0.57, 28.71)	132	4.94 (0.32)	4.15 (0.63, 24.90)
Vitamin E															
Total	1838	8.66 (0.08)	8.04 (1.47, 31.93)	53	8.86 (0.42)	8.38 (2.68, 18.44)	963	8.75 (0.11)	8.02 (1.66, 25.56)	597	8.70 (0.14)	8.10 (1.47, 31.93)	225	8.07 (0.17)	7.85 (3.16, 18.36)
Men	772	8.85 (0.13)	8.12 (2.68, 31.93)	30	9.38 (0.63)	8.62 (2.68, 18.44)	384	8.79 (0.19)	7.77 (3.28, 25.56)	265	9.05 (0.24)	8.44 (3.17, 31.93)	93	8.35 (0.27)	8.11 (4.43, 18.36)
Women	1066	8.51 (0.09) *	8.03 (1.47, 22.08)	23	8.18 (0.50)	8.29 (3.72, 15.77)	579	8.72 (0.14)	8.14 (1.66, 22.08)	332	8.43 (0.16) *	7.94 (1.47, 18.42)	132	7.87 (0.21)	7.77 (3.16, 15.32)

Table 2. Total, sex-, and age-specific consumption of dietary minerals for the participants of the MEAL study (*n* = 1838). * Denotes significant difference between sex (*p* <0.05).

		Total		<20 years			20-50			50-70			>70 years		
	n	Mean (SE)	Median (Range)	*n*	Mean (SE)	Median (Range)	*n*	Mean (SE)	Median (Range)	*n*	Mean (SE)	Median (Range)	*n*	Mean (SE)	Median (Range)
Calcium															
Total	1838	800.35 (7.76)	744.24 (169.71, 3109.93)	53	842.68 (44.12)	784.64 (304.55, 1662.32)	963	801.60 (11.36)	734.79 (171.97, 3109.93)	597	794.99 (12.61)	747.21 (169.71, 2294.96)	225	799.28 (20.69)	773.11 (172.00, 1839.22)
Men	772	803.27 (12.32)	736.34 (224.53, 3109.93)	30	900.43 (53.54)	861.76 (507.68, 1662.32)	384	801.99 (18.92)	705.14 (245.67, 3109.93)	265	785.15 (19.29)	733.38 (224.53, 2294.96)	93	828.90 (31.92)	813.83 (350.52, 1839.22)
Women	1066	798.23 (9.97)	753.51 (169.71, 2186.26)	23	767.36 (72.23)	653.88 (304.55, 1554.25)	579	801.34 (14.14)	749.32 (171.97, 2186.26)	332	802.84 (16.65)	767.44 (164.71, 2015.75)	132	778.40 (27.12)	733.15 (172.00, 1817.71)
Sodium															
Total	1838	2847.07 (25.59)	2689.04 (737.51, 9469.51)	53	2823.07 (139.92)	2940.59 (796.60, 5317.25)	963	2910.60 (36.70)	2774.86 (737.51, 9391.07)	597	2792.70 (45.29)	2625.90 (834.83, 9469.51)	225	2725.10 (58.25)	2613.69 (829.26, 5251.12)
Men	772	2924.44 (41.41)	2727.70 (796.60, 9469.51)	30	3027.64 (190.91)	3185.84 (796.60, 5317.25)	384	2970.18 (61.42)	2779.33 (870.96, 9273.63)	265	2906.64 (71.29)	2692.50 (903.42, 9469.51)	93	2753.03 (93.36)	2655.48 (1167.12, 5199.53)
Women	1066	2791.04 (32.28)	2661.75 (737.51, 9391.07)	23	2556.23 (195.57)	2453.54 (1097.52, 4746.73)	579	2871.08 (45.44)	2766.61 (737.51, 9391.07)	332	2701.76 (57.88)	2567.36 (834.83, 7747.75)	132	2705.43 (74.62)	2597.31 (829.26, 5251.12)
Iron															
Total	1838	15.27 (0.14)	14.40 (4.13, 84.06)	53	15.13 (0.66)	14.63 (4.13, 27.90)	963	15.41 (0.21)	14.33 (4.83, 84.06)	597	15.24 (0.22)	14.60 (4.42, 50.81)	225	14.76 (0.30)	13.96 (6.75, 31.34)
Men	772	15.59 (0.23)	14.53 (4.13, 84.06)	30	15.64 (0.86)	15.30 (4.13, 25.85)	384	15.68 (0.36)	14.45 (5.54, 84.06)	265	15.55 (0.37)	14.59 (4.42, 50.81)	93	15.34 (0.49)	14.47 (6.75, 31.34)
Women	1066	15.03 (0.17) *	14.26 (4.83, 41.09)	23	14.47 (1.02)	12.58 (7.72, 27.90)	579	15.23 (0.25)	14.26 (4.83, 41.09)	332	14.99 (0.27)	14.67 (5.25, 35.40)	132	14.35 (0.39)	13.50 (7.07, 29.99)

Table 2. *Cont.*

	Total			<20 years			20-50			50-70			>70 years		
	n	Mean (SE)	Median (Range)	n	Mean (SE)	Median (Range)	n	Mean (SE)	Median (Range)	n	Mean (SE)	Median (Range)	n	Mean (SE)	Median (Range)
Magnesium															
Total	1838	392.99 (3.24)	372.88 (109.85, 1665.14)	53	382.56 (16.08)	364.41 (109.85, 699.45)	963	395.21 (4.90)	369.92 (147.60, 1665.14)	597	394.57 (5.29)	385.25 (145.91, 1089.24)	225	381.76 (7.00)	366.82 (152.02, 709.36)
Men	772	396.92 (5.37)	374.76 (109.85, 1665.14)	30	392.08 (22.05)	379.14 (109.85, 699.45)	384	399.62 (8.48)	372.31 (158.30, 1665.14)	265	395.51 (8.63)	385.01 (149.58, 1089.24)	93	391.34 (10.68)	375.45 (162.04, 654.69)
Women	1066	390.15 (4.00)	372.08 (145.91, 954.25)	23	370.15 (23.67)	348.24 (246.17, 687.34)	579	392.29 (5.90)	369.29 (147.60, 954.25)	332	393.83 (6.57)	386.78 (145.91, 863.13)	132	375.01 (9.24)	363.79 (152.02, 709.36)
Potassium															
Total	1838	3653.59 (32.01)	3446.04 (812.57, 15693.66)	53	3731.19 (176.32)	3690.30 (812.57, 7852.01)	963	3674.80 (47.74)	3415.25 (872.14, 15693.66)	597	3669.38 (52.85)	3531.47 (1050.23, 11778.63)	225	3502.63 (72.19)	3375.85 (1206.18, 8039.89)
Men	772	3732.00 (53.25)	3509.80 (812.57, 15693.66)	30	3902.19 (228.04)	3795.98 (812.57, 7156.95)	384	3712.12 (82.16)	3455.26 (1574.87, 15693.66)	265	3757.96 (88.36)	3553.17 (1642.82, 11778.63)	93	3685.19 (109.37)	3626.06 (1775.75, 6615.19)
Women	1066	3596.81 (39.41) *	3397.32 (872.14, 9165.64)	23	3508.15 (275.53)	2998.47 (1799.16, 7852.01)	579	3650.05 (57.78)	3395.22 (872.14, 9165.64)	332	3598.68 (63.55)	3507.39 (1050.23, 8028.33)	132	3374.00 (94.69) *	3213.99 (1206.18, 8039.89)
Selenium															
Total	1838	103.07 (1.06)	93.91 (21.56, 431.10)	53	100.74 (5.07)	95.43 (21.56, 212.36)	963	102.42 (1.53)	91.61 (23.14, 431.10)	597	104.97 (1.82)	96.23 (24.69, 289.43)	225	101.36 (2.58)	94.02 (32.53, 239.21)
Men	772	103.96 (1.69)	91.84 (21.56, 431.10)	30	99.37 (5.90)	95.65 (21.56, 171.70)	384	104.57 (2.59)	90.70 (25.48, 431.10)	265	105.10 (2.78)	93.80 (24.69, 276.54)	93	99.69 (3.95)	90.64 (33.27, 239.21)
Women	1066	102.43 (1.35)	95.14 (23.14, 289.43)	23	102.52 (8.93)	95.43 (49.01, 212.36)	579	101.00 (1.88)	92.31 (23.14, 289.09)	332	104.86 (2.42)	97.79 (40.31, 289.43)	132	102.53 (3.42)	97.08 (32.53, 228.10)
Zinc															
Total	1838	12.19 (0.11)	11.48 (3.66, 54.05)	53	12.41 (0.51)	11.89 (5.15, 22.17)	963	12.27 (0.16)	11.37 (3.67, 54.05)	597	12.19 (0.18)	11.69 (3.67, 39.48)	225	11.76 (0.24)	11.30 (3.66, 22.69)
Men	772	12.40 (0.18)	11.51 (4.54, 54.05)	30	12.73 (0.68)	12.24 (5.15, 22.17)	384	12.39 (0.28)	11.28 (5.04, 54.05)	265	12.41 (0.28)	11.73 (4.54, 39.48)	93	12.26 (0.38)	11.62 (5.52, 22.69)
Women	1066	12.04 (0.13)	11.47 (3.66, 31.84)	23	12.00 (0.77)	11.47 (6.41, 20.56)	579	12.19 (0.18)	11.46 (3.67, 31.84)	332	12.02 (0.22)	11.61 (3.67, 26.81)	132	11.40 (0.30)	11.01 (3.66, 20.56)

Table 3 shows the percentage of individuals meeting the EFSA's dietary recommendations for vitamin and mineral intake. In our sample, 98% did not meet the criteria for vitamin D intake, both women and men. Among the highest adherence to the recommendations, the entire sample met the recommendations for vitamins B1 and B3, while more than half met those for iron, magnesium, selenium, zinc, and vitamins A, B2, B6, B9, B12, and C; however, more women than men met the recommendations for magnesium and zinc. In contrast, less than half of the total sample met the recommendations for calcium and potassium, while the majority did not meet the recommendations for vitamins D and E.

Table 3. Number and percentage of individuals meeting the European recommendations for mineral and vitamin intake in the MEAL study (*n* = 1838).

	Total (*n* = 1851)		Men (*n* = 772)		Women (*n* = 1066)	
	Yes, % (*n*)	No, % (*n*)	Yes, % (*n*)	No, % (*n*)	Yes, % (*n*)	No, % (*n*)
EFSA						
Calcium	24.0 (442)	76.0 (1397)	24.6 (190)	75.4 (582)	23.6 (252)	76.4 (814)
Iron	64.4 (1185)	35.6 (654)	79.1 (611)	20.9 (161)	53.8 (574)	46.2 (492)
Magnesium	66.9 (1230)	33.1 (609)	57.6 (445)	42.4 (327)	73.6 (785)	26.4 (281)
Potassium	48.1 (884)	51.9 (955)	50.5 (390)	49.5 (382)	46.3 (494)	53.7 (572)
Selenium	75.8 (1394)	24.2 (445)	76.3 (589)	23.7 (183)	75.4 (804)	24.6 (262)
Vitamin A	60.1 (1106)	39.9 (733)	53.6 (414)	46.4 (358)	64.9 (692)	35.1 (374)
Vitamin B1	100 (1839)	–	100 (772)	–	100 (1066)	–
Vitamin B2	77.0 (1416)	23.0 (423)	78.5 (606)	21.5 (166)	75.9 (809)	24.1 (257)
Vitamin B3	100 (1839)	–	100 (772)	–	100 (1066)	–
Vitamin B6	87.1 (1601)	12.9 (238)	86.8 (670)	13.2 (102)	87.3 (931)	12.7 (135)
Vitamin B9	60.3 (1108)	39.7 (731)	62.3 (481)	37.7 (291)	58.8 (627)	41.2 (439)
Vitamin B12	75.4 (1387)	24.6 (452)	77.2 (596)	22.8 (176)	74.1 (790)	25.9 (276)
Vitamin C	70.5 (1296)	29.5 (543)	65.5 (506)	34.5 (266)	74.1 (790)	25.9 (276)
Vitamin D	3.2 (58)	96.8 (1781)	3.2 (25)	96.8 (747)	3.1 (33)	96.9 (1033)
Vitamin E	14.8 (273)	85.2 (1566)	10.5 (81)	89.5 (691)	18.0 (192)	82.0 (874)
Zinc	62.4 (1148)	37.6 (691)	47.7 (368)	52.3 (404)	73.2 (780)	26.8 (286)

LARN's dietary recommendations reflected the European ones, with more than 80% of individuals meeting dietary recommendations for sodium, magnesium, selenium, and vitamin B complex (besides vitamin B9) (Table 4). However, some differences between the sexes in iron and zinc intake were found, with recommendations for the former met by a higher percentage of men and the latter by a higher percentage of women (Table 4).

Table 4. Number and percentage of individuals meeting the Italian recommendations for mineral and vitamin intake in the MEAL study (*n* = 1838).

	Total (*n* = 1851)		Men (*n* = 772)		Women (*n* = 1066)	
	Yes, % (*n*)	No, % (*n*)	Yes, % (*n*)	No, % (*n*)	Yes, % (*n*)	No, % (*n*)
LARN						
Calcium	20.1 (370)	79.9 (1469)	20.5 (158)	79.5 (614)	19.9 (212)	80.1 (854)
Sodium	91.6 (1695)	7.8 (144)	93.3 (720)	6.7 (102)	91.4 (974)	8.6 (92)
Iron	51.2 (942)	48.8 (897)	86.8 (670)	13.2 (102)	25.5 (272)	74.5 (794)
Magnesium	90.4 (1663)	9.6 (176)	90.4 (698)	9.6 (74)	90.5 (965)	9.5 (101)
Potassium	35.8 (658)	64.2 (1181)	37.6 (290)	62.4 (484)	34.5 (368)	65.5 (698)
Selenium	90.5 (1665)	9.5 (174)	92.1 (711)	7.9 (61)	89.4 (953)	10.6 (113)
Vitamin A	66.7 (1226)	33.3 (613)	61.4 (474)	38.6 (298)	70.5 (752)	29.5 (314)
Vitamin B1	80.2 (1475)	19.8 (364)	78.1 (603)	21.9 (169)	81.8 (872)	18.2 (194)
Vitamin B2	85.0 (1564)	15.0 (275)	78.5 (606)	21.5 (166)	89.9 (958)	10.1 (108)
Vitamin B3	68.7 (1264)	31.3 (575)	69.4 (536)	30.6 (236)	68.3 (728)	31.7 (338)
Vitamin B6	97.5 (1793)	2.5 (46)	97.9 (756)	2.1 (16)	97.2 (1036)	2.8 (30)
Vitamin B9	41.0 (754)	59.0 (1085)	43.0 (332)	57.0 (440)	39.6 (422)	60.4 (644)
Vitamin B12	96.2 (1770)	3.8 (69)	98.3 (759)	17 (13)	94.7 (1010)	5.3 (56)
Vitamin C	75.1 (1382)	24.9 (457)	68.8 (531)	31.2 (241)	79.8 (851)	20.2 (215)
Vitamin D	3.1 (57)	96.9 (1782)	3.2 (25)	96.8 (747)	3.0 (32)	97.0 (1034)
Vitamin E	11.5 (211)	88.5 (1628)	10.5 (81)	89.5 (691)	12.2 (130)	87.8 (936)
Zinc	62.9 (1157)	37.1 (682)	45.1 (348)	54.9 (424)	75.9 (809)	24.1 (257)

4. Discussion

This study provides an analysis of micronutrient intake in the MEAL cohort, a representative sample of adults from southern Italy. The main differences between sexes and age groups do not reflect any specific pattern of consumption and malnutrition among younger or older individuals. However, this report highlights alarming data concerning inadequate intake of vitamin D and vitamin E.

Women reported lower intake of certain micronutrients, which is in line with European data [40]. The same analysis showed a prevalence of inadequate micronutrient intake, especially vitamin D, vitamin C, vitamin B9, calcium, selenium, and iodine, in the adult and elderly population [40]. It has been estimated that 1 billion people worldwide suffer from vitamin D deficiency, including more than 40% of the population in the US and Europe, 82% in Italy, 70% in Korea, and about 70% in Malaysia [41]. Mean intake ranged from 1.3 mg/day in Spain to 3.5 mg/day in Poland and the Netherlands. The percentage of individuals with intake under the recommended amount ranged from 7% of men in the Netherlands to 95% of women in Spain. It is noteworthy that supplement intake or food fortification was not taken into account in this study. In other countries, especially in northern Europe, fortifying the food supply with vitamin D is mandatory [40]. Recently, it has been demonstrated by several studies that vitamin D is beneficial in the prevention of chronic diseases, and vitamin D intake is significantly associated with a reduction in the risk of osteoporosis, type 2 diabetes, cancer, multiple sclerosis, and rheumatoid arthritis [42,43]. When environmental, social, or physiological circumstances do not allow sufficient exposure to sunlight and adequate production of vitamin D, dietary compensation must occur to maintain serum 25(OH)D levels. Moreover, in those countries where the population has a diet poor in fatty fish (one of the major dietary sources of vitamin D), the only alternative way to increase exposure to natural or artificial UVB light is to fortify foods or use vitamin supplements. The major contributor of vitamin D in this sample was fish, thus the data presented in this study is further alarming, as fish consumption has been demonstrated to be high in this population [44]. Moreover, individuals with higher adherence to an overall Mediterranean diet have been reported to have higher intake of vitamin D, suggesting that other components may contribute to total vitamin D intake [45–47]. However, low consumption of dairy products and eggs (typical of the Mediterranean diet) does not allow reaching the suggested dietary intake of vitamin D [48]. Regardless of the high availability of sunshine, low serum 25(OH)D levels have been reported in adults in southern European countries, confirming the results obtained in the present study [49].

In the present study, there was a significant difference in vitamin E intake between men and women (the latter had lower intake), despite only a minority of both (about 10%) meeting the criteria for adequate consumption. At the European level, selenium deficiency was found in less than 10% of adults in northern Europe (Finland and the Netherlands), while in Italy and other northern European countries it was above 30% [40]. Vitamin E intake varied greatly in Europe, ranging from 6.5 mg/day and 7.2 mg/day in women and men, respectively, living in Denmark to 11.9 mg/day and 13.7 mg/day in women and men, respectively, living in the Netherlands; the contribution from dietary supplements among adults, particularly women, was substantial in the aforementioned countries [50]. Recent mechanistic studies indicate that vitamin E has unique antioxidant and anti-inflammatory properties that may play a role in preventing metabolic disorders and chronic diseases [51]. Consistently with mechanistic findings, animal and human studies show that vitamin E may be helpful in preventing inflammation-associated diseases [52]. From a clinical point of view, vitamin E may decrease the blood levels of C-reactive protein (CRP), a marker of chronic inflammation that has a major role in the etiology of chronic disease [53]. Major food contributors of vitamin E are fruits and vegetables, which have been reported to be highly consumed in this sample. However, certain sources of vitamin E (such as avocado, papaya, etc.) are not commonly consumed in the Mediterranean area. In contrast, almonds are the most typical nut consumed in Sicily that is high in vitamin E; moderate consumption of nuts may play a role in the benefits associated with the Mediterranean diet, as it has been suggested to be a healthy food for a successful aging [54,55].

Among the vitamin B group, it is noteworthy to underline that less than half of the sample met the recommendations for vitamin B9 (folic acid). These estimates are in line with most European countries, where vitamin B9 intake ranged from 376 mg/day in the UK to 212 mg/g in the Dutch adult male population; the distribution of average consumption among women was slightly different, as women in the UK consumed relatively low amounts of vitamin B9 compared to other countries, while the highest intake was reported for the Portuguese and Italian adult female population [40]. Vitamins involved in one-carbon metabolism (including B9 and B12) have a major role during pregnancy for the developing fetus, preventing major congenital malformations such as neural tube defects [56,57]. Vitamin B9 has been studied for cancer prevention, as gene-nutrient interaction between the genes in the folate metabolic pathway and dietary folate availability have shown that mutations in genes of folate metabolism alter individual susceptibility to certain childhood cancers [58]. Moreover, in vitro and in vivo studies have shown that vitamin B9 may improve vascular endothelial function, thereby preventing the progression of CVD in individuals with overt disease or elevated CVD risk [59]. Particularly for women of childbearing age, deficient intake can have serious consequences for their offspring. Although it was not significant, folate intake was higher in this group compared to the total group of women; however, women had lower intake than men in all age groups. There is evidence that vitamin B9 intake is generally low in adult populations and that food fortification and supplement intake are needed to reach the suggested amount [60]. As previously reported, the sample had a relatively high intake of vegetables and legumes, which may provide adequate amounts of several phytochemicals and antioxidant compounds [61,62]. However, concerning vitamin B9, common dietary intake seems to be insufficient to reach the amount suggested, and supplementation with synthetic products may not be equivalent to consuming natural sources of vitamins (including vitamin B9) [63]. Nevertheless, vitamin B9 supplementation has been demonstrated to protect against neural tube defects and CVD [64–66]. Thus, information campaigns on folic acid availability and preferences would be of benefit for the general population and have large-scale public health implications in preventing neurological conditions in newborns [67].

Regarding other micronutrients and minerals, in this study we found that the nutritional recommendations were adequate overall. However, less than half of the sample met the recommendations for sodium, potassium, and calcium intake. High consumption of sodium and low consumption of potassium is a widespread issue all around the world: the mean global sodium intake has been estimated to be 3.95 g/day, with the highest in Eastern (mean >4.2 g/day) and Central (3.9–4.2 g/day) Europe; among European countries, Italy scored among the highest [68]. Regarding potassium intake, lower consumption has been reported in Germany, Sweden, and southern Italy [69]. Similarly, Italy is among the countries with the lowest consumption of calcium, while Northern European countries have the highest [70]. The potential effects of calcium on human health are matched with the role of vitamin D and have been discussed previously. The relationship between sodium and potassium intake and blood pressure levels is relatively well understood, while further consideration should be given to their potential role in endothelial cell dysfunction, progression of albuminuria and kidney disease, and cardiovascular disease–related morbidity and mortality [71]. It has been demonstrated that globally, 1.65 million annual deaths related to CVD were attributed to inadequate sodium intake [72]. In contrast, potassium per se has been not univocally reported as beneficial for health [73], while sodium-to-potassium ratio has been suggested as a potential marker for metabolic risk disorders [74–76]. A major dietary source of sodium has been commonly reported to be "added salt" and processed meat [77], but in this sample we found that the main contributor of sodium intake was grains. This is not surprising, considering the cultural heritage of the southern Mediterranean population, with production of pasta and bread; the latter has to be considered responsible for the aforementioned sodium intake. However, a slow abandonment of traditional foods for a "Westernized" diet has also been noted in this area [78], and the introduction of salted snacks (such as baked goods and crackers) in the diet may significantly contribute to total sodium intake, especially among the younger population.

The results presented in this study should be considered in light of some limitations. Comparing between nutritional surveys is a great challenge due to different methodologies used in various studies. Specifically, studies using 24 h recalls usually reported lower micronutrient intake than studies using food diaries and FFQs. Moreover, the number of items included in the FFQs may also affect the reporting of micronutrient intake. It is clear that an ideal method does not exist and both methodologies are widely used in nutritional epidemiology. However, a certain degree of under- and overestimation for dietary recalls and FFQs, respectively, should be taken into account.

5. Conclusions

In conclusion, despite meeting the national and European dietary recommendations for most micronutrients and minerals, there is evidence that the southern Italian population has low intake of vitamin D and calcium, vitamin E, and potassium and high intake of sodium. Considering the body of evidence on the beneficial and detrimental effects of the aforementioned dietary components, population-wide campaigns are needed in order to raise awareness of such dietary issues as an essential public health effort to prevent chronic noncommunicable diseases.

Author Contributions: Conceptualization, Marina Marranzano; methodology, Dora Castiglione and Marina Marranzano; formal analysis, Armando Platania, Data Curation, Armando Platania and Alessandra Conti, and Mariagiovanna Falla; writing–original draft preparation, Dora Castiglione and Marina Marranzano; writing–review and editing, Marina Marranzano and Maurizio D'Urso; supervision, Marina Marranzano and Maurizio D'Urso.

Funding: This research received no external funding.

Conflicts of Interest: The authors declare no conflict of interest.

References

1. Grosso, G.; Bella, F.; Godos, J.; Sciacca, S.; Del Rio, D.; Ray, S.; Galvano, F.; Giovannucci, E.L. Possible role of diet in cancer: Systematic review and multiple meta-analyses of dietary patterns, lifestyle factors, and cancer risk. *Nutr. Rev.* **2017**, *75*, 405–419. [CrossRef] [PubMed]
2. Mozaffarian, D. Dietary and policy priorities for cardiovascular disease, diabetes, and obesity: A comprehensive review. *Circulation* **2016**, *133*, 187–225. [CrossRef] [PubMed]
3. Watson, J.; Lee, M.; Garcia-Casal, M.N. Consequences of inadequate intakes of vitamin A, vitamin B12, vitamin D, calcium, iron, and folate in older persons. *Curr. Geriatr. Rep.* **2018**, *7*, 103–113. [CrossRef] [PubMed]
4. Asensi-Fabado, M.A.; Munne-Bosch, S. Vitamins in plants: Occurrence, biosynthesis and antioxidant function. *Trends Plant Sci.* **2010**, *15*, 582–592. [CrossRef] [PubMed]
5. Yang, C.S.; Ho, C.T.; Zhang, J.; Wan, X.; Zhang, K.; Lim, J. Antioxidants: Differing meanings in food science and health science. *J. Agric. Food Chem.* **2018**, *66*, 3063–3068. [CrossRef] [PubMed]
6. Dennehy, C.; Tsourounis, C. A review of select vitamins and minerals used by postmenopausal women. *Maturitas* **2010**, *66*, 370–380. [CrossRef] [PubMed]
7. Haskell, M.J. The challenge to reach nutritional adequacy for vitamin A: Beta-carotene bioavailability and conversion—Evidence in humans. *Am. J. Clin. Nutr.* **2012**, *96*, 1193S–1203S. [CrossRef] [PubMed]
8. Chung, E.; Mo, H.; Wang, S.; Zu, Y.; Elfakhani, M.; Rios, S.R.; Chyu, M.C.; Yang, R.S.; Shen, C.L. Potential roles of vitamin E in age-related changes in skeletal muscle health. *Nutr. Res.* **2018**, *49*, 23–36. [CrossRef] [PubMed]
9. Grosso, G.; Bei, R.; Mistretta, A.; Marventano, S.; Calabrese, G.; Masuelli, L.; Giganti, M.G.; Modesti, A.; Galvano, F.; Gazzolo, D. Effects of vitamin C on health: A review of evidence. *Front. Biosci. (Landmark Ed.)* **2013**, *18*, 1017–1029. [PubMed]
10. Granger, M.; Eck, P. Dietary vitamin C in human health. *Adv. Food Nutr. Res.* **2018**, *83*, 281–310. [CrossRef] [PubMed]
11. Goltzman, D. Functions of vitamin D in bone. *Histochem. Cell. Biol.* **2018**, *149*, 305–312. [CrossRef] [PubMed]
12. Razzaque, M.S. Sunlight exposure: Do health benefits outweigh harm? *J. Steroid Biochem. Mol. Biol.* **2018**, *175*, 44–48. [CrossRef] [PubMed]

13. Issa, C.M. Vitamin d and type 2 diabetes mellitus. *Adv. Exp. Med. Biol.* **2017**, *996*, 193–205. [CrossRef] [PubMed]

14. Skaaby, T.; Thuesen, B.H.; Linneberg, A. Vitamin D, cardiovascular disease and risk factors. *Adv. Exp. Med. Biol.* **2017**, *996*, 221–230. [CrossRef] [PubMed]

15. Shui, I.; Giovannucci, E. Vitamin D status and cancer incidence and mortality. *Adv. Exp. Med. Biol.* **2014**, *810*, 33–51. [PubMed]

16. Abdou, E.; Hazell, A.S. Thiamine deficiency: An update of pathophysiologic mechanisms and future therapeutic considerations. *Neurochem. Res.* **2015**, *40*, 353–361. [CrossRef] [PubMed]

17. MacKay, D.; Hathcock, J.; Guarneri, E. Niacin: Chemical forms, bioavailability, and health effects. *Nutr. Rev.* **2012**, *70*, 357–366. [CrossRef] [PubMed]

18. Naderi, N.; House, J.D. Recent developments in folate nutrition. *Adv. Food Nutr. Res.* **2018**, *83*, 195–213. [CrossRef] [PubMed]

19. Saedisomeolia, A.; Ashoori, M. Riboflavin in human health: A review of current evidences. *Adv. Food Nutr. Res.* **2018**, *83*, 57–81. [CrossRef] [PubMed]

20. Hughes, C.F.; Ward, M.; Hoey, L.; McNulty, H. Vitamin B12 and ageing: Current issues and interaction with folate. *Ann. Clin. Biochem.* **2013**, *50*, 315–329. [CrossRef] [PubMed]

21. Farquhar, W.B.; Edwards, D.G.; Jurkovitz, C.T.; Weintraub, W.S. Dietary sodium and health: More than just blood pressure. *J. Am. Coll. Cardiol.* **2015**, *65*, 1042–1050. [CrossRef] [PubMed]

22. Weaver, C.M. Potassium and health. *Adv. Nutr.* **2013**, *4*, 368S–377S. [CrossRef] [PubMed]

23. Malavolta, M.; Piacenza, F.; Basso, A.; Giacconi, R.; Costarelli, L.; Mocchegiani, E. Serum copper to zinc ratio: Relationship with aging and health status. *Mech. Ageing Dev.* **2015**, *151*, 93–100. [CrossRef] [PubMed]

24. Brenneisen, P.; Steinbrenner, H.; Sies, H. Selenium, oxidative stress, and health aspects. *Mol. Asp. Med.* **2005**, *26*, 256–267. [CrossRef] [PubMed]

25. Galaris, D.; Pantopoulos, K. Oxidative stress and iron homeostasis: Mechanistic and health aspects. *Crit. Rev. Clin. Lab. Sci.* **2008**, *45*, 1–23. [CrossRef] [PubMed]

26. Samaniego-Vaesken, M.L.; Partearroyo, T.; Olza, J.; Aranceta-Bartrina, J.; Gil, A.; Gonzalez-Gross, M.; Ortega, R.M.; Serra-Majem, L.; Varela-Moreiras, G. Iron intake and dietary sources in the Spanish population: Findings from the anibes study. *Nutrients* **2017**, *9*. [CrossRef] [PubMed]

27. Ayuk, J.; Gittoes, N.J. Contemporary view of the clinical relevance of magnesium homeostasis. *Ann. Clin. Biochem.* **2014**, *51*, 179–188. [CrossRef] [PubMed]

28. Gröber, U.; Schmidt, J.; Kisters, K. Magnesium in prevention and therapy. *Nutrients* **2015**, *7*, 8199–8226. [CrossRef] [PubMed]

29. Rosique-Esteban, N.; Guasch-Ferre, M.; Hernandez-Alonso, P.; Salas-Salvado, J. Dietary magnesium and cardiovascular disease: A review with emphasis in epidemiological studies. *Nutrients* **2018**, *10*. [CrossRef] [PubMed]

30. Peterlik, M.; Grant, W.B.; Cross, H.S. Calcium, vitamin D and cancer. *Anticancer Res.* **2009**, *29*, 3687–3698. [PubMed]

31. Reid, I.R.; Birstow, S.M.; Bolland, M.J. Calcium and cardiovascular disease. *Endocrinol. Metab. (Seoul)* **2017**, *32*, 339–349. [CrossRef] [PubMed]

32. Tankeu, A.T.; Ndip Agbor, V.; Noubiap, J.J. Calcium supplementation and cardiovascular risk: A rising concern. *J. Clin. Hypertens. (Greenwich)* **2017**, *19*, 640–646. [CrossRef] [PubMed]

33. (SINU), S.I.d.N.U. Iv Revisione dei Livelli di Assunzione di Riferimento di Nutrienti ed Energia per la Popolazione Italiana (Larn). Available online: http://www.sinu.it/html/pag/tabelle_larn_2014_rev.asp (accessed on 1 April 2018).

34. (EFSA), E.F.S.A. Dietary Reference Values for Nutrients. Available online: https://www.efsa.europa.eu/sites/default/files/2017_09_DRVs_summary_report.pdf (accessed on 15 April 2018).

35. Grosso, G.; Marventano, S.; D'Urso, M.; Mistretta, A.; Galvano, F. The Mediterranean healthy eating, ageing, and lifestyle (meal) study: Rationale and study design. *Int. J. Food Sci. Nutr.* **2017**, *68*, 577–586. [CrossRef] [PubMed]

36. Mistretta, A.; Marventano, S.; Platania, A.; Godos, J.; Galvano, F.; Grosso, G. Metabolic profile of the Mediterranean healthy eating, lifestyle and aging (meal) study cohort. *Mediterr. J. Nutr. Metab.* **2017**, *10*, 131–140. [CrossRef]

37. Buscemi, S.; Rosafio, G.; Vasto, S.; Massenti, F.M.; Grosso, G.; Galvano, F.; Rini, N.; Barile, A.M.; Maniaci, V.; Cosentino, L.; et al. Validation of a food frequency questionnaire for use in Italian adults living in sicily. *Int. J. Food Sci. Nutr.* **2015**, *66*, 426–438. [CrossRef] [PubMed]

38. Marventano, S.; Mistretta, A.; Platania, A.; Galvano, F.; Grosso, G. Reliability and relative validity of a food frequency questionnaire for Italian adults living in sicily, southern Italy. *Int. J. Food Sci. Nutr.* **2016**, *67*, 857–864. [CrossRef] [PubMed]

39. Istituto Nazionale di Ricerca per gli Alimenti e la Nutrizione. Tabelle di composizione degli alimenti. 2009. Available online: http://nut.entecra.it/646/tabelle_di_composizione_degli_alimenti.html (accessed on 27 March 2018).

40. Roman Vinas, B.; Ribas Barba, L.; Ngo, J.; Gurinovic, M.; Novakovic, R.; Cavelaars, A.; de Groot, L.C.; van't Veer, P.; Matthys, C.; Serra Majem, L. Projected prevalence of inadequate nutrient intakes in Europe. *Ann. Nutr. Metab.* **2011**, *59*, 84–95. [CrossRef] [PubMed]

41. Majeed, F. Low levels of vitamin D an emerging risk for cardiovascular diseases: A review. *Int. J. Health Sci.* **2017**, *11*, 71–76.

42. Basit, S. Vitamin D in health and disease: A literature review. *Br. J. Biomed. Sci.* **2013**, *70*, 161–172. [CrossRef] [PubMed]

43. Theodoratou, E.; Tzoulaki, I.; Zgaga, L.; Ioannidis, J.P. Vitamin D and multiple health outcomes: Umbrella review of systematic reviews and meta-analyses of observational studies and randomised trials. *BMJ* **2014**, *348*, g2035. [CrossRef] [PubMed]

44. Mule, S.; Falla, M.; Conti, A.; Castiglione, D.; Blanco, I.; Platania, A.; D'Urso, M.; Marranzano, M. Macronutrient and major food group intake in a cohort of southern Italian adults. *Antioxidants (Basel)* **2018**, *7*. [CrossRef] [PubMed]

45. Grosso, G.; Marventano, S.; Giorgianni, G.; Raciti, T.; Galvano, F.; Mistretta, A. Mediterranean diet adherence rates in sicily, southern Italy. *Public Health Nutr.* **2014**, *17*, 2001–2009. [CrossRef] [PubMed]

46. Marventano, S.; Godos, J.; Platania, A.; Galvano, F.; Mistretta, A.; Grosso, G. Mediterranean diet adherence in the Mediterranean healthy eating, aging and lifestyle (meal) study cohort. *Int. J. Food Sci. Nutr.* **2018**, *69*, 100–107. [CrossRef] [PubMed]

47. Grosso, G.; Marventano, S.; Buscemi, S.; Scuderi, A.; Matalone, M.; Platania, A.; Giorgianni, G.; Rametta, S.; Nolfo, F.; Galvano, F.; et al. Factors associated with adherence to the Mediterranean diet among adolescents living in sicily, southern Italy. *Nutrients* **2013**, *5*, 4908–4923. [CrossRef] [PubMed]

48. Grosso, G.; Pajak, A.; Mistretta, A.; Marventano, S.; Raciti, T.; Buscemi, S.; Drago, F.; Scalfi, L.; Galvano, F. Protective role of the Mediterranean diet on several cardiovascular risk factors: Evidence from sicily, southern Italy. *Nutr. Metab. Cardiovasc. Dis.* **2014**, *24*, 370–377. [CrossRef] [PubMed]

49. Manios, Y.; Moschonis, G.; Lambrinou, C.P.; Tsoutsoulopoulou, K.; Binou, P.; Karachaliou, A.; Breidenassel, C.; Gonzalez-Gross, M.; Kiely, M.; Cashman, K.D. A systematic review of vitamin D status in southern European countries. *Eur. J. Nutr.* **2017**. [CrossRef] [PubMed]

50. Mensink, G.B.M.; Fletcher, R.; Gurinovic, M.; Huybrechts, I.; Lafay, L.; Serra-Majem, L.; Szponar, L.; Tetens, I.; Verkaik-Kloosterman, J.; Baka, A.; et al. Mapping low intake of micronutrients across Europe. *Br. J. Nutr.* **2013**, *110*, 755–773. [CrossRef] [PubMed]

51. Chen, G.; Ni, Y.; Nagata, N.; Xu, L.; Ota, T. Micronutrient antioxidants and nonalcoholic fatty liver disease. *Int. J. Mol. Sci.* **2016**, *17*. [CrossRef] [PubMed]

52. Jiang, Q. Natural forms of vitamin E: Metabolism, antioxidant, and anti-inflammatory activities and their role in disease prevention and therapy. *Free Radic. Biol. Med.* **2014**, *72*, 76–90. [CrossRef] [PubMed]

53. Saboori, S.; Shab-Bidar, S.; Speakman, J.R.; Yousefi Rad, E.; Djafarian, K. Effect of vitamin E supplementation on serum c-reactive protein level: A meta-analysis of randomized controlled trials. *Eur. J. Clin. Nutr.* **2015**, *69*, 867–873. [CrossRef] [PubMed]

54. de Souza, R.G.M.; Schincaglia, R.M.; Pimentel, G.D.; Mota, J.F. Nuts and human health outcomes: A systematic review. *Nutrients* **2017**, *9*. [CrossRef]

55. Grosso, G.; Estruch, R. Nut consumption and age-related disease. *Maturitas* **2016**, *84*, 11–16. [CrossRef] [PubMed]
56. Dhobale, M.; Joshi, S. Altered maternal micronutrients (folic acid, vitamin B(12)) and omega 3 fatty acids through oxidative stress may reduce neurotrophic factors in preterm pregnancy. *J. Matern. Fetal Neonatal Med.* **2012**, *25*, 317–323. [CrossRef] [PubMed]
57. Kancherla, V.; Black, R.E. Historical perspective on folic acid and challenges in estimating global prevalence of neural tube defects. *Ann. N. Y. Acad. Sci.* **2018**, *1414*, 20–30. [CrossRef] [PubMed]
58. Moulik, N.R.; Kumar, A.; Agrawal, S. Folic acid, one-carbon metabolism & childhood cancer. *Indian J. Med. Res.* **2017**, *146*, 163–174. [CrossRef] [PubMed]
59. Stanhewicz, A.E.; Kenney, W.L. Role of folic acid in nitric oxide bioavailability and vascular endothelial function. *Nutr. Rev.* **2017**, *75*, 61–70. [CrossRef] [PubMed]
60. Martiniak, Y.; Heuer, T.; Hoffmann, I. Intake of dietary folate and folic acid in Germany based on different scenarios for food fortification with folic acid. *Eur. J. Nutr.* **2015**, *54*, 1045–1054. [CrossRef] [PubMed]
61. Godos, J.; Marventano, S.; Mistretta, A.; Galvano, F.; Grosso, G. Dietary sources of polyphenols in the Mediterranean healthy eating, aging and lifestyle (meal) study cohort. *Int. J. Food Sci. Nutr.* **2017**, *68*, 750–756. [CrossRef] [PubMed]
62. Godos, J.; Rapisarda, G.; Marventano, S.; Galvano, F.; Mistretta, A.; Grosso, G. Association between polyphenol intake and adherence to the Mediterranean diet in sicily, southern Italy. *NFS J.* **2017**, *8*, 1–7. [CrossRef]
63. Boyles, A.L.; Yetley, E.A.; Thayer, K.A.; Coates, P.M. Safe use of high intakes of folic acid: Research challenges and paths forward. *Nutr. Rev.* **2016**, *74*, 469–474. [CrossRef] [PubMed]
64. Viswanathan, M.; Treiman, K.A.; Kish-Doto, J.; Middleton, J.C.; Coker-Schwimmer, E.J.; Nicholson, W.K. Folic acid supplementation for the prevention of neural tube defects: An updated evidence report and systematic review for the US preventive services task force. *JAMA* **2017**, *317*, 190–203. [CrossRef] [PubMed]
65. De-Regil, L.M.; Pena-Rosas, J.P.; Fernandez-Gaxiola, A.C.; Rayco-Solon, P. Effects and safety of periconceptional oral folate supplementation for preventing birth defects. *Cochrane Database Syst. Rev.* **2015**, CD007950. [CrossRef] [PubMed]
66. Li, Y.; Huang, T.; Zheng, Y.; Muka, T.; Troup, J.; Hu, F.B. Folic acid supplementation and the risk of cardiovascular diseases: A meta-analysis of randomized controlled trials. *J. Am. Heart Assoc.* **2016**, *5*. [CrossRef] [PubMed]
67. Herrera-Araujo, D. Folic acid advisories: A public health challenge? *Health Econ.* **2016**, *25*, 1104–1122. [CrossRef] [PubMed]
68. owles, J.; Fahimi, S.; Micha, R.; Khatibzadeh, S.; Shi, P.; Ezzati, M.; Engell, R.E.; Lim, S.S.; Danaei, G.; Mozaffarian, D.; et al. Global, regional and national sodium intakes in 1990 and 2010: A systematic analysis of 24 h urinary sodium excretion and dietary surveys worldwide. *BMJ Open* **2013**, *3*, e003733. [CrossRef]
69. Welch, A.A.; Fransen, H.; Jenab, M.; Boutron-Ruault, M.C.; Tumino, R.; Agnoli, C.; Ericson, U.; Johansson, I.; Ferrari, P.; Engeset, D.; et al. Variation in intakes of calcium, phosphorus, magnesium, iron and potassium in 10 countries in the European prospective investigation into cancer and nutrition study. *Eur. J. Clin. Nutr.* **2009**, *63* (Suppl. 4), S101–S121. [CrossRef] [PubMed]
70. Balk, E.M.; Adam, G.P.; Langberg, V.N.; Earley, A.; Clark, P.; Ebeling, P.R.; Mithal, A.; Rizzoli, R.; Zerbini, C.A.F.; Pierroz, D.D.; et al. Global dietary calcium intake among adults: A systematic review. *Osteoporos Int.* **2017**, *28*, 3315–3324. [CrossRef] [PubMed]
71. Aaron, K.J.; Sanders, P.W. Role of dietary salt and potassium intake in cardiovascular health and disease: A review of the evidence. *Mayo Clin. Proc.* **2013**, *88*, 987–995. [CrossRef] [PubMed]
72. Mozaffarian, D.; Fahimi, S.; Singh, G.M.; Micha, R.; Khatibzadeh, S.; Engell, R.E.; Lim, S.; Danaei, G.; Ezzati, M.; Powles, J.; et al. Global sodium consumption and death from cardiovascular causes. *N. Engl. J. Med.* **2014**, *371*, 624–634. [CrossRef] [PubMed]
73. Stone, M.S.; Martyn, L.; Weaver, C.M. Potassium intake, bioavailability, hypertension, and glucose control. *Nutrients* **2016**, *8*. [CrossRef] [PubMed]
74. Cai, X.; Li, X.; Fan, W.; Yu, W.; Wang, S.; Li, Z.; Scott, E.M.; Li, X. Potassium and obesity/metabolic syndrome: A systematic review and meta-analysis of the epidemiological evidence. *Nutrients* **2016**, *8*, 183. [CrossRef] [PubMed]

75. Iwahori, T.; Miura, K.; Ueshima, H. Time to consider use of the sodium-to-potassium ratio for practical sodium reduction and potassium increase. *Nutrients* **2017**, *9*. [CrossRef] [PubMed]

76. Binia, A.; Jaeger, J.; Hu, Y.; Singh, A.; Zimmermann, D. Daily potassium intake and sodium-to-potassium ratio in the reduction of blood pressure: A meta-analysis of randomized controlled trials. *J. Hypertens.* **2015**, *33*, 1509–1520. [CrossRef] [PubMed]

77. Kloss, L.; Meyer, J.D.; Graeve, L.; Vetter, W. Sodium intake and its reduction by food reformulation in the European union—A review. *NFS J.* **2015**, *1*, 9–19. [CrossRef]

78. Grosso, G.; Galvano, F. Mediterranean diet adherence in children and adolescents in southern European countries. *NFS J.* **2016**, *3*, 13–19. [CrossRef]

antioxidants

MDPI

Article

Macronutrient and Major Food Group Intake in a Cohort of Southern Italian Adults

Serena Mulè [1], Mariagiovanna Falla [1], Alessandra Conti [1], Dora Castiglione [1], Isabella Blanco [1], Armando Platania [1], Maurizio D'Urso [2] and Marina Marranzano [1,*]

[1] Department of Medical and Surgical Sciences and Advanced Technologies "G.F. Ingrassia", University of Catania, 95125 Catania, Italy; serenamule86@gmail.com (S.M.); mariagiovannafalla@yahoo.com (M.F.); alessandra_conti@ymail.com (A.C.); doracastiglione29@gmail.com (D.C.); dott.ssa.blanco.isabella@gmail.com (I.B.); armplt@hotmail.it (A.P.)

[2] Provincial Health Authority of Catania, 95127 Catania, Italy; ma.durso76@gmail.com

* Correspondence: marranz@unict.it; Tel.: +39-095-378-2180

Received: 11 March 2018; Accepted: 12 April 2018; Published: 15 April 2018

Abstract: Background: Dietary intake of macronutrient and foods is considered crucial to decrease the risk of diet-related non-communicable diseases. Methods: The aim of this study was to describe the intake of major food groups and macronutrients in a random sample of 1838 southern Italian adults. Results: No significant differences of macronutrient consumption between sexes were found. By contrast, younger individuals had significantly higher intake of animal protein than older ones. Men reported consuming significantly more total processed meats and less eggs than women; egg consumption significantly increased by age groups. Significantly lower intake of fruit in the younger age group compared to older ones was found. Various patterns of correlation between food groups were described. More than half of individuals reached the suggested recommendations for carbohydrate and fiber intake, and about two-thirds met the recommendations for total protein and cholesterol intake, while only a minority met for total fat intake. Total and plant protein, monounsaturated and omega-6 fatty acids, were significantly inversely related with BMI (body mass index), while trans fatty acids and cholesterol were directly correlated. A direct association with unprocessed meats and an inverse association with processed meats was also found. Conclusions: The overall findings suggest that relatively healthy dietary habits are common in southern Italy.

Keywords: macronutrients; food intake; body mass index; dietary recommendations; cohort

---

1. Introduction

Over the last decades, great efforts have been done to identify a nutritionally balanced diet that might help reduce the risk of chronic non-communicable diseases. There is convincing evidence that dietary factors, alongside with physical activity and abstinence from unhealthy lifestyle behaviors (such as smoking habits), play a crucial role in prolonging the lifespan and ameliorating human health [1,2]. Adequate nutritional requirements represent, nowadays, a key element of public health effort [3]; thus, assessment and knowledge of current populations' nutritional status is needed to design national recommendations [4]. Previous guidelines were mainly interested in macronutrient intake, but more recent dietary advice focused on food groups, in order to improve the understanding of the general population and facilitate public health educators and policymakers to better identify crucial priorities in the field [5,6].

Research in nutritional epidemiology produced over the last years investigated the association between macronutrients/major food groups, and the most common chronic non-communicable diseases [7]. As prevalence of metabolic disorders has increased in the last decades, major attention has been appointed to the risk of obesity, considered the potential lead mediating factor for many other conditions [8]. In contrast with the individual role of obesity as determinant of diet-related diseases, there is general agreement that calorie source matters, and that diet quality, intended as a proper ratio between macronutrients and individual food groups, constitutes an independent risk factor for negative outcomes [9]. As carbohydrates are generally the most common source of dietary energy, it is therefore intuitive to ascribe to them the major responsibility for higher risk of obesity. However, numerous studies failed in assessing such a relationship, making evidence on this matter difficult to understand [10]. In fact, whether carbohydrates come in the form of whole or refined grains, has been suggested to be relevant in the explanation for the uncertainty of the findings from the studies exploring the association between total carbohydrate intake and weight status [10]. Similar concerns regard dietary guidelines involving protein intake. In fact, there is adequate evidence (from randomized controlled trials, RCTs) showing that substitution of protein for carbohydrate may favorably affect weight management and improve cardiometabolic biomarkers [11,12]. However, the type of protein may have specific effects, and other studies reported that differences between animal and plant protein occur when exploring long-term association with metabolic disorders [13] and overall mortality risk [14,15]. Final important different effects have been recently associated with various dietary fats. The failure of "low-fat diets" in prolonging the lifespan [16] and the discovery of the beneficial effects of (relatively) "high-fat diets", such as the Mediterranean dietary pattern [17,18], underlined the need to better distinguish between dietary fats and their effects on health. There is evidence that mono- and polyunsaturated fatty acids (MUFA and PUFA, respectively), including omega-3 PUFA from fish and vegetable, may exert a number of beneficial effects compared to saturated or, even worse, trans-fatty acids [19,20]. However, evidence on unhealthy effects of saturated fatty acids, per se, is still controversial, and further research is needed, overall, to better distinguish between subgroups of macronutrients, as aforementioned.

National and international organizations are dealing with current evidence on the association between diet and health. Experts boards continuously draft and update dietary guidelines and recommendations in order to prevent, on a large population scale, common non-communicable diseases. However, data on actual food consumption in cohort studies is often underrated and scarcely described. The aim of the present study was to describe the intake of major food groups and macronutrients in a sample of southern Italian adults, and to analyze the differences in consumption between sexes and age groups. Additionally, the study aimed to explore the correlation between the variables investigated and the association with weight status of participants.

2. Materials and Methods

2.1. Study Population

A sample of 2044 men and women aged 18 or more was collected between 2014 and 2016 in the main districts of the city of Catania, southern Italy, to build the Mediterranean healthy eating, ageing, and lifestyle (MEAL) cohort. A detailed description of the study protocol is published elsewhere [21]. Briefly, the theoretical sample size was set at 1500 individuals to provide a specific relative precision of 5% (type I error, 0.05; type II error, 0.10), taking into account an anticipated 70% participation rate. The sampling technique included stratification by municipality area, age, and sex of inhabitants, and randomization into subgroups, with randomly selected general practitioners being the sampling units, and individuals registered to them comprising the final sample units. Out of 2405 individuals invited, the final sample size was 2044 participants (response rate of 85%). All participants were informed about the aims of the study and provided a written informed consent. All the study procedures were carried out in accordance with the Declaration of Helsinki (1989) of the World Medical Association. The study protocol has been approved by the concerning ethical committee (protocol number: 802/23 December 2014).

2.2. Data Collection

Data was collected by a face-to-face computer-assisted personal interview using tablet computers. In order to visualize the response options, participants were provided of a paper copy of the questionnaire, however, final answers were filled in by the interviewer directly on the digital device (tablet computer). The demographic and anthropometric data were collected according to standard procedures [22]. Regarding anthropometric measurements, height was measured to the nearest 0.5 cm without shoes, with the back square against the wall tape, eyes looking straight ahead, with a right-angle triangle resting on the scalp and against the wall. Weight was measured with a lever balance to the nearest 100 g without shoes and with light undergarments. Body mass index (BMI) was finally calculated [23].

2.3. Dietary Assessment

A food frequency questionnaire (FFQ) previously validated for the Sicilian population was administered to collect information on food consumption [24,25]. The long version of the FFQ used to retrieve the dietary estimates presented in this study consisted of 110 food items; intake of seasonal foods referred to consumption during the period in which the food was available, and then adjusted by its proportional intake in one year. Following the identification of the food frequency consumption, the estimated intakes were converted into daily intake (g/day) and were used to calculate energy and macronutrient content based on online food composition databases (such as the Research Center for Foods and Nutrition CREA—*Consiglio per la ricerca in agricoltura e l'analisi dell'economia agraria*) [26]. Nutrient intake was finally adjusted for total energy intake (kcal/day) using the residual method [27]. FFQs with unreliable intakes (we arbitrarily considered <1000 or >6000 kcal/day as realistically unreliable energy intake; $n = 107$) as well as missing items for the purposes of this study ($n = 99$) were excluded from the analyses, leaving a total of 1838 individuals included in the analysis.

2.4. Dietary Recommendations

To investigate agreement with dietary recommendations, we used the European proposed values for macronutrient intake of the European Food Safety Agency (EFSA) [28] and those proposed by the Italian Society of Human Nutrition "*Livelli di Assunzione di Riferimento di Nutrienti*" (LARN) [29], while for major food groups we used the World Health Organization (WHO) recommendations [30].

2.5. Statistical Analysis

Frequencies are presented as absolute numbers and percentages; continuous variables are presented as means and standard errors, medians and ranges. Differences between groups for continuous variables were compared with Student's *t* test and ANOVA for continuous variables distributed normally, and Mann–Whitney U test and Kruskal–Wallis test for variables not normally distributed. Correlations among major food groups were tested through calculation of Pearson's or Spearman's correlation coefficients, depending on the distribution of the variable. Linear association between variables of interest and BMI levels were tested through linear regression analyses. All reported *P* values were based on two-sided tests and compared to a significance level of 5%. SPSS 17 (SPSS Inc., Chicago, IL, USA) software was used for all the statistical calculations.

3. Results

Table 1 shows the distribution of total energy, macronutrient and fiber intake in the study cohort, by sex and age groups. No significant differences of mean consumption of macronutrients between sexes were found. All macronutrients were mostly equally distributed, and even though men had slightly higher intake of cholesterol and total protein, the difference was not significant compared to women. In contrast, younger individuals consumed significantly more animal protein than older ones.

The description of consumption of major food group from animal source is shown in Table 2. Regarding differences between sex, men reported consuming significantly more total processed meats and less eggs than women (18.00 g/day vs. 14.54 g/day and 2.04 g/day vs. 2.62 g/day, respectively). Difference in egg consumption was also found between age groups, as intake significantly increased with age. Table 3 describes distribution of intake of plant food groups between sex and age groups. No significant differences were evident between sexes, but a significantly lower intake of fruit in the younger age group compared to older ones was found.

The correlation between intake of all major food groups is shown in Table 4. A correlation between fruit, vegetables, legumes, and seafood was found; however, the latter were also correlated with all other animal products, including cheese, eggs, and processed and unprocessed red meats. Whole and refined grain intake was correlated with yoghurt, while nuts and seeds were correlated with both meat and vegetable product intake. However, most of the significant associations were very weak and arguably negligible.

Table 5 describes the percentage of individuals meeting recommendations from LARN, EFSA, and WHO on macronutrients and food group intake. Generally, more than half of individuals reached the suggested recommendations for carbohydrate and fiber intake, while the proportion of adherent individuals was even higher for total protein and cholesterol intake recommendations. By contrast, only a minority met the recommendations for total fat intake.

Tables 6 and 7 describe the association between macronutrient and major food group intake and BMI levels in the investigated population, total and by sex. Total protein, and specifically plant protein, monounsaturated fatty acids and omega-6 fatty acids were significantly inversely related with BMI, while trans fatty acids and cholesterol were directly correlated (Table 6). However, no significant results were found for major food groups, with the exception of a direct association with unprocessed meats and an inverse association with processed meats (Table 7). It was noteworthy that the magnitude of the latter associations and of proteins, in general, were very small compared to those of dietary fats.

Table 1. Total, sex, and age group-specific consumption of macronutrients and fiber in the study participants of the Mediterranean healthy eating, ageing, and lifestyle (MEAL) study (*n* = 1838). * denotes *p* < 0.05.

	Total			<20 years			20 < years < 50			50 < years < 70			>70 years		
	n	Mean (SE)	Median (Range)	*n*	Mean (SE)	Median (Range)	*n*	Mean (SE)	Median (Range)	*n*	Mean (SE)	Median (Range)	*n*	Mean (SE)	Median (Range)
Total Energy															
Total	1838	2022.80 (15.30)	1927.63 (1000.80, 4974.01)	53	2037.36 (79.24)	1986.05 (1025.75, 3728.68)	963	2027.51 (21.91)	1930.64 (1000.80, 4865.38)	597	2038.93 (27.00)	1920.70 (1015.33, 4974.01)	225	1956.43 (36.66)	1923.88 (1010.31, 3379.32)
Men	772	2054.16 (25.38)	1939.18 (1012.84, 4974.01)	30	2101.73 (111.24)	2048.38 (1025.75, 3728.68)	384	2047.09 (37.99)	1907.56 (1012.84, 4865.38)	265	2076.45 (43.57)	1938.57 (1015.33, 4974.01)	93	2004.48 (56.27)	2025.94 (1019.88, 3334.93)
Women	1066	2000.09 (18.90)	1915.60 (1000.80, 3915.46)	23	1953.41 (111.15)	1873.20 (1113.85, 3547.77)	579	2014.52 (26.35)	1942.98 (1000.80, 3915.46)	332	2008.99 (33.84)	1902.58 (1016.84, 3911.26)	132	1922.58 (48.27)	1887.88 (1010.31, 3379.32)
Saturated fat															
Total	1838	23.58 (0.23)	22.17 (6.49, 80.45)	53	25.55 (1.29)	26.51 (11.85, 55.77)	963	23.71 (0.33)	22.11 (6.49, 80.45)	597	23.52 (0.41)	21.88 (6.91, 74.26)	225	22.69 (0.55)	22.87 (6.60, 45.05)
Men	772	23.86 (0.38)	22.09 (6.60, 80.45)	30	27.89 (1.80)	26.80 (14.82, 55.77)	384	23.45 (0.55)	20.88 (8.61, 80.45)	265	24.04 (0.64)	22.20 (9.47, 74.26)	93	23.75 (0.88)	24.47 (6.60, 44.82)
Women	1066	23.37 (0.29)	22.33 (6.49, 62.82)	23	22.50 (1.64)	21.09 (11.85, 34.77)	579	23.88 (0.40)	23.00 (6.49, 62.82)	332	23.11 (0.52)	21.46 (6.91, 61.79)	132	21.94 (0.69)	21.65 (6.97, 45.05)

Table 1. *Cont.*

		Total		<20 years			20 < years < 50			50 < years < 70			>70 years		
	n	Mean (SE)	Median (Range)	n	Mean (SE)	Median (Range)	n	Mean (SE)	Median (Range)	n	Mean (SE)	Median (Range)	n	Mean (SE)	Median (Range)
Monounsaturated fat															
Total	1838	25.30 (0.21)	24.13 (7.29, 93.85)	53	27.41 (1.23)	25.73 (13.19, 63.32)	963	25.39 (0.29)	24.04 (7.29, 80.32)	597	25.28 (0.38)	23.99 (11.01, 93.85)	225	24.48 (0.49)	24.07 (7.61, 47.14)
Men	772	25.62 (0.36)	23.94 (7.61, 93.85)	30	29.12 (1.83)	26.56 (15.20, 63.32)	384	25.12 (0.51)	23.19 (11.71, 80.32)	265	25.98 (0.66)	24.26 (11.77, 93.85)	93	25.56 (0.80)	26.43 (7.61, 47.14)
Women	1066	25.07 (0.25)	24.19 (7.29, 62.38)	23	25.18 (1.45)	24.10 (13.19, 41.50)	579	25.57 (0.35)	24.74 (7.29, 62.38)	332	24.72 (0.43)	23.67 (11.01, 55.05)	132	23.71 (0.60)	23.04 (11.03, 44.77)
Total omega-6 fatty acids															
Total	1838	9.92 (0.10)	9.21 (3.08, 55.51)	53	10.33 (0.61)	8.94 (3.36, 27.60)	963	9.97 (0.14)	9.21 (3.08, 30.64)	597	10.03 (0.19)	9.22 (3.29, 55.51)	225	9.29 (0.21)	9.13 (3.64, 20.44)
Men	772	9.99 (0.16)	9.09 (3.08, 55.51)	30	10.49 (0.86)	9.43 (3.36, 27.60)	384	9.93 (0.23)	8.96 (3.08, 29.06)	265	10.19 (0.31)	9.18 (3.29, 55.51)	93	9.54 (0.30)	9.43 (4.01, 16.09)
Women	1066	9.87 (0.12)	9.27 (3.61, 32.07)	23	10.11 (0.89)	8.92 (4.89, 23.60)	579	10.00 (0.17)	9.37 (3.61, 30.64)	332	9.91 (0.23)	9.24 (3.69, 32.07)	132	9.12 (0.29)	9.03 (3.64, 20.44)

Table 1. *Cont.*

	Total			<20 years			20 < years < 50			50 < years < 70			>70 years		
	n	Mean (SE)	Median (Range)	n	Mean (SE)	Median (Range)	n	Mean (SE)	Median (Range)	n	Mean (SE)	Median (Range)	n	Mean (SE)	Median (Range)
Seafood omega-3 fat															
Total	1838	0.53 (0.01)	0.38 (0.00, 5.24)	53	0.55 (0.05)	0.42 (0.00, 1.56)	963	0.52 (0.02)	0.36 (0.00, 5.24)	597	0.56 (0.02)	0.41 (0.01, 5.06)	225	0.53 (0.03)	0.43 (0.03, 3.26)
Men	772	0.53 (0.02)	0.38 (0.00, 5.24)	30	0.49 (0.07)	0.36 (0.00, 1.34)	384	0.50 (0.03)	0.34 (0.00, 5.24)	265	0.58 (0.04)	0.44 (0.05, 5.06)	93	0.56 (0.06)	0.42 (0.05, 3.26)
Women	1066	0.53 (0.02)	0.39 (0.00, 3.97)	23	0.63 (0.08)	0.56 (0.17, 1.56)	579	0.53 (0.02)	0.38 (0.00, 3.97)	332	0.54 (0.03)	0.38 (0.01, 3.05)	132	0.51 (0.04)	0.43 (0.03, 2.98)
Plant omega-3 fat															
Total	1838	1.17 (0.01)	1.06 (0.39, 5.66)	53	1.23 (0.08)	1.08 (0.51, 3.78)	963	1.17 (0.02)	1.06 (0.41, 5.51)	597	1.20 (0.02)	1.06 (0.42, 5.66)	225	1.10 (0.03)	1.05 (0.39, 2.79)
Men	772	1.17 (0.02)	1.06 (0.41, 4.48)	30	1.25 (0.11)	1.09 (0.60, 3.78)	384	1.17 (0.03)	1.02 (0.41, 3.93)	265	1.18 (0.03)	1.07 (0.46, 4.48)	93	1.14 (0.04)	1.13 (0.46, 2.32)
Women	1066	1.18 (0.02)	1.06 (0.39, 5.66)	23	1.20 (0.12)	0.90 (0.51, 2.82)	579	1.18 (0.02)	1.09 (0.42, 5.51)	332	1.21 (0.03)	1.05 (0.42, 5.66)	132	1.08 (0.04)	1.02 (0.39, 2.79)

Table 1. *Cont.*

	Total			<20 years			20 < years < 50			50 < years < 70			>70 years		
	n	Mean (SE)	Median (Range)	*n*	Mean (SE)	Median (Range)	*n*	Mean (SE)	Median (Range)	*n*	Mean (SE)	Median (Range)	*n*	Mean (SE)	Median (Range)
Trans fatty acid															
Total	1838	32.31 (0.28)	30.83 (10.30, 135.12)	53	34.60 (1.69)	31.68 (16.68, 84.81)	963	32.38 (0.39)	30.82 (10.30, 100.42)	597	32.46 (0.50)	31.02 (12.26, 135.12)	225	31.09 (0.62)	30.07 (10.99, 59.82)
Men	772	32.63 (0.47)	30.63 (10.99, 135.12)	30	36.18 (2.50)	32.79 (17.03, 84.81)	384	32.05 (0.66)	29.66 (11.97, 100.42)	265	33.19 (0.88)	31.42 (12.26, 135.12)	93	32.26 (1.00)	32.83 (10.99, 59.82)
Women	1066	32.08 (0.33)	30.87 (10.30, 84.75)	23	32.55 (2.12)	30.14 (16.68, 59.93)	579	32.60 (0.47)	31.38 (10.30, 84.75)	332	31.88 (0.58)	30.35 (14.88, 68.99)	132	30.28 (0.78)	29.62 (13.84, 55.51)
Dietary cholesterol															
Total	1838	187.55 (1.93)	175.00 (17.29, 921.07)	53	198.15 (9.17)	191.00 (87.71, 371.79)	963	187.15 (2.74)	173.89 (17.29, 921.07)	597	188.24 (3.44)	172.62 (59.94, 876.81)	225	184.84 (4.92)	180.36 (42.92, 521.31)
Men	772	191.35 (3.22)	174.09 (42.92, 921.07)	30	206.26 (12.70)	206.42 (102.85, 371.79)	384	186.61 (4.64)	164.24 (56.99, 921.07)	265	195.15 (5.78)	176.70 (63.53, 876.81)	93	195.29 (7.83)	189.94 (42.92, 521.31)
Women	1066	184.80 (2.37)	176.30 (17.29, 594.94)	23	187.57 (13.09)	177.00 (87.71, 333.42)	579	187.51 (3.35)	181.67 (17.29, 594.94)	332	182.73 (4.10)	164.69 (59.94, 487.38)	132	177.65 (6.26)	169.40 (56.35, 475.25)

Table 1. *Cont.*

	Total			<20 years			20 < years < 50			50 < years < 70			>70 years		
	n	Mean (SE)	Median (Range)	*n*	Mean (SE)	Median (Range)	*n*	Mean (SE)	Median (Range)	*n*	Mean (SE)	Median (Range)	*n*	Mean (SE)	Median (Range)
Total protein															
Total	1838	83.98 (0.66)	80.02 (29.27, 332.66)	53	86.24 (3.14)	83.52 (33.65, 138.33)	963	84.27 (0.96)	79.31 (29.27, 332.66)	597	83.89 (1.14)	80.18 (29.35, 303.04)	225	82.41 (1.63)	80.23 (29.35, 185.58)
Men	772	85.22 (1.11)	79.67 (33.65, 332.66)	30	88.27 (4.03)	89.30 (33.65, 137.17)	384	85.01 (1.63)	79.21 (43.56, 332.66)	265	85.59 (1.96)	79.54 (37.92, 303.04)	93	84.07 (2.56)	80.73 (39.51, 185.58)
Women	1066	83.07 (0.81)	80.23 (29.27, 215.96)	23	83.59 (5.01)	78.86 (50.69, 138.33)	579	83.78 (1.17)	80.38 (29.27, 215.96)	332	82.53 (1.32)	80.24 (29.35, 185.93)	132	81.25 (2.12)	79.47 (29.35, 156.18)
Animal protein															
Total	1838	25.65 (0.44)	22.75 (0.00, 449.25)	53	30.68 (1.83)	30.97 (6.63, 68.99)	963	26.03 (0.68)	23.08 (0.00, 449.25)	597	25.73 (0.74)	22.63 (0.00, 238.85)	225	22.67 (0.77) *	18.98 (3.05, 61.61)
Men	772	26.46 (0.67)	23.07 (0.00, 238.85)	30	31.26 (2.85)	29.70 (6.63, 68.99)	384	26.88 (0.91)	23.55 (0.00, 161.09)	265	26.46 (1.32)	23.20 (0.00, 238.85)	93	23.13 (1.09)	20.69 (6.23, 54.31)
Women	1066	25.07 (0.59)	22.38 (0.00, 449.25)	23	29.93 (2.08)	31.34 (7.14, 47.55)	579	25.46 (0.95)	22.91 (0.00, 449.25)	332	25.15 (0.81)	22.19 (5.25, 148.42)	132	22.35 (1.07)	18.71 (3.05, 61.61)

Table 1. *Cont.*

		Total		<20 years			20 < years < 50			50 < years < 70			>70 years		
	n	Mean (SE)	Median (Range)	n	Mean (SE)	Median (Range)	n	Mean (SE)	Median (Range)	n	Mean (SE)	Median (Range)	n	Mean (SE)	Median (Range)
Dairy protein															
Total	1838	14.01 (0.21)	12.24 (0.00, 67.63)	53	14.84 (1.02)	12.69 (0.00, 28.48)	963	14.19 (0.30)	11.78 (0.00, 67.63)	597	13.75 (0.34)	12.85 (0.00, 52.81)	225	13.69 (0.53)	13.09 (0.00, 50.39)
Men	772	13.59 (0.32)	11.60 (0.00, 67.63)	30	13.69 (1.55)	11.91 (0.00, 28.48)	384	13.70 (0.52)	10.76 (0.00, 67.63)	265	13.17 (0.48)	13.08 (0.00, 44.02)	93	14.28 (0.78)	15.25 (0.00, 42.42)
Women	1066	14.31 (0.27)	12.71 (0.00, 54.33)	23	16.35 (1.17)	15.53 (8.52, 27.50)	579	14.52 (0.37)	12.77 (0.00, 54.33)	332	14.21 (0.47)	12.39 (0.00, 52.81)	132	13.27 (0.72)	12.04 (0.00, 50.39)
Plant protein															
Total	1838	44.71 (0.41)	41.90 (6.90, 178.86)	53	44.13 (1.89)	43.91 (6.90, 86.56)	963	45.11 (0.61)	41.92 (13.67, 178.86)	597	44.69 (0.70)	41.88 (11.96, 140.10)	225	43.16 (0.99)	40.57 (15.11, 83.01)
Men	772	45.35 (0.67)	42.14 (6.90, 178.86)	30	44.75 (2.53)	45.04 (6.90, 70.78)	384	45.93 (1.02)	42.46 (13.67, 178.86)	265	45.19 (1.11)	41.79 (16.35, 140.10)	93	43.64 (1.55)	40.55 (17.71, 82.61)
Women	1066	44.24 (0.52)	41.80 (11.96, 117.03)	23	43.32 (2.91)	40.89 (25.91, 86.56)	579	44.57 (0.74)	41.69 (16.36, 117.03)	332	44.29 (0.89)	42.31 (11.96, 100.77)	132	42.82 (1.30)	40.62 (15.11, 83.01)

Table 1. *Cont.*

	Total			<20 years			20 < years < 50			50 < years < 70			>70 years		
	n	Mean (SE)	Median (Range)	n	Mean (SE)	Median (Range)	n	Mean (SE)	Median (Range)	n	Mean (SE)	Median (Range)	n	Mean (SE)	Median (Range)
Total carbohydrates															
Total	1838	296.02 (2.56)	274.18 (100.18, 897.76)	53	289.04 (13.47)	275.29 (119.11, 590.50)	963	296.17 (3.69)	271.23 (109.87, 897.76)	597	300.52 (4.44)	278.50 (100.18, 673.97)	225	285.14 (6.31)	268.17 (114.52, 560.82)
Men	772	300.69 (4.13)	278.62 (109.87, 897.76)	30	290.46 (17.43)	275.52 (119.11, 482.17)	384	302.36 (6.33)	276.32 (109.87, 897.76)	265	303.96 (6.78)	286.63 (126.94, 673.97)	93	287.83 (9.23)	269.57 (132.74, 504.86)
Women	1066	292.64 (3.25)	270.39 (100.18, 670.72)	23	287.19 (21.57)	270.44 (137.86, 590.50)	579	292.06 (4.47)	270.50 (112.30, 670.72)	332	297.77 (5.86)	272.22 (100.18, 608.56)	132	283.24 (8.60)	265.18 (114.52, 560.82)
Fiber															
Total	1838	31.69 (0.33)	29.30 (2.81, 150.50)	53	29.77 (1.55)	27.98 (2.81, 57.65)	963	31.81 (0.50)	29.11 (6.63, 150.50)	597	32.09 (0.54)	30.43 (5.46, 100.03)	225	30.53 (0.79)	29.26 (8.31, 81.77)
Men	772	32.25 (0.54)	29.52 (2.81, 150.50)	30	30.87 (2.18)	28.23 (2.81, 57.65)	384	32.23 (0.86)	28.34 (9.48, 150.50)	265	32.65 (0.86)	30.39 (8.83, 100.03)	93	31.65 (1.14)	30.36 (11.01, 57.97)
Women	1066	31.28 (0.42)	29.23 (5.46, 85.11)	23	28.34 (2.17)	26.42 (12.77, 56.68)	579	31.54 (0.60)	29.28 (6.63, 85.11)	332	31.63 (0.68)	30.46 (5.46, 78.36)	132	29.73 (1.08)	27.75 (8.31, 81.77)

Table 2. Total, sex, and age group-specific consumption of animal food groups in the study participants of the MEAL study ($n = 1838$). * denotes $p < 0.05$, ** denotes $p < 0.001$.

	Total			<20 years			20 < years < 50			50 < years < 70			>70 years		
	n	Mean (SE)	Median (Range)	n	Mean (SE)	Median (Range)	n	Mean (SE)	Median (Range)	n	Mean (SE)	Median (Range)	n	Mean (SE)	Median (Range)
Total processed meats															
Total	1838	15.99 (0.43) **	11.50 (0.00, 168.00)	53	16.82 (1.92)	17.05 (0.00, 53.00)	963	17.57 (0.58)	11.50 (0.00, 129.50)	597	14.54 (0.82)	7.00 (0.00, 168.00)	225	12.85 (0.99)	7.00 (0.00, 157.00)
Men	772	18.00 (0.75)	11.50 (0.00, 168.00)	30	18.68 (2.67)	18.00 (0.00, 53.00)	384	19.12 (1.02)	11.50 (0.00, 129.50)	265	17.63 (1.49)	11.50 (0.00, 168.00)	93	14.24 (1.36)	7.85 (0.00, 50.00)
Women	1066	14.52 (0.49)	7.42 (0.00, 157.00)	23	14.40 (2.73)	7.00 (1.50, 53.00)	579	16.54 (0.68)	11.50 (0.00, 99.35)	332	12.08 (0.84)	7.00 (0.00, 129.50)	132	11.87 (1.39)	7.00 (0.00, 157.00)
Unprocessed meats															
Total	1838	33.78 (0.59)	28.00 (0.00, 286.00)	53	38.01 (4.06)	28.00 (0.00, 114.00)	963	33.58 (0.76)	28.00 (0.00, 136.00)	597	33.78 (1.10)	28.00 (0.00, 286.00)	225	33.67 (1.85)	28.00 (0.00, 164.00)
Men	772	34.65 (0.65)	28.00 (0.00, 286.00)	30	35.06 (5.18)	28.00 (0.00, 100.00)	384	33.58 (1.19)	28.00 (0.00, 128.00)	265	36.23 (1.80)	28.00 (0.00, 286.00)	93	34.40 (2.96)	28.00 (3.00, 164.00)
Women	1066	33.16 (0.75)	28.00 (0.00, 164.00)	23	41.86 (6.51)	28.00 (0.00, 114.00)	579	33.58 (0.99)	28.00 (0.00, 136.00)	332	31.82 (1.34)	28.00 (0.00, 136.00)	132	33.16 (2.36)	28.00 (0.00, 164.00)

Table 2. *Cont.*

	Total			<20 years			20 < years < 50			50 < years < 70			>70 years		
	n	Mean (SE)	Median (Range)	*n*	Mean (SE)	Median (Range)	*n*	Mean (SE)	Median (Range)	*n*	Mean (SE)	Median (Range)	*n*	Mean (SE)	Median (Range)
Total seafood															
Total	1838	60.81 (1.28)	47.40 (0.00, 784.70)	53	60.35 (5.30)	54.70 (0.00, 145.00)	963	59.17 (1.85)	45.00 (0.00, 784.70)	597	63.71 (2.19)	50.40 (0.00, 442.00)	225	60.26 (3.45)	48.10 (0.00, 448.00)
Men	772	61.07 (2.18)	46.80 (0.00, 784.70)	30	56.56 (7.47)	46.50 (0.00, 142.00)	384	58.41 (3.28)	43.30 (0.00, 784.70)	265	65.84 (3.55)	54.10 (3.00, 442.00)	93	59.91 (6.03)	45.00 (6.00, 448.00)
Women	1066	60.62 (1.55)	48.00 (0.00, 408.00)	23	65.29 (7.40)	58.40 (12.70, 145.00)	579	59.68 (2.18)	47.40 (0.00, 408.00)	332	62.01 (2.72)	47.70 (0.00, 373.70)	132	60.50 (4.09)	50.45 (0.00, 250.00)
Eggs															
Total	1838	2.38 (0.11) *	0.77 (0.00, 24.75)	53	1.84 (0.38)	0.77 (0.00, 13.75)	963	1.92 (0.12)	0.77 (0.00, 24.75)	597	2.80 (0.22)	0.77 (0.00, 24.75)	225	3.30 (0.38) **	0.77 (0.00, 24.75)
Men	772	2.04 (0.14)	0.77 (0.00, 24.75)	30	1.72 (0.48)	0.77 (0.00, 13.75)	384	1.42 (0.11)	0.77 (0.00, 24.75)	265	2.27 (0.27)	0.77 (0.00, 24.75)	93	4.08 (0.67)	1.98 (0.00, 24.75)
Women	1066	2.62 (0.16)	0.77 (0.00, 24.75)	23	2.00 (0.61)	0.77 (0.16, 13.75)	579	2.26 (0.19)	0.77 (0.00, 24.75)	332	3.23 (0.32)	0.77 (0.00, 24.75)	132	2.75 (0.44)	0.77 (0.00, 24.75)
Cheese															
Total	1838	53.45 (0.80)	46.70 (0.00, 328.01)	53	56.29 (4.27)	50.20 (15.51, 147.47)	963	53.68 (1.13)	46.82 (0.00, 310.01)	597	53.49 (1.43)	46.08 (0.00, 328.01)	225	51.74 (2.02)	46.33 (0.00, 231.88)

Table 2. Cont.

	Total			<20 years			20 < years < 50			50 < years < 70			>70 years		
	n	Mean (SE)	Median (Range)	n	Mean (SE)	Median (Range)	n	Mean (SE)	Median (Range)	n	Mean (SE)	Median (Range)	n	Mean (SE)	Median (Range)
Men	772	55.16 (1.30)	47.53 (0.00, 328.01)	30	64.43 (5.02)	52.67 (26.48, 147.47)	384	53.49 (1.86)	44.85 (1.50, 310.01)	265	56.61 (2.35)	48.63 (0.00, 328.01)	93	54.98 (3.13)	52.08 (0.00, 123.20)
Women	1066	52.22 (1.01)	45.82 (0.00, 296.01)	23	45.69 (6.84)	31.43 (15.51, 138.88)	579	53.80 (1.42)	47.50 (0.00, 296.01)	332	51.00 (1.74)	43.82 (0.00, 213.71)	132	49.46 (2.64)	45.72 (0.00, 231.88)
Yoghurt															
Total	1838	28.79 (1.07)	8.38 (0.00, 312.50)	53	37.23 (9.20)	8.38 (0.00, 312.50)	963	26.83 (1.36)	8.38 (0.00, 312.50)	597	29.71 (1.88)	8.38 (0.00, 312.50)	225	32.77 (3.57)	8.38 (0.00, 312.50)
Men	772	28.27 (1.66)	8.38 (0.00, 312.50)	30	50.05 (14.85)	17.50 (0.00, 312.50)	384	26.31 (2.09)	8.38 (0.00, 312.50)	265	27.70 (2.79)	8.38 (0.00, 312.50)	93	30.96 (5.28)	8.38 (0.00, 312.50)
Women	1066	29.17 (1.40)	8.38 (0.00, 312.50)	23	20.51 (7.72)	0.00 (0.00, 125.00)	579	27.18 (1.79)	8.38 (0.00, 312.50)	332	31.32 (2.55)	8.38 (0.00, 312.50)	132	34.04 (4.83)	8.38 (0.00, 312.50)
Reduced fat milk															
Total	1838	124.71 (3.68)	90.00 (0.00, 1125.00)	53	129.83 (19.42)	90.00 (0.00, 625.00)	963	127.22 (5.16)	90.00 (0.00, 1125.00)	597	118.01 (6.28)	90.00 (0.00, 1125.00)	225	130.56 (10.82)	90.00 (0.00, 625.00)
Men	772	121.70 (5.48)	90.00 (0.00, 1125.00)	30	125.20 (20.49)	90.00 (0.00, 250.00)	384	130.70 (8.31)	90.00 (0.00, 1125.00)	265	107.36 (8.60)	35.00 (0.00, 625.00)	93	124.31 (15.55)	90.00 (0.00, 625.00)
Women	1066	126.89 (4.95)	90.00 (0.00, 1125.00)	23	135.87 (36.48)	90.00 (0.00, 625.00)	579	124.91 (6.59)	90.00 (0.00, 625.00)	332	126.52 (8.94)	90.00 (0.00, 1125.00)	132	134.96 (14.88)	90.00 (0.00, 625.00)

Table 3. Total, sex, and age group-specific consumption of plant food groups in the study participants of the MEAL study (*n* = 1838). * denotes $p < 0.05$.

	Total			<20 years			20 < years < 50			50 < years < 70			>70 years		
	n	Mean (SE)	Median (Range)	*n*	Mean (SE)	Median (Range)	*n*	Mean (SE)	Median (Range)	*n*	Mean (SE)	Median (Range)	*n*	Mean (SE)	Median (Range)
Fruits															
Total	1838	395.92 (7.43)	295.13 (0.00, 2801.47)	53	335.51 (28.08)	303.58 (0.00, 951.57)	963	402.02 (11.09)	295.09 (0.00, 2801.47)	597	412.67 (12.52)	318.32 (0.00, 1822.92)	225	339.64 (16.42)	268.78 (0.00, 1545.11)
Men	772	410.55 (11.81)	305.19 (0.00, 2801.47)	30	375.82 (44.50)	326.02 (0.00, 951.57)	384	408.13 (18.20)	302.11 (0.00, 2801.47)	265	430.97 (19.18)	308.70 (0.60, 1822.92)	93	373.56 (27.59)	289.75 (18.08, 1207.35)
Women	1066	385.33 (9.54)	285.57 (0.00, 2305.08)	23	282.93 (25.86)	257.78 (72.56, 541.09)	579	397.96 (13.95)	291.35 (0.00, 2305.08)	332	398.06 (16.50)	320.91 (0.00, 1791.90)	132	315.74 (19.98)	253.81 (0.00, 1545.11)
Non-starchy vegetables															
Total	1838	219.48 (3.22)	195.86 (0.00, 1506.75)	53	250.47 (28.28)	192.78 (1.13, 1236.37)	963	214.51 (4.54)	189.54 (0.00, 1506.75)	597	222.71 (5.29)	199.63 (0.00, 1254.12)	225	224.87 (8.58)	217.68 (0.00, 1268.28)
Men	772	221.43 (5.04)	195.80 (0.00, 1506.75)	30	254.14 (31.47)	204.14 (1.13, 799.94)	384	211.29 (7.58)	182.75 (0.00, 1506.75)	265	227.06 (7.88)	203.94 (1.50, 709.37)	93	236.69 (12.51)	235.67 (36.69, 567.68)
Women	1066	218.07 (4.19)	195.86 (0.00, 1268.28)	23	245.68 (51.48)	183.20 (33.78, 1236.37)	579	216.65 (5.63)	192.75 (0.00, 1146.28)	332	219.23 (7.15)	197.86 (0.00, 1254.12)	132	216.54 (11.56)	208.32 (0.0, 1268.28)

Table 3. *Cont.*

	Total			<20 years			20 < years < 50			50 < years < 70			>70 years		
	n	Mean (SE)	Median (Range)	n	Mean (SE)	Median (Range)	n	Mean (SE)	Median (Range)	n	Mean (SE)	Median (Range)	n	Mean (SE)	Median (Range)
Other starchy vegetables															
Total	1838	16.33 (0.44)	14.00 (0.00, 450.90)	53	15.88 (2.06)	14.00 (0.00, 66.00)	963	17.24 (0.72)	14.00 (0.00, 450.90)	597	15.22 (0.57)	14.00 (0.00, 130.00)	225	15.46 (0.87)	14.00 (0.00, 66.00)
Men	772	17.07 (0.80)	14.00 (0.00, 450.90)	30	17.08 (2.94)	14.45 (0.00, 66.00)	384	18.04 (1.43)	14.00 (0.00, 450.90)	265	16.03 (0.93)	14.00 (0.00, 130.00)	93	16.00 (1.34)	14.00 (0.00, 66.00)
Women	1066	15.80 (0.48)	14.00 (0.00, 130.00)	23	14.31 (2.85)	8.71 (0.00, 46.80)	579	16.72 (0.73)	14.00 (0.00, 130.00)	332	14.57 (0.72)	14.00 (0.00, 74.80)	132	15.08 (1.14)	14.00 (0.00, 66.00)
Beans and legumes															
Total	1838	35.61 (0.88)	23.70 (0.00, 655.33)	53	35.35 (4.78)	23.10 (0.00, 130.23)	963	36.69 (1.32)	24.00 (0.00, 655.33)	597	33.69 (1.33)	23.70 (0.00, 325.33)	225	36.15 (2.48)	22.33 (0.00, 184.00)
Men	772	36.33 (1.48)	23.85 (0.00, 655.33)	30	36.18 (6.39)	27.17 (0.00, 129.00)	384	37.19 (2.33)	24.10 (0.00, 655.33)	265	34.65 (2.15)	24.70 (0.00, 325.33)	93	37.58 (4.09)	22.33 (3.00, 179.00)
Women	1066	35.09 (1.08)	23.40 (0.00, 210.70)	23	34.27 (7.35)	22.33 (5.23, 130.23)	579	36.36 (1.56)	23.70 (0.00, 210.70)	332	32.91 (1.67)	23.40 (0.00, 210.70)	132	35.14 (3.10)	22.28 (0.00, 184.00)

Table 3. *Cont.*

	Total			<20 years			20 < years < 50			50 < years < 70			>70 years		
	n	Mean (SE)	Median (Range)	*n*	Mean (SE)	Median (Range)	*n*	Mean (SE)	Median (Range)	*n*	Mean (SE)	Median (Range)	*n*	Mean (SE)	Median (Range)
Nuts and seeds															
Total	1838	20.30 (0.73)	11.52 (0.00, 408.40)	53	19.77 (4.48)	9.05 (0.00, 190.00)	963	19.87 (0.98)	10.35 (0.00, 408.40)	597	21.89 (1.42)	12.75 (0.00, 408.40)	225	18.10 (1.60)	10.05 (0.00, 153.40)
Men	772	20.91 (1.21)	10.40 (0.00, 408.40)	30	23.29 (7.54)	7.71 (0.00, 190.00)	384	19.42 (1.92)	7.94 (0.00, 408.40)	265	22.69 (1.70)	13.40 (0.00, 190.00)	93	21.20 (3.05)	10.35 (0.00, 153.40)
Women	1066	19.87 (0.90)	11.70 (0.00, 408.40)	23	15.17 (3.16)	11.70 (0.00, 68.80)	579	20.16 (1.02)	12.73 (0.00, 190.00)	332	21.26 (2.17)	11.52 (0.00, 408.40)	132	15.91 (1.66)	9.92 (0.00, 101.48)
Potatoes															
Total	1838	25.52 (0.58)	17.75 (0.00, 450.75)	53	31.09 (5.13)	24.20 (0.00, 253.00)	963	25.86 (0.85)	17.75 (0.00, 450.75)	597	24.98 (0.95)	17.00 (0.00, 169.20)	225	24.21 (1.30)	17.00 (0.00, 106.70)
Men	772	26.48 (0.87)	17.75 (0.00, 180.00)	30	30.72 (4.27)	24.20 (0.00, 100.00)	384	26.17 (1.22)	17.50 (0.00, 180.00)	265	26.56 (1.61)	17.00 (0.00, 169.20)	93	26.23 (2.06)	20.70 (0.00, 103.00)
Women	1066	24.82 (0.78)	17.50 (0.00, 450.75)	23	31.56 (10.58)	20.70 (0.00, 253.00)	579	25.65 (1.16)	18.68 (0.00, 450.75)	332	23.73 (1.12)	17.00 (0.00, 136.00)	132	22.78 (1.68)	16.34 (0.00, 106.70)

Table 3. *Cont.*

	Total			<20 years			20 < years < 50			50 < years < 70			>70 years		
	n	Mean (SE)	Median (Range)	*n*	Mean (SE)	Median (Range)	*n*	Mean (SE)	Median (Range)	*n*	Mean (SE)	Median (Range)	*n*	Mean (SE)	Median (Range)
Whole grains															
Total	1838	27.38 (1.19)	3.00 (0.00, 330.00)	53	26.21 (5.42)	1.01 (0.00, 151.20)	963	29.34 (1.74)	3.00 (0.00, 330.00)	597	25.90 (1.99)	3.00 (0.00, 298.70)	225	23.21 (3.01)	2.10 (0.00, 270.36)
Men	772	26.56 (1.77)	3.00 (0.00, 330.00)	30	37.73 (8.69)	9.00 (0.00, 151.20)	384	30.52 (2.78)	5.40 (0.00, 330.00)	265	20.54 (2.58)	3.00 (0.00, 298.70)	93	23.75 (4.59)	3.00 (0.00, 252.85)
Women	1066	27.97 (1.59)	2.10 (0.00, 330.00)	23	11.20 3.50	0.45 (0.00, 46.80)	579	28.55 (2.23)	3.00 (0.00, 330.00)	332	30.17 (2.91)	3.00 (0.00, 298.70)	132	22.84 (4.00)	0.73 (0.00, 270.36)
Refined grains															
Total	1838	214.10 (3.04)	184.15 (3.00, 909.26)	53	197.89 (17.80)	174.05 (3.00, 576.71)	963	210.54 (4.15)	180.89 (4.50, 909.26)	597	220.89 (5.51)	189.00 (6.70, 909.26)	225	215.19 (8.41)	185.05 (11.28, 630.03)
Men	772	217.24 (4.60)	187.39 (3.00, 909.26)	30	169.00 (20.22)	173.83 (3.00, 420.10)	384	217.19 (6.47)	186.51 (12.60, 589.85)	265	226.87 (8.25)	196.70 (11.28, 909.26)	93	205.57 (11.86)	180.31 (25.20, 541.06)
Women	1066	211.83 (4.06)	182.40 (4.50, 909.26)	23	235.58 (30.14)	173.83 (3.00, 420.10)	579	206.12 (5.41)	179.66 (4.50, 909.26)	332	216.12 (7.41)	182.60 (6.70, 696.60)	132	221.98 (11.65)	187.62 (11.28, 630.03)

Table 4. Pearson/Spearman correlation coefficients between major food groups intake. * denotes $p < 0.05$, ** denotes $p < 0.001$.

	Total Processed Meats	Unprocessed Red Meats	Total Seafood	Eggs	Cheese	Yoghurt	Fruits	Non-Starchy Vegetables	Potatoes	Other Starchy Vegetables	Beans and Legumes	Nuts and Seeds	Refined Grains	Whole Grains
Total processed meats	1	-	-	-	-	-	-	-	-	-	-	-	-	-
Unprocessed red meats	0.217 **	1	-	-	-	-	-	-	-	-	-	-	-	-
Total seafood	0.162 **	0.072 **	1	-	-	-	-	-	-	-	-	-	-	-
Eggs	0.003	0.179 **	0.093 **	1	-	-	-	-	-	-	-	-	-	-
Cheese	0.251 **	0.200 **	0.189 **	0.094 **	1	-	-	-	-	-	-	-	-	-
Yoghurt	−0.010	−0.032	0.145 **	0.034	0.0123 **	1	-	-	-	-	-	-	-	-
Fruits	0.004	−0.004	0.121 **	−0.010	0.074 **	0.113 **	1	-	-	-	-	-	-	-
Non-starchy vegetables	−0.002	−0.037	0.209 **	0.016	0.167 **	0.138 **	0.297 **	1	-	-	-	-	-	-
Potatoes	0.316 **	0.085 **	0.151 **	0.084 **	0.305 **	0.063 **	0.073 **	0.075 **	1	-	-	-	-	-
Other starchy vegetables	0.075 **	−0.029	0.203 **	0.018	0.138 **	0.073 **	0.258 **	0.399 **	0.144 **	1	-	-	-	-
Beans and legumes	0.044	0.000	0.268 **	0.041	0.108 **	0.114 **	0.203 **	0.370 **	0.052 *	0.211 **	1	-	-	-
Nuts and seeds	0.098 **	0.069 **	0.048 *	0.036	0.084 **	0.005	−0.037	0.060 **	0.080 **	−0.002	0.071 **	1	-	-
Refined grains	0.052 *	0.189 **	−0.055 *	0.154 **	0.197 **	−0.152 **	0.034	−0.029	0.055 *	−0.032	−0.017	−0.021	1	-
Whole grains	0.058 *	−0.057 *	0.109 **	−0.065 **	0.079 **	0.190 **	0.151 **	0.196 **	0.008	0.070 **	0.090 **	−0.042	−0.119 **	1

Table 5. Percentage of study population meeting various recommendations for macronutrients (EFSA, LARN) and selected food groups (WHO).

	Total (*n* = 1839)		Men (*n* = 772)		Women (*n* = 1066)	
	Yes, % (*n*)	No, % (*n*)	Yes, % (n)	No, % (*n*)	Yes, % (n)	No, % (*n*)
EFSA						
Total carbohydrate (45–60%E)	56.3 (1035)	43.7 (804)	56.6 (437)	43.4 (335)	56.0 (597)	44.0 (469)
Total protein (>0.83 g/kg/day)	89.7 (1649)	10.3 (190)	85.5 (660)	14.5 (112)	92.8 (989)	7.2 (77)
Total fat (20–35%E)	17.1 (315)	82.9 (1524)	17.6 (136)	82.4 (636)	16.8 (179)	83.2 (887)
Fiber (>25 g/day)	62.8 (1154)	37.2 (685)	62.7 (484)	37.3 (288)	62.9 (670)	37.1 (396)
LARN						
Total carbohydrates (40–60%E)	59.4 (1092)	40.6 (747)	59.5 (459)	40.5 (313)	59.3 (632)	40.7 (434)
Total protein (>0.90 g/kg/day)	83.9 (1543)	16.1 (296)	78.6 (607)	21.4 (165)	87.8 (936)	12.2 (130)
Total fat (20–35%E)	17.1 (315)	82.9 (1524)	17.6 (136)	82.4 (636)	16.8 (179)	83.2 (887)
Cholesterol (<300 mg/day)	91.8 (1688)	8.2 (151)	91.3 (705)	8.7 (67)	92.1 (982)	7.9 (84)
Fiber (12.6–16.7 g/1000 kcal/day)	53.5 (983)	46.5 (856)	54.3 (419)	45.7 (353)	52.9 (564)	47.1 (502)
WHO						
Fruit and vegetable (>400 g/day)	74.6 (1371)	25.4 (468)	76.4 (590)	23.6 (182)	73.2 (780)	26.8 (286)
Pulses and nuts (>30 g/day)	81.7 (1503)	18.3 (336)	82.5 (637)	17.5 (135)	81.2 (866)	18.8 (200)
Total meat (<70 g/day)	77.0 (1416)	23.0 (423)	75.4 (582)	24.6 (190)	78.1 (833)	21.9 (233)

Table 6. Linear association between macronutrient intake and BMI levels in the study participants of the MEAL study (*n* = 1838). * denotes $p < 0.05$, ** denotes $p < 0.001$.

	Total	Men	Women
Total carbohydrates	0.000 (0.006)	−0.002 (0.008)	−0.001 (0.008)
Total protein	−0.035 (0.016) *	−0.022 (0.025)	−0.046 (0.022) *
Animal protein	0.000 (0.006)	−0.002 (0.010)	0.002 (0.008)
Dairy protein	−0.013 (0.013)	0.002 (0.020)	−0.023 (0.018)
Plant protein	−0.049 (0.022) *	0.010 (0.033)	−0.081 (0.031) **
Saturated fat	−0.023 (0.041)	−0.108 (0.064)	0.028 (0.053)
Monounsaturated fat	−0.707 (0.138) **	−0.594 (0.212) **	−0.838 (0.184) **
Total omega-6 fatty acids	−0.657 (0.173) **	−0.722 (0.251) **	−0.647 (0.242) **
Seafood omega-3 fat	0.024 (0.356)	0.272 (0.548)	−0.204 (0.471)
Plant omega-3 fat	0.654 (0.522)	2.178 (0.879) *	−0.213 (0.748)
Trans fatty acid	0.666 (0.135) **	0.520 (0.202) *	0.807 (0.184) **
Dietary cholesterol	0.021 (0.004) **	0.015 (0.007) *	0.026 (0.006) **
Fiber	−0.013 (0.016)	−0.054 (0.024) *	0.014 (0.021)

Table 7. Linear association between major food group intake and BMI levels in the study participants of the MEAL study (*n* = 1838). * denotes $p < 0.05$, ** denotes $p < 0.001$.

	Total	Men	Women
Total processed meats	−0.023 (0.007) **	−0.031 (0.010) **	−0.015 (0.010)
Unprocessed meats	0.017 (0.005) **	0.014 (0.007) *	0.019 (0.006) **
Total seafood	0.004 (0.002)	0.007 (0.003) *	0.002 (0.003)
Eggs	0.039 (0.024)	0.037 (0.043)	0.042 (0.031)
Cheese	0.004 (0.004)	−0.001 (0.007)	0.008 (0.006)
Yoghurt	−0.002 (0.003)	−0.010 (0.004) *	0.003 (0.003)
Fruits	−0.001 (0.000)	−0.001 (0.001)	−0.001 (0.001)
Non-starchy vegetables	0.001 (0.001)	0.002 (0.002)	0.001 (0.001)
Potatoes	−0.005 (0.005)	−0.003 (0.009)	−0.007 (0.006)
Other starchy vegetables	0.002 (0.007)	0.004 (0.009)	−0.005 (0.011)
Beans and legumes	0.001 (0.003)	−0.008 (0.005)	0.007 (0.005)
Nuts and seeds	0.002 (0.004)	0.002 (0.006)	0.000 (0.006)
Refined grains	0.002 (0.002)	0.001 (0.002)	0.002 (0.002)
Whole grains	−0.004 (0.003)	−0.004 (0.004)	−0.003 (0.003)

4. Discussion

The present study provided updated information on intake of major food groups and macronutrients and their association with weight status in a sample of southern Italian adults. We found that a large proportion of individuals had adequate intake of protein, fiber, fats, fruit and vegetable, meat, and pulses according to national and international recommendations. These results suggest that the investigated population has generally healthy dietary choices; however, investigating major food group consumption and comparison with other reports is crucial to better understand dietary priorities for future strategies to improve dietary habits and overall health.

Despite the importance of monitoring dietary intakes at population level, previous studies investigating macronutrient and food consumption are scarce. A recent report of Global Burden of Diseases Nutrition and Chronic Diseases Expert Group aimed to describe consumption of major food groups worldwide and at national level [31]. Despite that the report showed standardized intake to the same isocaloric diet (2000 kcal/day), our data are comparable, due to similar average total energy intake in both men and women. In 2010, mean global fruit consumption in adults has been reported to be 81.3 g/day, with the highest intake in Greece, and no clear pattern of variation of consumption worldwide. In this study, we reported a much higher fruit intake (about 400 g/day) only comparable with reports from Jamaica and Malaysia. However, two Italian surveys [32,33] showed an average national consumption of fruit closer to those reported in the present study (about 200–300 g/day); our estimates might be higher, due to the higher availability of fruit and lower prices in the regional territory [34] (taking into account that none of the previous reports included the municipality of Catania for sampling), or represent an overestimation, due to potential limitation of this type of recall studies (i.e., higher number of food items coding for "fruit" compared to other FFQs). Mean vegetable and legume consumption in our study was more in line with worldwide average intake (about 250 g/day versus 208.8 g/day, respectively) and those reported in the other Italian studies [32,33]. Moreover, fruit and vegetable consumption were strongly intercorrelated, reflecting a global trend. Consumption of nuts, seeds, and wholegrain is relatively low worldwide (around 10 g/day and 40 g/day), with the highest consumption in Southeast Asian nations and the lowest in Central European nations. Our reports were similar to worldwide average regarding whole-grain consumption, but much higher concerning nut intake (about 20 g/day); again, this can be the result of increased intake due to local production of certain nut subtypes (i.e., pistachios), which might be easier available and at lower price, or an overestimation due to various questions on nut-subtypes in our FFQs. Regarding animal products, in our cohort we found a higher consumption of seafood (about 60 g/day versus 28 g/day), similar of processed meat (about 16 g/day versus 14 g/day), and slightly lower of unprocessed meat (about 34 g/day versus 42 g/day) compared to worldwide reports. However, the higher seafood intake was evident in Pacific Island nations, the Mediterranean Basin, South Korea, and Japan, consistently with historical cultures and local availability. Also, the other Italian report showed similar intake of processed and unprocessed meat products than those reported in the present study, while consumption of fish was lower [32,33]. According to the Italian National Institute of Statistics (ISTAT), the mean expenditure for major food groups in the Italian islands (including Sicily) does not substantially differ from the national average, with the exception of higher purchase of seafood, thus reflecting a regional preference in consuming such products. Interestingly, we have found that seafood intake was weakly correlated with most of the other food groups investigated, suggesting that preference for fish might be common, and related with either healthy or unhealthy food groups.

Global and national reports on macronutrient intake have underlined dramatic diversity across nations and the need for inform policies to improve global health. Our estimates for dietary fats are slightly "healthier" than those previously reported in the Italian population (i.e., lower cholesterol and saturated fatty acid intake) [35]; however, no previous data on specific subgroup of fats (i.e., omega-6, omega-3, etc.) or protein (plant protein, animal protein) has been reported for the Italian population. When comparing our data to global consumption of fat, we reported lower intake of dietary cholesterol

(187 mg/day vs. 228 mg/day), higher of seafood omega-3 (0.53 g/day vs. 0.16 g/day) and similar of plant omega-3 (1.17 g/day vs. 1.37 g/day) [36]. Comparative data on type of protein is harder to retrieve. By roughly converting our estimates as percentage of total energy (%E), we may consider that the population investigated in this study consumed an average 5%E of animal protein (not including dairy protein) and about 9%E of plant protein: cohort studies conducted in the United States reported animal and plant protein intake of about 14% and 6%, respectively [15]; another Australian cohort reported slightly lower median intake of animal protein (about 10% of total energy) and similar of plant protein (about 6.5%) [37,38]. Thus, despite that studies to compare our reports to are scarce, we found a pattern of protein source intake healthier than in the aforementioned countries. These data on macronutrients, together with the aforementioned findings on major food groups, reflects the other findings on adherence to dietary recommendations. Various studies across the globe have reported an overall poor adherence to dietary guidelines of adult populations. Recent reports showed that diet quality of Americans, measured as agreement with dietary recommendations listed in the Healthy Eating Index (HEI), were far from optimal, regardless of socioeconomic status and race. Similarly, comparable trends have been observed in European countries. In Spain, there is a general low adherence to dietary guidelines, and these trends are particularly evident in individuals with overweight and obesity [39]. Nutrition surveys from France [40] and Germany [41] reported that consumption of fruit and vegetable does not meet dietary recommendations: similar findings were showed in other studies, where only about 30% of people living in United Kingdom [42] and 10% in Italy [43] reported eating the recommended five portions of fruits and vegetables per day. A report from Eastern European countries showed that roughly half individuals met WHO criteria for fruit and vegetable consumption, but only a minority met those for pulses and nut consumption [44]. By contrast, we found that half to two-thirds of the participants in our cohort met dietary guidelines on macronutrient and food group consumption, with the exception for total fats. However, despite that most of the individuals were under or, most likely, over the recommended intake, the results are not necessarily alarming, as we reported a higher intake of healthy rather than unhealthy fats. It has been shown that food sources of fat, such as olive oil, fish, and nuts, are associated with positive outcomes for health and a general recommendation in limiting total dietary fats may not entirely reflect a proper advice [45–47].

In this study, we found a correlation between certain macronutrients and food groups with BMI levels of the participants. Mostly in line with expectations, among dietary proteins, only plant protein intake was inversely correlated with BMI, while among dietary fats, monounsaturated and omega-6 fatty acids were inversely correlated, whereas trans fatty acids and cholesterol were directly correlated. However, these results did not entirely fit with correlations obtained with major food groups, as processed and unprocessed meats were indirectly and directly correlated with BMI levels, respectively. A possible reason for such unexpected findings may be the relative good quality of processed meat in southern Italy, which according the results of individual questions of the FFQ, we found it mostly referred to cured meat rather than fast foods (data not shown). Another explanation is the relatively low magnitude of the correlation for protein and meat products, which in fact might be spurious. Regarding the findings on dietary fats, we hypothesize that a major contributor to monounsaturated fatty acid intake was olive oil, highly consumed in this cohort as reported in previous studies [48]. General high levels of adherence to the Mediterranean diet has been previously shown in this cohort, as well as the association with lower likelihood of being obese and other metabolic conditions for those participants highly adherent to this dietary pattern; however, the association was not driven by olive oil or any other of the components of the score [49–51]. These findings corroborate the results of several other studies and suggest that the overall dietary pattern was more descriptive for a healthy nutritional alternative associated with better metabolic health [52–54]. Possible mediating factors have been hypothesized to be dietary polyphenols, which have been reported to exert potential beneficial effects on health [55,56]. With special regards to metabolic outcomes, dietary polyphenols have been shown to mediate, at least in part, the observed association with better metabolic health in this cohort [57–59]. Further studies are needed to investigate whether such compounds may explain, from a mechanistic point of view, the beneficial effects of healthy dietary pattern rich in fruit and vegetable, and other features typical of the Mediterranean diet.

The results presented in this study should be considered in light of methodological limitations. The use of FFQs is a widely-consolidated methodology, but they are also known to only provide estimates and not true intake, as they are subject to recall bias and over- and underestimation, depending on the number of food items included and social desirability bias, respectively. However, comparative reports used similar methodology and results are generally in line with literature and expected findings.

5. Conclusions

In conclusion, the present study provided updated information on macronutrient and major food group intake in a southern Italian adult population, taking into account specific subgroup of macronutrients rarely reported in current literature. The overall findings suggest that relatively healthy dietary habits are common in southern Italy, in up to two-thirds of the sample investigated. Further in-depth studies are needed to better understand whether findings related to foods may translate in equally adequate micronutrient intake in this cohort. However, further efforts should be made to improve diet quality of the remaining population in order to prevent non-communicable diseases.

Author Contributions: M.M. conceived and designed the experiments; A.P. performed the experiments; M.F., A.C., D.C. and I.B. analyzed the data; S.M. and M.M. and M.D'U. wrote the paper.

Conflicts of Interest: The authors declare no conflict of interest.

References

1. Lim, S.S.; Vos, T.; Flaxman, A.D.; Danaei, G.; Shibuya, K.; Adair-Rohani, H.; Amann, M.; Anderson, H.R.; Andrews, K.G.; Aryee, M.; et al. A comparative risk assessment of burden of disease and injury attributable to 67 risk factors and risk factor clusters in 21 regions, 1990–2010: A systematic analysis for the global burden of disease study 2010. *Lancet* **2012**, *380*, 2224–2260. [CrossRef]

2. Grosso, G.; Bella, F.; Godos, J.; Sciacca, S.; Del Rio, D.; Ray, S.; Galvano, F.; Giovannucci, E.L. Possible role of diet in cancer: Systematic review and multiple meta-analyses of dietary patterns, lifestyle factors, and cancer risk. *Nutr. Rev.* **2017**, *75*, 405–419. [CrossRef] [PubMed]

3. Mozaffarian, D. Dietary and policy priorities for cardiovascular disease, diabetes, and obesity: A comprehensive review. *Circulation* **2016**, *133*, 187–225. [CrossRef] [PubMed]

4. Khatibzadeh, S.; Saheb Kashaf, M.; Micha, R.; Fahimi, S.; Shi, P.; Elmadfa, I.; Kalantarian, S.; Wirojratana, P.; Ezzati, M.; Powles, J.; et al. A global database of food and nutrient consumption. *Bull. World Health Organ.* **2016**, *94*, 931–934. [CrossRef] [PubMed]

5. Morgan, P.J. Back to the future: The changing frontiers of nutrition research and its relationship to policy. *Proc. Nutr. Soc.* **2012**, *71*, 190–197. [CrossRef] [PubMed]

6. Mozaffarian, D.; Ludwig, D.S. Dietary guidelines in the 21st century—A time for food. *JAMA* **2010**, *304*, 681–682. [CrossRef] [PubMed]

7. Ezzati, M.; Riboli, E. Behavioral and dietary risk factors for noncommunicable diseases. *N. Engl. J. Med.* **2013**, *369*, 954–964. [CrossRef] [PubMed]

8. Nishida, C.; Uauy, R.; Kumanyika, S.; Shetty, P. The joint WHO/FAO expert consultation on diet, nutrition and the prevention of chronic diseases: Process, product and policy implications. *Public Health Nutr.* **2004**, *7*, 245–250. [CrossRef] [PubMed]

9. Mozaffarian, D. Foods, obesity, and diabetes-are all calories created equal? *Nutr. Rev.* **2017**, *75*, 19–31. [CrossRef] [PubMed]

10. Jebb, S.A. Carbohydrates and obesity: From evidence to policy in the UK. *Proc. Nutr. Soc.* **2015**, *74*, 215–220. [CrossRef] [PubMed]

11. Santesso, N.; Akl, E.A.; Bianchi, M.; Mente, A.; Mustafa, R.; Heels-Ansdell, D.; Schunemann, H.J. Effects of higher-versus lower-protein diets on health outcomes: A systematic review and meta-analysis. *Eur. J. Clin. Nutr.* **2012**, *66*, 780–788. [CrossRef] [PubMed]

12. Wycherley, T.P.; Moran, L.J.; Clifton, P.M.; Noakes, M.; Brinkworth, G.D. Effects of energy-restricted high-protein, low-fat compared with standard-protein, low-fat diets: A meta-analysis of randomized controlled trials. *Am. J. Clin. Nutr.* **2012**, *96*, 1281–1298. [CrossRef] [PubMed]

13. Tian, S.; Xu, Q.; Jiang, R.; Han, T.; Sun, C.; Na, L. Dietary protein consumption and the risk of type 2 diabetes: A systematic review and meta-analysis of cohort studies. *Nutrients* **2017**, *9*, 982. [CrossRef] [PubMed]

14. Farvid, M.S.; Malekshah, A.F.; Pourshams, A.; Poustchi, H.; Sepanlou, S.G.; Sharafkhah, M.; Khoshnia, M.; Farvid, M.; Abnet, C.C.; Kamangar, F.; et al. Dietary protein sources and all-cause and cause-specific mortality: The golestan cohort study in Iran. *Am. J. Prev. Med.* **2017**, *52*, 237–248. [CrossRef] [PubMed]

15. Song, M.; Fung, T.T.; Hu, F.B.; Willett, W.C.; Longo, V.D.; Chan, A.T.; Giovannucci, E.L. Association of animal and plant protein intake with all-cause and cause-specific mortality. *JAMA Intern. Med.* **2016**, *176*, 1453–1463. [CrossRef] [PubMed]

16. Dalen, J.E.; Devries, S. Diets to prevent coronary heart disease 1957–2013: What have we learned? *Am. J. Med.* **2014**, *127*, 364–369. [CrossRef] [PubMed]

17. Bloomfield, H.E.; Koeller, E.; Greer, N.; MacDonald, R.; Kane, R.; Wilt, T.J. Effects on health outcomes of a mediterranean diet with no restriction on fat intake: A systematic review and meta-analysis. *Ann. Intern. Med.* **2016**, *165*, 491–500. [CrossRef] [PubMed]

18. Grosso, G.; Marventano, S.; Yang, J.; Micek, A.; Pajak, A.; Scalfi, L.; Galvano, F.; Kales, S.N. A comprehensive meta-analysis on evidence of mediterranean diet and cardiovascular disease: Are individual components equal? *Crit. Rev. Food Sci. Nutr.* **2017**, *57*, 3218–3232. [CrossRef] [PubMed]

19. Liu, A.G.; Ford, N.A.; Hu, F.B.; Zelman, K.M.; Mozaffarian, D.; Kris-Etherton, P.M. A healthy approach to dietary fats: Understanding the science and taking action to reduce consumer confusion. *Nutr. J.* **2017**, *16*, 53. [CrossRef] [PubMed]

20. Marventano, S.; Kolacz, P.; Castellano, S.; Galvano, F.; Buscemi, S.; Mistretta, A.; Grosso, G. A review of recent evidence in human studies of n-3 and n-6 PUFA intake on cardiovascular disease, cancer, and depressive disorders: Does the ratio really matter? *Int. J. Food Sci. Nutr.* **2015**, *66*, 611–622. [CrossRef] [PubMed]

21. Grosso, G.; Marventano, S.; D'Urso, M.; Mistretta, A.; Galvano, F. The mediterranean healthy eating, ageing, and lifestyle (MEAL) study: Rationale and study design. *Int. J. Food Sci. Nutr.* **2017**, *68*, 577–586. [CrossRef] [PubMed]

22. Mistretta, A.; Marventano, S.; Platania, A.; Godos, J.; Galvano, F.; Grosso, G. Metabolic profile of the mediterranean healthy eating, lifestyle and aging (MEAL) study cohort. *Mediterr. J. Nutr. Metab.* **2017**, *10*, 131–140. [CrossRef]

23. World Health Organization. *Obesity: Preventing and Managing the Global Epidemic*; Report of a Who Consultation Presented at the World Health Organization, 3–5 June 1997; World Health Organization: Geneva, Switzerland, 1997.

24. Buscemi, S.; Rosafio, G.; Vasto, S.; Massenti, F.M.; Grosso, G.; Galvano, F.; Rini, N.; Barile, A.M.; Maniaci, V.; Cosentino, L.; et al. Validation of a food frequency questionnaire for use in Italian adults living in sicily. *Int. J. Food Sci. Nutr.* **2015**, *66*, 426–438. [CrossRef] [PubMed]

25. Marventano, S.; Mistretta, A.; Platania, A.; Galvano, F.; Grosso, G. Reliability and relative validity of a food frequency questionnaire for Italian adults living in Sicily, Southern Italy. *Int. J. Food Sci. Nutr.* **2016**, *67*, 857–864. [CrossRef] [PubMed]

26. Istituto Nazionale di Ricerca per gli Alimenti e la Nutrizione. *Tabelle di Composizione Degli Alimenti*; Istituto Nazionale di Ricerca per gli Alimenti e la Nutrizione: Rome, Italy, 2009.

27. Willett, W.C.; Lenart, E. Reproducibility and validity of food frequency questionnaire. In *Nutritional Epidemiology*, 2nd ed.; Oxford University Press: Oxford, UK, 1998.

28. EFSA. Dietary Reference Values for Nutrients. Available online: https://www.efsa.europa.eu/sites/default/files/2017_09_DRVs_summary_report.pdf (accessed on 18 March 2018).

29. Società Italiana di Nutrizione Umana (SINU). Iv Revisione dei Livelli di Assunzione di Riferimento di Nutrienti ed Energia per la Popolazione Italiana (LARN). Available online: http://www.sinu.it/html/pag/tabelle_larn_2014_rev.asp (accessed on 18 Marh 2018).

30. WHO. Diet, Nutrition and the Prevention of Chronic Diseases. Available online: http://apps.who.int/iris/bitstream/handle/10665/42665/WHO_TRS_916.pdf;jsessionid= B7E43124EB39AE47E19C06B0A876C0FD?sequence=1 (accessed on 18 March 2018).

31. Micha, R.; Khatibzadeh, S.; Shi, P.; Andrews, K.G.; Engell, R.E.; Mozaffarian, D.; Global Burden of Diseases Nutrition; Chronic Diseases Expert Group. Global, regional and national consumption of major food groups in 1990 and 2010: A systematic analysis including 266 country-specific nutrition surveys worldwide. *BMJ Open* **2015**, *5*, e008705. [CrossRef] [PubMed]

32. Pounis, G.; Bonanni, A.; Ruggiero, E.; Di Castelnuovo, A.; Costanzo, S.; Persichillo, M.; Bonaccio, M.; Cerletti, C.; Riccardi, G.; Donati, M.B.; et al. Food group consumption in an Italian population using the updated food classification system FoodEx2: Results from the Italian Nutrition & Health Survey (INHES) study. *Nutr. Metab. Cardiovasc. Dis.* **2017**, *27*, 307–328. [CrossRef] [PubMed]

33. Leclercq, C.; Arcella, D.; Piccinelli, R.; Sette, S.; Le Donne, C.; Turrini, A.; INRAN-SCAI 2005–06 Study Group. The Italian national food consumption survey INRAN-SCAI 2005–06: Main results in terms of food consumption. *Public Health Nutr.* **2009**, *12*, 2504–2532. [CrossRef] [PubMed]

34. EUROSTAT. Comparative Price Levels for Food, Beverages and Tobacco. Available online: http://ec.europa.eu/eurostat/statistics-explained/index.php/Comparative_price_levels_for_food,_beverages_ and_tobacco-Bread_and_cereals.2C_meat.2C_fish_and_dairy_products (accessed on 26 January 2018).

35. Sette, S.; Le Donne, C.; Piccinelli, R.; Arcella, D.; Turrini, A.; Leclercq, C.; Group, I.-S.S. The third Italian national food consumption survey, INRAN-SCAI 2005–06—Part 1: Nutrient intakes in Italy. *Nutr. Metab. Cardiovasc. Dis.* **2011**, *21*, 922–932. [CrossRef] [PubMed]

36. Micha, R.; Khatibzadeh, S.; Shi, P.; Fahimi, S.; Lim, S.; Andrews, K.G.; Engell, R.E.; Powles, J.; Ezzati, M.; Mozaffarian, D.; et al. Global, regional, and national consumption levels of dietary fats and oils in 1990 and 2010: A systematic analysis including 266 country-specific nutrition surveys. *BMJ* **2014**, *348*, g2272. [CrossRef] [PubMed]

37. Shang, X.; Scott, D.; Hodge, A.; English, D.R.; Giles, G.G.; Ebeling, P.R.; Sanders, K.M. Dietary protein from different food sources, incident metabolic syndrome and changes in its components: An 11-year longitudinal study in healthy community-dwelling adults. *Clin. Nutr.* **2017**, *36*, 1540–1548. [CrossRef] [PubMed]

38. Shang, X.; Scott, D.; Hodge, A.M.; English, D.R.; Giles, G.G.; Ebeling, P.R.; Sanders, K.M. Dietary protein intake and risk of type 2 diabetes: Results from the melbourne collaborative cohort study and a meta-analysis of prospective studies. *Am. J. Clin. Nutr.* **2016**, *104*, 1352–1365. [CrossRef] [PubMed]

39. Rodriguez-Rodriguez, E.; Aparicio, A.; Aranceta-Bartrina, J.; Gil, A.; Gonzalez-Gross, M.; Serra-Majem, L.; Varela-Moreiras, G.; Ortega, R.M. Low adherence to dietary guidelines in Spain, especially in the overweight/obese population: The ANIBES study. *J. Am. Coll. Nutr.* **2017**, *36*, 240–247. [CrossRef] [PubMed]

40. Castetbon, K.; Vernay, M.; Malon, A.; Salanave, B.; Deschamps, V.; Roudier, C.; Oleko, A.; Szego, E.; Hercberg, S. Dietary intake, physical activity and nutritional status in adults: The French nutrition and health survey (ENNS, 2006–2007). *Br. J. Nutr.* **2009**, *102*, 733–743. [CrossRef] [PubMed]

41. Gose, M.; Krems, C.; Heuer, T.; Hoffmann, I. Trends in food consumption and nutrient intake in Germany between 2006 and 2012: Results of the german national nutrition monitoring (NEMONIT). *Br. J. Nutr.* **2016**, *115*, 1498–1507. [CrossRef] [PubMed]

42. Whitton, C.; Nicholson, S.K.; Roberts, C.; Prynne, C.J.; Pot, G.K.; Olson, A.; Fitt, E.; Cole, D.; Teucher, B.; Bates, B.; et al. National diet and nutrition survey: UK food consumption and nutrient intakes from the first year of the rolling programme and comparisons with previous surveys. *Br. J. Nutr.* **2011**, *106*, 1899–1914. [CrossRef] [PubMed]

43. Istituto Superiore di Sanità. Rapporto Nazionale Passi 2012: Consumo di Frutta e Verdura. Available online: http://www.epicentro.iss.it/passi/rapporto2012/5ADay.asp (accessed on 23 February 2018).

44. Boylan, S.; Welch, A.; Pikhart, H.; Malyutina, S.; Pajak, A.; Kubinova, R.; Bragina, O.; Simonova, G.; Stepaniak, U.; Gilis-Januszewska, A.; et al. Dietary habits in three central and eastern european countries: The hapiee study. *BMC Public Health* **2009**, *9*, 439. [CrossRef] [PubMed]

45. Grosso, G.; Estruch, R. Nut consumption and age-related disease. *Maturitas* **2016**, *84*, 11–16. [CrossRef] [PubMed]

46. Lopez-Miranda, J.; Perez-Jimenez, F.; Ros, E.; De Caterina, R.; Badimon, L.; Covas, M.I.; Escrich, E.; Ordovas, J.M.; Soriguer, F.; Abia, R.; et al. Olive oil and health: Summary of the II international conference on olive oil and health consensus report, Jaen and Cordoba (Spain) 2008. *Nutr. Metab. Cardiovasc. Dis.* **2010**, *20*, 284–294. [CrossRef] [PubMed]

47. Mozaffarian, D.; Rimm, E.B. Fish intake, contaminants, and human health: Evaluating the risks and the benefits. *JAMA* **2006**, *296*, 1885–1899. [CrossRef] [PubMed]

48. Marventano, S.; Godos, J.; Platania, A.; Galvano, F.; Mistretta, A.; Grosso, G. Mediterranean diet adherence in the mediterranean healthy eating, aging and lifestyle (MEAL) study cohort. *Int. J. Food Sci. Nutr.* **2018**, *69*, 100–107. [CrossRef] [PubMed]

49. Zappala, G.; Buscemi, S.; Mule, S.; La Verde, M.; D'Urso, M.; Corleo, D.; Marranzano, M. High adherence to mediterranean diet, but not individual foods or nutrients, is associated with lower likelihood of being obese in a mediterranean cohort. *Eat. Weight Disord.* **2017**, 1–10. [CrossRef] [PubMed]

50. La Verde, M.; Mule, S.; Zappala, G.; Privitera, G.; Maugeri, G.; Pecora, F.; Marranzano, M. Higher adherence to the mediterranean diet is inversely associated with having hypertension: Is low salt intake a mediating factor? *Int. J. Food Sci. Nutr.* **2018**, *69*, 235–244. [CrossRef] [PubMed]

51. Platania, A.; Zappala, G.; Mirabella, M.U.; Gullo, C.; Mellini, G.; Beneventano, G.; Maugeri, G.; Marranzano, M. Association between mediterranean diet adherence and dyslipidaemia in a cohort of adults living in the Mediterranean area. *Int. J. Food Sci. Nutr.* **2017**, 1–11. [CrossRef] [PubMed]

52. Mancini, J.G.; Filion, K.B.; Atallah, R.; Eisenberg, M.J. Systematic review of the mediterranean diet for long-term weight loss. *Am. J. Med.* **2016**, *129*, 407–415.e4. [CrossRef] [PubMed]

53. Grosso, G.; Mistretta, A.; Marventano, S.; Purrello, A.; Vitaglione, P.; Calabrese, G.; Drago, F.; Galvano, F. Beneficial effects of the mediterranean diet on metabolic syndrome. *Curr. Pharm. Des.* **2014**, *20*, 5039–5044. [CrossRef] [PubMed]

54. Godos, J.; Zappala, G.; Bernardini, S.; Giambini, I.; Bes-Rastrollo, M.; Martinez-Gonzalez, M. Adherence to the mediterranean diet is inversely associated with metabolic syndrome occurrence: A meta-analysis of observational studies. *Int. J. Food Sci. Nutr.* **2017**, *68*, 138–148. [CrossRef] [PubMed]

55. Grosso, G.; Godos, J.; Lamuela-Raventos, R.; Ray, S.; Micek, A.; Pajak, A.; Sciacca, S.; D'Orazio, N.; Del Rio, D.; Galvano, F. A comprehensive meta-analysis on dietary flavonoid and lignan intake and cancer risk: Level of evidence and limitations. *Mol. Nutr. Food Res.* **2017**, *61*. [CrossRef] [PubMed]

56. Grosso, G.; Micek, A.; Godos, J.; Pajak, A.; Sciacca, S.; Galvano, F.; Giovannucci, E.L. Dietary flavonoid and lignan intake and mortality in prospective cohort studies: Systematic review and dose-response meta-analysis. *Am. J. Epidemiol.* **2017**, *185*, 1304–1316. [CrossRef] [PubMed]

57. Godos, J.; Bergante, S.; Satriano, A.; Pluchinotta, F.R.; Marranzano, M. Dietary phytoestrogen intake is inversely associated with hypertension in a cohort of adults living in the Mediterranean area. *Molecules* **2018**, *23*, 368. [CrossRef] [PubMed]

58. Godos, J.; Sinatra, D.; Blanco, I.; Mule, S.; La Verde, M.; Marranzano, M. Association between dietary phenolic acids and hypertension in a mediterranean cohort. *Nutrients* **2017**, *9*, 69. [CrossRef]

59. Marranzano, M.; Ray, S.; Godos, J.; Galvano, F. Association between dietary flavonoids intake and obesity in a cohort of adults living in the Mediterranean area. *Int. J. Food Sci. Nutr.* **2018**, 1–10. [CrossRef] [PubMed]

antioxidants

MDPI

Article

Fluid Intake and Beverage Consumption Description and Their Association with Dietary Vitamins and Antioxidant Compounds in Italian Adults from the Mediterranean Healthy Eating, Aging and Lifestyles (MEAL) Study

Armando Platania [1], Dora Castiglione [1], Dario Sinatra [1], Maurizio D' Urso [2] and Marina Marranzano [1,*]

[1] Department of Medical and Surgical Sciences and Advanced Technologies "G.F. Ingrassia", University of Catania, 95123 Catania, Italy; armplt@hotmail.it (A.P.); doracastiglione29@gmail.com (D.C.); sinatradario@gmail.com (D.S.)

[2] Provincial Health Authority of Catania, 95127 Catania, Italy; ma.durso76@gmail.com

* Correspondence: marranz@unict.it; Tel.: +39-095-378-2180

Received: 28 February 2018; Accepted: 4 April 2018; Published: 9 April 2018

Abstract: The aim of the present study was to investigate the total water intake (TWI) from drinks and foods and to evaluate the correlation between the different types of drinks on energy and antioxidant intake. The cohort comprised 1602 individuals from the city of Catania in Southern Italy. A food frequency questionnaire was administered to assess dietary and water intake. The mean total water intake was 2.7 L; more than about two thirds of the sample met the European recommendations for water intake. Water and espresso coffee were the most consumed drinks. Alcohol beverages contributed about 3.0% of total energy intake, and sugar sweetened beverages contributed about 1.4%. All antioxidant vitamins were significantly correlated with TWI. However, a higher correlation was found for water from food rather than water from beverages, suggesting that major food contributors to antioxidant vitamin intake might be fruits and vegetables, rather than beverages other than water. A mild correlation was found between fruit juices and vitamin C; coffee, tea and alcohol, and niacin and polyphenols; and milk and vitamin B12. The findings from the present study show that our sample population has an adequate intake of TWI and that there is a healthy association between beverages and dietary antioxidants.

Keywords: beverages; antioxidants; water; coffee; alcohol

1. Introduction

Water comprises from 75% body weight in infants to 55% in the elderly and is an essential nutrient that is involved in all body functions—cellular homeostasis, thermoregulation and chemical reactions to name just a few. Hydration status is regulated by a complex sensitive network of physiological controls, which aims to balance water intake and loss [1]. Proper hydration status is essential in order to preserve physical and mental functions; as a matter of fact, even slight dehydration can alter these two functions [2,3]. The importance of an adequate water intake has long been recognized, and guidelines have been established by the European Food Safety Authority (EFSA). Total water intake from all sources for men and women should be about 2.5 L and 2.0 L, respectively [4]. However, it must be stated that water requirements can change according to climatic state, salt intake and physical activity as well as health status, metabolism and age. However, it has been reported that the percentage of the population with an inadequate water intake may vary from 5% to 35% among European countries [5].

Fluid intake in all age groups depends not only on water, but also on a variety of beverages, such as tea, coffee, milk, sugar-sweetened soft drinks, juice and alcoholic drinks, that may contribute, to a various extent, to the level of hydration of an individual. The percentage of water intake in a typical Western diet has been estimated to be nearly 20–30% from food and 70–80% from fluids (including water and the aforementioned beverages) [5]. Moreover, there is plenty of scientific literature regarding the investigation of the role of specific beverages on health conditions, for instance, the adverse impact of sugar-sweetened soft drinks on metabolic disorders [6–8], the potential beneficial effects of milk on cardiovascular diseases [9,10], the positive impact on cardio-metabolic outcomes and decreased risk of certain cancers associated with coffee [11,12] and tea consumption [13–15], and the increased risk of cancer related to excessive alcohol drinking [16,17]. In addition, understanding the pattern of water contribution from food and beverages as well as beverages is essential for estimating individual or population-level adherence to recommended water intakes and the appropriateness of beverage habits. In fact, it is desirable to develop specific recommendations that take into account the hydration capability and potential effects related to energy and antioxidant content of beverages [18]. A previous report [19] attempted to summarize current knowledge regarding the relationship between water intake and human health, but the authors emphasized the scarcity of scientific literature on this matter, especially with respect to how rarely fluid intake is collected in epidemiological studies. The aim of the present study was to quantify the total water and beverage intake in a cohort of Southern Italian adults in order to test their compliance with the international guidelines of the European Food Safety Authority (EFSA). Correlations between beverages, energy and dietary antioxidant intake were further explored.

2. Materials and Methods

2.1. Study Design and Population

This study was a cross-sectional investigation of fluid intake and beverage patterns in the Mediterranean healthy eating, ageing, and lifestyle (MEAL) cohort. The total sample consisted of 2044 men and women, aged 18 to >70 years old, living in the city of Catania in Southern Italy. The sample was stratified by gender and 10-year age groups. The survey was conducted between the years 2014 and 2015; subjects and their clinical information were extracted randomly from a pool made up of general practitioners' databases. More detailed information on the study protocol has been published elsewhere [20]. The aim of the study was explained to all participants, and a written informed consent was filled in by the subjects involved. The conduction of the study and all of its procedures were performed in accordance with the Declaration of Helsinki (1989) of the World Medical Association. The study protocol has been approved by the ethical committee of the University of Catania (protocol number: 802/23 December 2014).

2.2. Data Collection

The survey consisted of a face-to-face computer-assisted personal interview with tablet computers. Participants were also provided with a paper copy of the survey to read the response options, and the interviewer filled in the digital copy of the survey on the tablet computer with the final answers from the interviewee. A section of the survey collected demographic data such as gender, age at recruitment, educational status, occupation (the main source of employment during the year before the survey was collected in more detail) or last occupation before retirement, and marital status. Participants were categorized into 3 age groups: (i) 18–35 years; (ii) 35–50 years; and (iii) >50 years. Educational status was categorized in 3 groups: (i) low (primary/secondary); (ii) medium (high school); and (iii) high (university). Occupational status was categorized in 4 groups: (i) unemployed; (ii) low (unskilled workers); (iii) medium (partially skilled workers); and (iv) high (skilled workers). Physical activity status was assessed through the International Physical Activity Questionnaires (IPAQ) [21]. This is a set of questionnaires (5 domains) collecting information about the time spent practicing physical activity in the last 7 days; following the IPAQ guidelines, final scores were used to categorize physical activity level in 3 groups: (i) low; (ii) moderate; and (iii) high. Smoking status was categorized in 3 groups: (i) non-smoker; (ii) ex-smoker; and (iii) current smoker. Alcohol consumption was categorized in 3 groups: (i) none; (ii) moderate drinker (0.1–12 g/day) and (iii) regular drinker (>12 g/day). Anthropometric measurements were collected using standardized methods [22]. Height was measured to the nearest 0.5 cm without shoes, with the back square against the wall tape, eyes looking straight ahead, and a right-angle triangle resting on the scalp and against the wall. Body mass index (BMI) was calculated, and patients were categorized as under/normal weight (BMI < 25 kg/m^2), overweight (BMI 25 to 29.9 kg/m^2), and obese (BMI > 30 kg/m^2) [23].

2.3. Dietary Assessment

Two food frequency questionnaires (a long and a short version) were administered to assess dietary and water intakes. These questionnaires have been previously tested for validity and reliability in the Sicilian population [24,25]. Beverages were combined into eight categories for further analysis: (i) alcohol beverages; (ii) sugar-sweetened beverages (including sugar-sweetened fruit juices); (iii) tea; (iv) coffee; (v) fruit juices (only 100% fruit juices); (vi) milk (all types of milk without separation by fat percentage); (vii) soy milk; (viii) water (including tap water and bottled water). Seasonal food consumption was registered during the period when the food was available; then it was adjusted by its proportional intake in one year. Food composition tables from the Research Center for Foods and Nutrition (CREA; *Consiglio per la ricerca in agricoltura e l'analisi dell'economia agraria*) [26] were used to assess the water content of food and beverages as well as energy and nutrient intakes. Total polyphenol and phytoestrogen (isoflavones and lignans) intake was calculated through a comparison with the Phenol-Explorer database (www.phenol-explorer.eu) [27]. The process of identification and calculation of the polyphenol content in foods is described in detail elsewhere [28]. Finally, intakes were adjusted for total energy intake (kcal/day) using the residual method in line with the Global Nutrition and Policy Consortium guidelines [29]. FFQs (Food Frequency Questionnaires) with unreliable intakes (<1000 or >6000 kcal/day) (*n* = 107) as well as missing items for the purposes of this study (*n* = 335) were excluded from the analyses, leaving a total of 1602 individuals included in the analysis. Total water intake (TWI) was compared with the EFSA Dietary Reference Values (DRV) regarding the Adequate Intake (AI) of water for men and women (2.5 L and 2.0 L). However, Nordic countries consider inadequate to be an intake of less than 1 g of water per 1 kcal of energy. For this reason, a combined classification was used to provide a more comprehensive evaluation of water intake: a classification based on the EFSA AI value (criterion 1), a ratio between water intake in g and energy intake in kcal (criterion 2), and a combination of criterion 1 and criterion 2 (criterion 3).

2.4. Statistical Analysis

Absolute numbers and percentages were used to show all frequencies, while means and standard errors were used to show continuous variables. Chi-square tests were used to compare categorical variables between groups, Student's *t*-test and ANOVA were used to compare continuous variables distributed normally, and the Mann–Whitney *U*-test and Kruskall–Wallis test were used to compare variables that were not normally distributed. An analysis of partial correlations, adjusted for age, gender, body weight, and physical activity, was performed between water intake, energy intake, beverage consumption, vitamins and polyphenols. All analyses were based on two-sided tests and significance was set to a *p*-value of 0.05. SPSS 17 (SPSS Inc., Chicago, IL, USA) software was used for all the statistical computations.

3. Results

Our sample comprised 1602 participants, ranging from 18 to >70 years old, with more women than men (Table 1). One quarter were current smokers, with men being more physically active than women. On average, BMI levels suggested a certain degree of overweight individuals in the sample, though women had slightly lower BMI levels. When considering BMI in categories, there was a higher prevalence of overweight men than women, while the opposite was observed for obesity prevalence.

Table 1. Descriptive statistics of the Mediterranean healthy eating, ageing, and lifestyle (MEAL) sample.

	Male (*n* = 669) (41.8% column)	Female (*n* = 933) (58.2% column)	Total (*n* = 1602)
Age groups, *n* (%)			
18–35	210 (31.4)	260 (27.9)	470 (29.3)
35–50	199 (29.7)	273 (29.3)	472 (29.5)
>50	260 (38.9)	400 (42.9)	660 (41.2)
Education, *n* (%)			
Elementary or Middle school	157 (23.5)	314 (33.7)	471 (29.4)
Diploma	308 (46.0)	341 (36.5)	649 (40.5)
Graduate	204 (30.5)	278 (29.8)	482 (30.1)
Occupation, *n* (%)			
Unemployed	66 (11.7)	299 (37.8)	365 (26.9)
Low (unskilled workers)	96 (17.1)	122 (15.4)	218 (16.1)
Medium (partially skilled workers)	170 (30.2)	153 (19.3)	323 (23.8)
High (skilled workers)	231 (41.0)	218 (27.5)	449 (33.1)
Smoking, *n* (%)			
Yes	173 (25.9)	225 (24.1)	398 (24.8)
No	374 (55.9)	648 (69.5)	1022 (63.8)
Ex-smoker	122 (18.2)	60 (6.4)	182 (11.4)
Body Mass Index (BMI) (kg/m^2) mean (SE)	26.3 (3.7)	25.1 (4.9)	25.6 (4.5)
BMI category (kg/m^2) *n* (%)			
Underweight/Normal (<18.5–24.9)	237 (39.4)	486 (55.7)	723 (49.1)
Overweight (25–29.9)	280 (46.5)	231 (26.5)	511 (34.7)
Obese (>29.9)	85 (14.1)	155 (17.8)	240 (15.0)
Physical activity (scale) *n* (%)			
Low	76 (11.4)	201 (21.5)	277 (17.3)
Moderate	317 (47.4)	481 (51.6)	798 (49.8)
High	276 (41.3)	247 (26.5)	523 (32.6)
Total Water Intake (L) mean (SE)	2729 (25.4)	2682 (23.6)	2701 (17.4)

The frequency distribution of total water intake (TWI) showed an average of 2.7 L for both men and women (Figure 1).

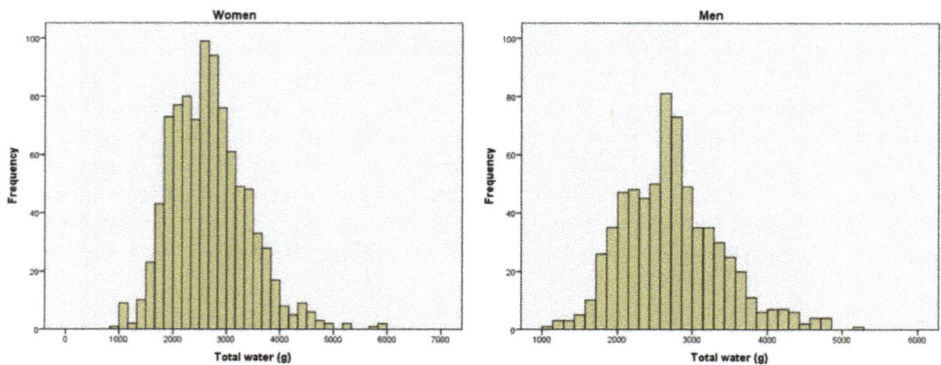

Figure 1. The frequency distribution of total water intake (TWI).

The combined classification including the EFSA AI recommendations was evaluated for both sexes (Table 2). Sixty-four percent of men and 83% of women met EFSA criterion 1 (total water intake >2.5 L in men, >2.0 L in women), while 88% of men and 86% of women met EFSA criterion 2 (water/energy intake >1).

Table 2. Combined classification for total water intake (TWI) following established criteria.

Criteria Classification	Men (*n* = 669)	Women (*n* = 933)
Criterion 1: (%)	63.7	82.7
Criterion 2: (%)	88.3	86.4
Criterion 3: (%)	59.8	75.1

EFSA: European Food Safety Authority. (1) Criterion 1: TWI >2.5 L men, >2 L women; (2) Criterion 2: ratio of total water intake and total energy >1; (3) Criterion 3: both criteria 1 and 2.

The major contribution to TWI was given by total beverages (73%), for which preference of consumption was assessed in both men and women (Figure 2). Water was consumed by all participants, followed by coffee (91% by both genders), alcohol beverages (85% by men, 79% by women) and sugar-sweetened beverages (83% by men, 75% by women). Tea and soy milk beverages were the less preferred, being consumed, respectively, by 35% of men and 30% of women (tea), and 12% of men and 15% of women (soy milk).

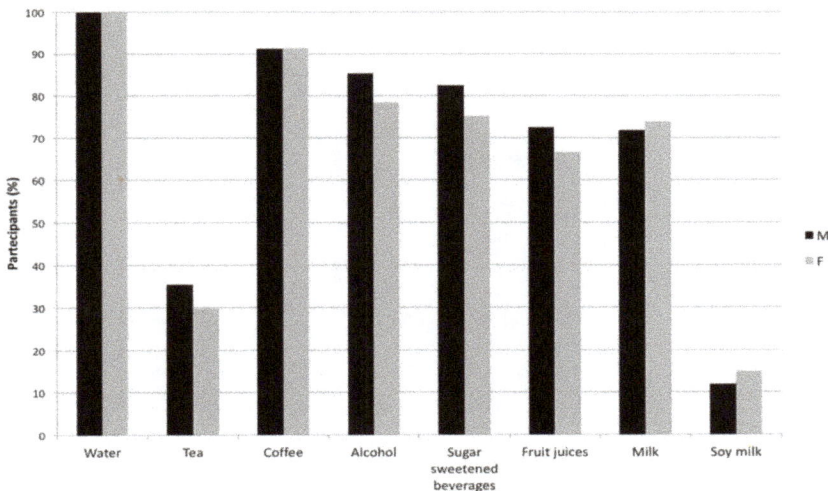

Figure 2. Prevalence of beverage consumption (% consumption by males and females).

Percentages of total weight (food and beverages) consumed (g/day), daily energy intake (kcal/day), and water intake (g/day) were calculated (Table 3). Regarding TWI, 73% came from beverages and 27% from food. These percentages met the EFSA recommendations regarding the percentage of water intake that should be provided by beverages (70–80%) and food (20–30%). Furthermore, water was the main source of TWI with a score of 55%. The mean total energy intake was 1951 kcal/day, with a relative contribution of 9% from beverages. Of these, alcohol beverages contributed 3.0% for men and 2.7% for women, while sugar-sweetened beverages contributed 1.4% for men and 1.3% for women. Little difference was found between the total weight percentage of alcohol beverages consumed by men and women (respectively, 3.5% and 3.0%).

Table 3. Contribution of food and beverages to total water and energy intake.

	Total Weight Consumed (g/day)			Contribution to Energy Intake (kcal/day)			Contribution to Water Intake (g/day)		
	Total	Men	Women	Total	Men	Women	Total	Men	Women
All food and drink, mean (SE)	3072.70 (19.76)	3102.36 (28.07)	3051.43 (27.31)	1950.53 (16.89)	1975.22 (24.81)	1932.83 (22.90)	2701.71 (17.35)	2729.08 (25.40)	2682.08 (23.57)
Food only, mean (SE)	36.0% (0.3%)	35.3% (0.4%)	36.5% (0.3%)	91.3% (0.1%)	91.4% (0.2%)	91.2% (0.2%)	27.4% (0.3%)	26.6% (0.4%)	28.0% (0.3%)
Beverages only, mean (SE)	64.0% (0.3%)	64.7% (0.4%)	63.5% (0.3%)	8.7% (0.1%)	8.6% (0.2%)	8.8% (0.2%)	72.6% (0.3%)	73.4% (0.4%)	72.0% (0.4%)
Alcohol beverages, mean (SE)	3.2% (0.1%)	3.5% (0.2%)	3.0% (0.2%)	2.8% (0.1%)	3.0% (0.1%)	2.7% (0.1%)	3.7% (0.1%)	4.0% (0.2%)	3.5% (0.2%)
Sugar-sweetened beverages, mean (SE)	2.1% (0.1%)	2.2% (0.1%)	2.0% (0.1%)	1.3% (0.1%)	1.4% (0.1%)	1.3% (0.1%)	2.5% (0.1%)	2.6% (0.2%)	2.4% (0.2%)
Tea, mean (SE)	2.4% (0.1%)	2.0% (0.2%)	2.7% (0.1%)	0.0% (0.0%)	0.0% (0.0%)	0.0% (0.0%)	2.7% (0.1%)	2.3% (0.2%)	3.0% (0.2%)
Coffee, mean (SE)	1.9% (0.0%)	1.9% (0.1%)	1.9% (0.0%)	0.3% (0.0%)	0.3% (0.0%)	0.3% (0.0%)	2.1% (0.0%)	2.2% (0.1%)	2.1% (0.1%)
Fruit juices, mean (SE)	1.1% (0.0%)	1.2% (0.1%)	1.1% (0.1%)	0.7% (0.0%)	0.8% (0.1%)	0.7% (0.0%)	1.3% (0.1%)	1.4% (0.1%)	1.2% (0.1%)
Milk, mean (SE)	4.3% (0.1%)	4.0% (0.2%)	4.5% (0.2%)	3.2% (0.1%)	2.9% (0.1%)	3.4% (0.1%)	4.9% (0.1%)	4.5% (0.2%)	5.2% (0.2%)
Soy milk, mean (SE)	0.5% (0.0%)	0.3% (0.0%)	0.6% (0.1%)	0.3% (0.0%)	0.2% (0.0%)	0.4% (0.0%)	0.5% (0.0%)	0.3% (0.0%)	0.7% (0.1%)
Water, mean (SE)	48.4% (0.3%)	49.5% (0.4%)	47.7% (0.4%)	/ /	/ /	/ /	54.9% (0.3%)	56.2% (0.5%)	54.0% (0.4%)

Water intake was also calculated separately by age group (Table 4).

Table 4. Total water intake and beverage consumption (g/day) by age group.

	Men				Women			
	Age Group			*p*	Age Group			*p*
	1	2	3		1	2	3	
Total water intake from food and beverages, mean (SE)	2804 53	2676 43	2709 37	0.12	2686 43	2739 44	2640 36	0.22
Water from food, mean (SE)	745 30	720 24	738 19	0.77	770 23	772 30	755 18	0.83
Water from beverages, mean (SE)	2059 40	1956 32	1971 29	0.07	1916 32	1967 32	1885 26	0.11
Alcoholic beverages, mean (SE)	112 9	91 7	127 12	0.05	93 8	78 7	98 8	0.20
Sugar-sweetened beverages, mean (SE)	84 9	65 8	66 7	0.18	76 9	71 8	55 6	0.13
Tea, mean (SE)	84 13	60 11	67 10	0.34	74 7	102 12	84 7	0.10
Coffee, mean (SE)	52 3	57 3	61 3	0.07	53 3	50 2	56 2	0.14
Fruit juices, mean (SE)	39 4	32 3	44 5	0.10	36 4	41 5	31 3	0.19
Milk, mean (SE)	126 10	117 11	114 9	0.68	129 10	144 11	132 8	0.52
Soy milk, mean (SE)	11 3	9 3	7 2	0.48	27 5	25 5	16 3	0.10
Water, mean (SE)	1552 31	1525 29	1485 26	0.22	1428 27	1456 27	1413 24	0.48

p-values were obtained through ANOVA tests.

The amount of water consumed in beverages did not change with age, while alcohol beverage consumption was higher in the elderly, followed by younger participants and then middle-aged ones. TWI correlated highly with the consumption of water from beverages ($r = 0.82$) and mildly with food weight ($r = 0.68$) (Table 5).

Table 5. Partial correlations between water intake, energy intake and beverage consumption.

	Total Water	Water from Beverages	Water from Food	Food Weight	Total Energy	Energy from Beverages	Energy from Food
Total water	1.00	0.82 **	0.66 **	0.68 **	0.53 **	0.37 **	0.49 **
Water from beverages	0.82 **	1.00	0.12 **	0.18 **	0.25 **	0.40 **	0.19 **
Water from food	0.66 **	0.12 **	1.00	0.96 **	0.60 **	0.11 **	0.61 **
Food weight	0.68 **	0.18 **	0.96 **	1.00	0.73 **	0.22 **	0.73 **
Total energy	0.53 **	0.25 **	0.60 **	0.73 **	1.00	0.39 **	0.98 **
Energy from beverages	0.37 **	0.40 **	0.11 **	0.22 **	0.39 **	1.00	0.22 **
Energy from food	0.49 **	0.19 **	0.61 **	0.73 **	0.98 **	0.22 **	1.00
Alcohol beverages	0.22 **	0.23 **	0.08 **	0.10 **	0.21 **	0.64 **	0.10 **
Sugar-sweetened beverages	0.20 **	0.24 **	0.05 *	0.20 **	0.35 **	0.47 **	0.28 **
Tea	0.34 **	0.23 **	0.16 **	0.15 **	0.03	0.06 *	0.02
Coffee	0.08 **	0.06 *	0.05 *	0.03	0.07 **	0.14 **	0.05 *
Fruit juices	0.29 **	0.21 **	0.22 **	0.27 **	0.18 **	0.28 **	0.14 **
Milk	0.12 **	0.20 **	−0.04	0.00	0.12 **	0.51 **	0.03
Soy milk	0.16 **	0.14 **	0.09 **	0.10 **	0.05 *	0.10 **	0.04
Water	0.61 **	0.81 **	0.02	0.02	0.03	−0.14 **	0.06 *

** Correlation is significant at the 0.01 level (bilateral); * Correlation is significant at the 0.05 level (bilateral).

Alcohol beverages, milk and sugar-sweetened beverages had the highest correlations with total energy from beverages ($r = 0.64$, $r = 0.51$, and $r = 0.47$, respectively), while the scores were low for the other beverages and near zero for water and tea. All antioxidant vitamins were significantly correlated with TWI (Table 6).

However, a higher correlation with TWI was found for water from food than for water from beverages, suggesting that the major food contributors to antioxidant vitamin intake might be fruits and vegetables, rather than beverages other than water. Another strong correlation was found between total phytoestrogen intake and soy milk intake ($r = 0.78$), suggesting that soy milk is a major contributor to phytoestrogen intake in this cohort. A mild correlation was found between fruit juices and vitamin C ($r = 0.42$), and tea and polyphenols ($r = 0.37$), and other correlations were found between coffee, and niacin ($r = 0.31$) and polyphenols ($r = 0.20$); milk and vitamin B12 ($r = 0.32$); and alcohol and polyphenols ($r = 0.31$).

Table 6. Partial correlations between water intake, energy intake, beverage consumption, vitamins and polyphenols.

Title	Total Water	Water from Beverages	Water from Food	Food Weight	Total Energy	Energy from Beverages	Energy from Food	Alcohol Beverages	Sugar-Sweetened Beverages	Tea	Coffee	Fruit Juices	Milk	Soy Milk	Water
Vitamin A	0.60 **	0.21 **	0.75 **	0.74 **	0.54 **	0.25 **	0.52 **	0.11 **	0.01	0.23 **	0.05	0.21 **	0.01	0.15 **	0.03
Vitamin B1	0.48 **	0.22 **	0.54 **	0.56 **	0.76 **	0.27 **	0.75 **	0.15 **	0.23 **	0.15 **	0.05	0.09 **	0.08 **	0.11 **	0.07 **
Vitamin B2	0.51 **	0.27 **	0.52 **	0.53 **	0.72 **	0.39 **	0.69 **	0.17 **	0.19 **	0.18 **	0.12 **	0.06 *	0.27 **	0.17 **	0.04
Niacin	0.54 **	0.27 **	0.60 **	0.69 **	0.77 **	0.25 **	0.76 **	0.12 **	0.16 **	0.11 **	0.31 **	0.10 **	0.00	0.14 **	0.10 **
Vitamin B6	0.66 **	0.29 **	0.73 **	0.82 **	0.83 **	0.30 **	0.82 **	0.08 **	0.12 **	0.25 **	0.02	0.17 **	0.05 *	0.18 **	0.08 **
Folate	0.59 **	0.23 **	0.73 **	0.79 **	0.69 **	0.25 **	0.68 **	0.12 **	0.06 *	0.20 **	0.06 *	0.23 **	0.01	0.21 **	0.06 *
Vitamin B12	0.35 **	0.21 **	0.35 **	0.41 **	0.44 **	0.14 **	0.44 **	0.06 *	0.27 **	0.11 **	−0.03	0.11 **	0.32 **	0.04	0.20 **
Vitamin C	0.62 **	0.18 **	0.84 **	0.85 **	0.50 **	0.29 **	0.47 **	0.12 **	0.06 *	0.18 **	0.02	0.42 **	0.00	0.13 **	−0.03
Vitamin D	0.31 **	0.17 **	0.32 **	0.36 **	0.35 **	0.04	0.37 **	−0.02	0.15 **	0.14 **	−0.05	0.14 **	0.06 *	0.10 **	0.20 **
Vitamin E	0.57 **	0.21 **	0.71 **	0.74 **	0.76 **	0.28 **	0.76 **	0.15 **	0.14 **	0.19 **	0.01	0.15 **	−0.05 *	0.19 **	0.02
Total polyphenols	0.46 **	0.24 **	0.43 **	0.44 **	0.42 **	0.32 **	0.39 **	0.31 **	−0.07 **	0.37 **	0.20 **	0.13 **	−0.11 **	0.14 **	0.04
Total phytoestrogens	0.34 **	0.14 **	0.40 **	0.39 **	0.38 **	0.05 *	−0.05 *	0.15	0.01	0.12 *	−0.05 *	0.11 **	−0.08 **	0.78 **	0.04

** Correlation is significant at the 0.01 level (bilateral); * Correlation is significant at the 0.05 level (bilateral).

4. Discussion

This study analysed the total water intake from both food and beverages among the MEAL cohort, a representative sample of Southern Italian adults. Our study is one of the few attempts to describe the consumption of water and beverages in Italy, the relationship between energy intake and beverages, as well as being a reference for the role of some antioxidant vitamins and polyphenols contained in beverages. The EFSA Scientific Opinion on Dietary Reference Values for Water states that the dietary reference intake values for water should include water from drinking water (tap or bottled), all kinds of beverages, and food moisture. The adequate total water intake suggested by EFSA guidelines (2 L for women and 2.5 L for men) was met by the majority of our sample population.

Comparisons with similar studies are not easy to perform due to the use of different populations with different dietary habits, as well as different tools administered. Nonetheless, it can be stated that participants from two previous studies on hydration (Spanish and Italian) did not reach the EFSA guidelines recommended intake for water, while participants in another one (English) did reach the recommended intake [30–32]. In our study, the main contributor to TWI was plain water. Our results are only partially in line with previous studies in which the main source of TWI was either food (Italy and Spain) or hot beverages (UK); this can be explained by the hotter seasonal temperatures in Southern Italy, which stimulate thirst induction, and the preference for plain water over sugar-sweetened beverages due to the cultural heritage of adults living in Southern Italy who are still adherent, at least in part, to traditional dietary patterns [33,34]. The main benefit from a hydration pattern which mainly focuses on plain water intake is the fulfillment of its purpose without adding caloric intake, while providing mineral micronutrients. In line with other studies [30–32], water intake from beverages decreased with age, underlining a common trend which sees the elderly drinking less. Evidence from studies conducted in Mediterranean populations shows that adherence to traditional dietary patterns (including beverage consumption) might be beneficial for body composition in young, adult and elderly individuals [35–37]. However, more studies are needed to evaluate whether this trend could be linked with a risk of dehydration in the elderly or if it is a harmless manifestation of the lower physiological water needs of the metabolism in late years.

The second most popular beverage was coffee, namely as espresso coffee, which was drunk by 91% of our sample, but still had a low contribution to TWI (2%) due to the small quantity contained in a serving. In our analysis, we found a statistically significant correlation between coffee and niacin and total polyphenols. Coffee contains around 5 mg of niacin and thus can be an important source of this vitamin [38]. Moreover, previous studies have demonstrated that coffee is among the highest source of polyphenols in individuals living in Northern and Eastern European populations [39,40] due to the large intake compared with individuals living in Southern European countries, such as France, Spain and Italy [41–43]. Coffee drinking has been associated with positive health outcomes [44]. Regarding espresso coffee, despite it being reported to provide an important antioxidant capacity within the context of a common diet [45,46], results regarding its effects on health are varied, ranging from positive [47–49] to negative [50–52]. In general, scientific literature on espresso coffee is scarce, and further research is needed to test the impact of this beverage on human health.

Regarding alcoholic beverage consumption, women almost reached the same intake as men (3.0% vs. 3.5%) in contrast with previous studies in which men had, in some cases, double the consumption of women [32]. Regular and moderate daily alcohol consumption is a typical pattern of the Mediterranean diet, with wine being the most consumed type of alcoholic beverage [53]. This habit is especially rooted in the elderly, as confirmed by our analysis stratified by age groups. Furthermore, a small correlation has been found between alcohol beverages and polyphenols. The caloric contribution of alcohol beverages was 3%; the downside of the non-negligible caloric intake of alcoholic beverages could be softened by the benefits of moderate alcohol use. In fact, there is evidence that dietary polyphenols may exert beneficial effect toward human health [54,55]. It has been established that alcohol has a protective effect at low doses, in particular, for red wine in regard to cardiovascular diseases, and the presence of polyphenols contained in such beverages has been proposed as one of the main contributors to this effect [56,57]. Another source of polyphenols is tea, as remarked by our correlation analysis. However, its consumption was relatively low in our sample, comprising 2% of total water weight consumed, with a popularity of only 30% of the population. Finally, the least popular beverage was soy milk, which was highly correlated with phytoestrogen intake. Soy foods and beverages are hardly consumed in the Mediterranean area compared to Asian countries [58], but evidence suggests that phytoestrogens, such as isoflavones, may exert beneficial effects on health [59]. A previous study involving the present cohort showed certain positive associations between phytoestrogen consumption and hypertension despite the low consumption [34], suggesting that this beverage has the potential for dietary recommendation in Mediterranean countries and could be taken into account for future dietary strategies. Besides polyphenols, antioxidant vitamins were correlated with total water intake, and specifically with water from food. This suggests that foods with a high content of water, like fruits and some kinds of vegetables, are the main source of these substances [60–62]. As expected, correlations between vitamin C and fruit juices, and vitamin B12 and milk were found, suggesting that such beverages are among the major sources of these vitamins.

Beverages' contribution to total energy intake was 9%, less than the 10% suggested by EFSA and WHO (World Health Organization) [4,63]. It has been recently discussed worldwide whether energy intake from beverages could have a role in metabolic disorders such as obesity [64–67]. According to recent scientific literature, there is a wide difference in beverage energy contribution between countries, ranging from around 5–10% in Italy and Spain, to 16% in the UK and 21% in the US [30–32,68]. The habit of drinking sugar-sweetened beverages is not common in the context of a Mediterranean diet pattern, as reported in our results (sugar-sweetened beverages accounted for only 1.3% of total caloric intake), as well as findings from other Italian and Spanish studies (0.4–2%) [30,31]. There is evidence from this cohort that traditional dietary patterns may be associated with a lower likelihood of having metabolic disorders [69–71]. In contrast, studies conducted in the UK and the US reported alarming trends in sugar-sweetened beverage consumption, which have been ascribed as being potentially responsible for the rise in metabolic disorders in these areas [32,68].

The main limitation of this study was the lack of a hydration biomarker that could definitely correlate fluid intake with the hydration levels of the individuals involved in the survey. Second, the cross-sectional nature of the study design and the use of a FFQ to estimate dietary information may lead to recall bias, with some overestimations (for instance on water intakes) and underestimations (alcohol beverages).

5. Conclusions

In conclusion, the results of the present study show that our sample population had adequate intake of TWI in accordance with EFSA reference values. Caloric beverages did not have a big impact on total caloric intake among the sample population, entering the boundaries of EFSA and WHO recommendation limits. Some beverages, like fruit juices, can cause a significant intake of certain vitamins and could be included in specific recommendations to avoid nutritional deficiencies. A small number of studies have investigated hydration status on a population level, thus further investigations are needed to identify trends and issues that may require specific health and nutrition policies or targeted interventions.

Author Contributions: Marina Marranzano conceived and designed the experiments; Armando Platania performed the experiments; Armando Platania and Dario Sinatra analyzed the data; Armando Platania wrote the paper; Dora Castiglione and Maurizio D' Urso provided data and clinical revision.

Conflicts of Interest: The authors declare no conflict of interest.

References

1. Arnaud, M.J. *Hydration Throughout Life: International Conference, Vittel, France, June 9–12, 1998*; John Libbey Eurotext: Montrouge, France, 1998.
2. Watson, P.; Whale, A.; Mears, S.A.; Reyner, L.A.; Maughan, R.J. Mild hypohydration increases the frequency of driver errors during a prolonged, monotonous driving task. *Physiol .Behav.* **2015**, *147*, 313–318. [CrossRef] [PubMed]
3. Maughan, R.J. Impact of mild dehydration on wellness and on exercise performance. *Eur. J. Clin. Nutr.* **2003**, *57* (Suppl. 2), S19–S23. [CrossRef] [PubMed]
4. European Food Safety Authority (EFSA) Panel on Dietetic Products, Nutrition and Allergies. Scientific opinion on dietary reference values for water. *EFSA J.* **2010**, *8*, 1459.
5. Nissensohn, M.; Castro-Quezada, I.; Serra-Majem, L. Beverage and water intake of healthy adults in some european countries. *Int. J. Food Sci. Nutr.* **2013**, *64*, 801–805. [CrossRef] [PubMed]
6. Imamura, F.; O'Connor, L.; Ye, Z.; Mursu, J.; Hayashino, Y.; Bhupathiraju, S.N.; Forouhi, N.G. Consumption of sugar sweetened beverages, artificially sweetened beverages, and fruit juice and incidence of type 2 diabetes: Systematic review, meta-analysis, and estimation of population attributable fraction. *BMJ* **2015**, *351*, 3576. [CrossRef] [PubMed]
7. Xi, B.; Huang, Y.; Reilly, K.H.; Li, S.; Zheng, R.; Barrio-Lopez, M.T.; Martinez-Gonzalez, M.A.; Zhou, D. Sugar-sweetened beverages and risk of hypertension and CVD: A dose-response meta-analysis. *Br. J. Nutr.* **2015**, *113*, 709–717. [CrossRef] [PubMed]
8. Ruanpeng, D.; Thongprayoon, C.; Cheungpasitporn, W.; Harindhanavudhi, T. Sugar and artificially sweetened beverages linked to obesity: A systematic review and meta-analysis. *QJM* **2017**, *110*, 513–520. [CrossRef] [PubMed]
9. Larsson, S.C.; Crippa, A.; Orsini, N.; Wolk, A.; Michaelsson, K. Milk consumption and mortality from all causes, cardiovascular disease, and cancer: A systematic review and meta-analysis. *Nutrients* **2015**, *7*, 7749–7763. [CrossRef] [PubMed]
10. Guo, J.; Astrup, A.; Lovegrove, J.A.; Gijsbers, L.; Givens, D.I.; Soedamah-Muthu, S.S. Milk and dairy consumption and risk of cardiovascular diseases and all-cause mortality: Dose-response meta-analysis of prospective cohort studies. *Eur. J. Epidemiol.* **2017**, *32*, 269–287. [CrossRef] [PubMed]
11. Grosso, G.; Micek, A.; Godos, J.; Sciacca, S.; Pajak, A.; Martinez-Gonzalez, M.A.; Giovannucci, E.L.; Galvano, F. Coffee consumption and risk of all-cause, cardiovascular, and cancer mortality in smokers and non-smokers: A dose-response meta-analysis. *Eur. J. Epidemiol.* **2016**, *31*, 1191–1205. [CrossRef] [PubMed]
12. Marventano, S.; Salomone, F.; Godos, J.; Pluchinotta, F.; Del Rio, D.; Mistretta, A.; Grosso, G. Coffee and tea consumption in relation with non-alcoholic fatty liver and metabolic syndrome: A systematic review and meta-analysis of observational studies. *Clin. Nutr.* **2016**, *35*, 1269–1281. [CrossRef] [PubMed]
13. Yang, J.; Mao, Q.X.; Xu, H.X.; Ma, X.; Zeng, C.Y. Tea consumption and risk of type 2 diabetes mellitus: A systematic review and meta-analysis update. *BMJ Open* **2014**, *4*, e005632. [CrossRef] [PubMed]
14. Tang, J.; Zheng, J.S.; Fang, L.; Jin, Y.; Cai, W.; Li, D. Tea consumption and mortality of all cancers, CVD and all causes: A meta-analysis of eighteen prospective cohort studies. *Br. J. Nutr.* **2015**, *114*, 673–683. [CrossRef] [PubMed]
15. Zhang, C.; Qin, Y.Y.; Wei, X.; Yu, F.F.; Zhou, Y.H.; He, J. Tea consumption and risk of cardiovascular outcomes and total mortality: A systematic review and meta-analysis of prospective observational studies. *Eur. J. Epidemiol.* **2015**, *30*, 103–113. [CrossRef] [PubMed]
16. Bagnardi, V.; Rota, M.; Botteri, E.; Tramacere, I.; Islami, F.; Fedirko, V.; Scotti, L.; Jenab, M.; Turati, F.; Pasquali, E.; et al. Alcohol consumption and site-specific cancer risk: A comprehensive dose-response meta-analysis. *Br. J. Cancer* **2015**, *112*, 580–593. [CrossRef] [PubMed]
17. Praud, D.; Rota, M.; Rehm, J.; Shield, K.; Zatonski, W.; Hashibe, M.; La Vecchia, C.; Boffetta, P. Cancer incidence and mortality attributable to alcohol consumption. *Int. J. Cancer* **2016**, *138*, 1380–1387. [CrossRef] [PubMed]
18. Rosado, C.I.; Marin, A.L.V.; Martinez, J.A.; Cabrerizo, L.; Gargallo, M.; Lorenzo, H.; Quiles, J.; Planas, M.; Polanco, I.; de Avila, D.R.; et al. Importance of water in the hydration of the spanish population: Fesnad 2010 document. *Nutr. Hosp.* **2011**, *26*, 27–36.

19. Popkin, B.M.; D'Anci, K.E.; Rosenberg, I.H. Water, hydration, and health. *Nutr. Rev.* **2010**, *68*, 439–458. [CrossRef] [PubMed]

20. Grosso, G.; Marventano, S.; D'Urso, M.; Mistretta, A.; Galvano, F. The mediterranean healthy eating, ageing, and lifestyle (meal) study: Rationale and study design. *Int. J. Food Sci. Nutr.* **2017**, *68*, 577–586. [CrossRef] [PubMed]

21. Craig, C.L.; Marshall, A.L.; Sjostrom, M.; Bauman, A.E.; Booth, M.L.; Ainsworth, B.E.; Pratt, M.; Ekelund, U.; Yngve, A.; Sallis, J.F.; et al. International physical activity questionnaire: 12-country reliability and validity. *Med. Sci. Sports Exerc.* **2003**, *35*, 1381–1395. [CrossRef] [PubMed]

22. Mistretta, A.; Marventano, S.; Platania, A.; Godos, J.; Galvano, F.; Grosso, G. Metabolic profile of the mediterranean healthy eating, lifestyle and aging (meal) study cohort. *Mediterr. J. Nutr. Metab.* **2017**, *10*, 131–140. [CrossRef]

23. World Health Organization. *Obesity: Preventing and Managing the Global Epidemic*; Report of a Who Consultation Presented at the World Health Organization; June 3–5, 1997; World Health Organization: Geneva, Switzerland, 1997.

24. Buscemi, S.; Rosafio, G.; Vasto, S.; Massenti, F.M.; Grosso, G.; Galvano, F.; Rini, N.; Barile, A.M.; Maniaci, V.; Cosentino, L.; et al. Validation of a food frequency questionnaire for use in italian adults living in sicily. *Int. J. Food Sci. Nutr.* **2015**, *66*, 426–438. [CrossRef] [PubMed]

25. Marventano, S.; Mistretta, A.; Platania, A.; Galvano, F.; Grosso, G. Reliability and relative validity of a food frequency questionnaire for italian adults living in sicily, southern italy. *Int. J. Food Sci. Nutr.* **2016**, *67*, 857–864. [CrossRef] [PubMed]

26. Marletta, L.; Carnovale, E. *Tabelle di Composizione Degli Alimenti*; Istituto Nazionale di Ricerca per gli Alimenti e la Nutrizione: Milano, Italy, 2000.

27. Neveu, V.; Perez-Jiménez, J.; Vos, F.; Crespy, V.; Du Chaffaut, L.; Mennen, L.; Knox, C.; Eisner, R.; Cruz, J.; Wishart, D.; et al. Phenol-explorer: An online comprehensive database on polyphenol contents in foods. *Database* **2010**. [CrossRef] [PubMed]

28. Godos, J.; Marventano, S.; Mistretta, A.; Galvano, F.; Grosso, G. Dietary sources of polyphenols in the mediterranean healthy eating, aging and lifestyle (meal) study cohort. *Int. J. Food Sci. Nutr.* **2017**, 1–7. [CrossRef] [PubMed]

29. Willett, W.C. Reproducibility and validity of food frequency questionnaire. In *Nutritional Epidemiology*, 2nd ed.; Oxford University Press: Oxford, UK, 1998.

30. Nissensohn, M.; Sanchez-Villegas, A.; Ortega, R.M.; Aranceta-Bartrina, J.; Gil, A.; Gonzalez-Gross, M.; Varela-Moreiras, G.; Serra-Majem, L. Beverage consumption habits and association with total water and energy intakes in the spanish population: Findings of the anibes study. *Nutrients* **2016**, *8*, 232. [CrossRef] [PubMed]

31. Mistura, L.; D'Addezio, L.; Turrini, A. Beverage consumption habits in italian population: Association with total water intake and energy intake. *Nutrients* **2016**, *8*. [CrossRef] [PubMed]

32. Gibson, S.; Shirreffs, S.M. Beverage consumption habits "24/7" among british adults: Association with total water intake and energy intake. *Nutr. J.* **2013**, *12*, 9. [CrossRef] [PubMed]

33. Grosso, G.; Marventano, S.; Giorgianni, G.; Raciti, T.; Galvano, F.; Mistretta, A. Mediterranean diet adherence rates in sicily, southern italy. *Public Health Nutr.* **2014**, *17*, 2001–2009. [CrossRef] [PubMed]

34. Marventano, S.; Godos, J.; Platania, A.; Galvano, F.; Mistretta, A.; Grosso, G. Mediterranean diet adherence in the mediterranean healthy eating, aging and lifestyle (meal) study cohort. *Int. J. Food Sci. Nutr.* **2018**, *69*, 100–107. [CrossRef] [PubMed]

35. Bonaccorsi, G.; Lorini, C.; Santomauro, F.; Sofi, F.; Vannetti, F.; Pasquini, G.; Macchi, C.; Mugello Study, G. Adherence to mediterranean diet and nutritional status in a sample of nonagenarians. *Exp. Gerontol.* **2018**, *103*, 57–62. [CrossRef] [PubMed]

36. De Lorenzo, A.; Noce, A.; Bigioni, M.; Calabrese, V.; Della Rocca, D.G.; Di Daniele, N.; Tozzo, C.; Di Renzo, L. The effects of italian mediterranean organic diet (IMOD) on health status. *Curr. Pharm. Des.* **2010**, *16*, 814–824. [CrossRef] [PubMed]

37. Mistretta, A.; Marventano, S.; Antoci, M.; Cagnetti, A.; Giogianni, G.; Nolfo, F.; Rametta, S.; Pecora, G.; Marranzano, M. Mediterranean diet adherence and body composition among southern italian adolescents. *Obes. Res. Clin. Pract.* **2017**, *11*, 215–226. [CrossRef] [PubMed]

38. Kremer, J.I.; Gompel, K.; Bakuradze, T.; Eisenbrand, G.; Richling, E. Urinary excretion of niacin metabolites in humans after coffee consumption. *Mol. Nutr. Food Res.* **2018**, e1700735. [CrossRef] [PubMed]

39. Grosso, G.; Stepaniak, U.; Topor-Madry, R.; Szafraniec, K.; Pajak, A. Estimated dietary intake and major food sources of polyphenols in the polish arm of the hapiee study. *Nutrition* **2014**, *30*, 1398–1403. [CrossRef] [PubMed]

40. Ovaskainen, M.L.; Torronen, R.; Koponen, J.M.; Sinkko, H.; Hellstrom, J.; Reinivuo, H.; Mattila, P. Dietary intake and major food sources of polyphenols in finnish adults. *J. Nutr.* **2008**, *138*, 562–566. [CrossRef] [PubMed]

41. Tresserra-Rimbau, A.; Medina-Remon, A.; Perez-Jimenez, J.; Martinez-Gonzalez, M.A.; Covas, M.I.; Corella, D.; Salas-Salvado, J.; Gomez-Gracia, E.; Lapetra, J.; Aros, F.; et al. Dietary intake and major food sources of polyphenols in a spanish population at high cardiovascular risk: The predimed study. *Nutr. Metab. Cardiovasc. Dis.* **2013**, *23*, 953–959. [CrossRef] [PubMed]

42. Perez-Jimenez, J.; Neveu, V.; Vos, F.; Scalbert, A. Identification of the 100 richest dietary sources of polyphenols: An application of the phenol-explorer database. *Eur. J. Clin. Nutr.* **2010**, *64* (Suppl. 3), S112–S120. [CrossRef]

43. Godos, J.; Rapisarda, G.; Marventano, S.; Galvano, F.; Mistretta, A.; Grosso, G. Association between polyphenol intake and adherence to the mediterranean diet in sicily, southern Italy. *NFS J.* **2017**, *8*, 1–7. [CrossRef]

44. Grosso, G.; Godos, J.; Galvano, F.; Giovannucci, E.L. Coffee, caffeine, and health outcomes: An umbrella review. *Annu. Rev. Nutr.* **2017**, *37*, 131–156. [CrossRef] [PubMed]

45. Caprioli, G.; Cortese, M.; Maggi, F.; Minnetti, C.; Odello, L.; Sagratini, G.; Vittori, S. Quantification of caffeine, trigonelline and nicotinic acid in espresso coffee: The influence of espresso machines and coffee cultivars. *Int. J. Food Sci. Nutr.* **2014**, *65*, 465–469. [CrossRef] [PubMed]

46. Caprioli, G.; Cortese, M.; Sagratini, G.; Vittori, S. The influence of different types of preparation (espresso and brew) on coffee aroma and main bioactive constituents. *Int. J. Food Sci. Nutr.* **2015**, *66*, 505–513. [CrossRef] [PubMed]

47. Godos, J.; Sinatra, D.; Blanco, I.; Mule, S.; La Verde, M.; Marranzano, M. Association between dietary phenolic acids and hypertension in a mediterranean cohort. *Nutrients* **2017**, *9*. [CrossRef]

48. Pounis, G.; Tabolacci, C.; Costanzo, S.; Cordella, M.; Bonaccio, M.; Rago, L.; D'Arcangelo, D.; Di Castelnuovo, A.F.; de Gaetano, G.; Donati, M.B.; et al. Reduction by coffee consumption of prostate cancer risk: Evidence from the Moli-sani cohort and cellular models. *Int. J. Cancer* **2017**, *141*, 72–82. [CrossRef] [PubMed]

49. Solfrizzi, V.; Panza, F.; Imbimbo, B.P.; D'Introno, A.; Galluzzo, L.; Gandin, C.; Misciagna, G.; Guerra, V.; Osella, A.; Baldereschi, M.; et al. Coffee consumption habits and the risk of mild cognitive impairment: The Italian longitudinal study on aging. *J. Alzheimers Dis.* **2015**, *47*, 889–899. [CrossRef] [PubMed]

50. Parodi, S.; Merlo, F.D.; Stagnaro, E. Working Group for the Epidemiology of Hematolymphopoietic Malignancies in, I. Coffee consumption and risk of non-hodgkin's lymphoma: Evidence from the Italian multicentre case-control study. *Cancer Causes Control* **2017**, *28*, 867–876. [CrossRef] [PubMed]

51. Grioni, S.; Agnoli, C.; Sieri, S.; Pala, V.; Ricceri, F.; Masala, G.; Saieva, C.; Panico, S.; Mattiello, A.; Chiodini, P.; et al. Espresso coffee consumption and risk of coronary heart disease in a large Italian cohort. *PLoS ONE* **2015**, *10*, e0126550. [CrossRef] [PubMed]

52. Filiberti, R.A.; Fontana, V.; De Ceglie, A.; Blanchi, S.; Grossi, E.; Della Casa, D.; Lacchin, T.; De Matthaeis, M.; Ignomirelli, O.; Cappiello, R.; et al. Association between coffee or tea drinking and barrett's esophagus or esophagitis: An Italian study. *Eur. J. Clin. Nutr.* **2017**, *71*, 980–986. [CrossRef] [PubMed]

53. Giacosa, A.; Barale, R.; Bavaresco, L.; Faliva, M.A.; Gerbi, V.; La Vecchia, C.; Negri, E.; Opizzi, A.; Perna, S.; Pezzotti, M.; et al. Mediterranean way of drinking and longevity. *Crit. Rev. Food Sci. Nutr.* **2016**, *56*, 635–640. [CrossRef] [PubMed]

54. Grosso, G.; Godos, J.; Lamuela-Raventos, R.; Ray, S.; Micek, A.; Pajak, A.; Sciacca, S.; D'Orazio, N.; Del Rio, D.; Galvano, F. A comprehensive meta-analysis on dietary flavonoid and lignan intake and cancer risk: Level of evidence and limitations. *Mol. Nutr. Food Res.* **2017**, *61*. [CrossRef] [PubMed]

55. Grosso, G.; Micek, A.; Godos, J.; Pajak, A.; Sciacca, S.; Galvano, F.; Giovannucci, E.L. Dietary flavonoid and lignan intake and mortality in prospective cohort studies: Systematic review and dose-response meta-analysis. *Am. J. Epidemiol.* **2017**, *185*, 1304–1316. [CrossRef] [PubMed]

56. Chiva-Blanch, G.; Arranz, S.; Lamuela-Raventos, R.M.; Estruch, R. Effects of wine, alcohol and polyphenols on cardiovascular disease risk factors: Evidences from human studies. *Alcohol Alcohol.* **2013**, *48*, 270–277. [CrossRef] [PubMed]

57. Arranz, S.; Chiva-Blanch, G.; Valderas-Martinez, P.; Medina-Remon, A.; Lamuela-Raventos, R.M.; Estruch, R. Wine, beer, alcohol and polyphenols on cardiovascular disease and cancer. *Nutrients* **2012**, *4*, 759–781. [CrossRef] [PubMed]

58. Mocciaro, G.; Ziauddeen, N.; Godos, J.; Marranzano, M.; Chan, M.Y.; Ray, S. Does a mediterranean-type dietary pattern exert a cardio-protective effect outside the mediterranean region? A review of current evidence. *Int. J. Food Sci. Nutr.* **2017**, 1–12. [CrossRef] [PubMed]

59. Godos, J.; Bergante, S.; Satriano, A.; Pluchinotta, F.R.; Marranzano, M. Dietary phytoestrogen intake is inversely associated with hypertension in a cohort of adults living in the mediterranean area. *Molecules* **2018**, *23*. [CrossRef] [PubMed]

60. Olza, J.; Aranceta-Bartrina, J.; Gonzalez-Gross, M.; Ortega, R.M.; Serra-Majem, L.; Varela-Moreiras, G.; Gil, A. Reported dietary intake and food sources of zinc, selenium, and vitamins A, E and C in the spanish population: Findings from the anibes study. *Nutrients* **2017**, *9*. [CrossRef] [PubMed]

61. Dominguez-Perles, R.; Mena, P.; Garcia-Viguera, C.; Moreno, D.A. Brassica foods as a dietary source of vitamin C: A review. *Crit. Rev. Food Sci. Nutr.* **2014**, *54*, 1076–1091. [CrossRef] [PubMed]

62. Marranzano, M.; Godos, J.; Sumantra, R.; Galvano, F. Association between dietary flavonoids intake and obesity in a cohort of adults living in the Mediterranean area. *Int. J. Food Sci. Nutr.* **2018**, ahead of print. [CrossRef] [PubMed]

63. World Health Organization. *Diet, Nutrition and the Prevention of Chronic Diseases*; Report No. 916; World Health Organization: Geneva, Switzerland, 2002.

64. Lustig, R.H.; Schmidt, L.A.; Brindis, C.D. Public health: The toxic truth about sugar. *Nature* **2012**, *482*, 27–29. [CrossRef] [PubMed]

65. Johnson, R.J.; Segal, M.S.; Sautin, Y.; Nakagawa, T.; Feig, D.I.; Kang, D.H.; Gersch, M.S.; Benner, S.; Sanchez-Lozada, L.G. Potential role of sugar (fructose) in the epidemic of hypertension, obesity and the metabolic syndrome, diabetes, kidney disease, and cardiovascular disease. *Am. J. Clin. Nutr.* **2007**, *86*, 899–906. [PubMed]

66. Johnson, R.K.; Appel, L.J.; Brands, M.; Howard, B.V.; Lefevre, M.; Lustig, R.H.; Sacks, F.; Steffen, L.M.; Wylie-Rosett, J. Dietary sugars intake and cardiovascular health: A scientific statement from the American Heart Association. *Circulation* **2009**, *120*, 1011–1020. [CrossRef] [PubMed]

67. Malik, V.S.; Popkin, B.M.; Bray, G.A.; Despres, J.P.; Hu, F.B. Sugar-sweetened beverages, obesity, type 2 diabetes mellitus, and cardiovascular disease risk. *Circulation* **2010**, *121*, 1356–1364. [CrossRef] [PubMed]

68. Drewnowski, A.; Rehm, C.D.; Constant, F. Water and beverage consumption among adults in the United States: Cross-sectional study using data from NHANES 2005–2010. *BMC Public Health* **2013**, *13*, 1068. [CrossRef] [PubMed]

69. La Verde, M.; Mule, S.; Zappala, G.; Privitera, G.; Maugeri, G.; Pecora, F.; Marranzano, M. Higher adherence to the Mediterranean diet is inversely associated with having hypertension: Is low salt intake a mediating factor? *Int. J. Food Sci. Nutr.* **2018**, *69*, 235–244. [CrossRef] [PubMed]

70. Platania, A.; Zappala, G.; Mirabella, M.U.; Gullo, C.; Mellini, G.; Beneventano, G.; Maugeri, G.; Marranzano, M. Association between mediterranean diet adherence and dyslipidaemia in a cohort of adults living in the Mediterranean area. *Int. J. Food Sci. Nutr.* **2017**, 1–11. [CrossRef] [PubMed]

71. Zappala, G.; Buscemi, S.; Mule, S.; La Verde, M.; D'Urso, M.; Corleo, D.; Marranzano, M. High adherence to Mediterranean diet, but not individual foods or nutrients, is associated with lower likelihood of being obese in a Mediterranean cohort. *Eat. Weight Disord.* **2017**. [CrossRef] [PubMed]

antioxidants

MDPI

Review

Phytochemical Profile of Brown Rice and Its Nutrigenomic Implications

Keneswary Ravichanthiran [1,†], Zheng Feei Ma [2,3,*,†], Hongxia Zhang [4,†], Yang Cao [5], Chee Woon Wang [6], Shahzad Muhammad [7], Elom K. Aglago [8], Yihe Zhang [9], Yifan Jin [2] and Binyu Pan [10]

[1] Faculty of Food Science and Nutrition, Universiti Malaysia Sabah, Kota Kinabalu 2073, Sabah, Malaysia; rkeneswary@yahoo.com
[2] Department of Public Health, Xi'an Jiaotong-Liverpool University, Suzhou 215123, China; Yifan.Jin14@student.xjtlu.edu.cn
[3] School of Medical Sciences, Universiti Sains Malaysia, Kota Bharu 15200, Kelantan, Malaysia
[4] Department of Food Science, University of Otago, Dunedin 9016, New Zealand; zhanghongxia326@hotmail.com
[5] Department of Health Promotion, Pudong Maternal and Child Health Care Institution, Shanghai 201399, China; evacaoyang@163.com
[6] Department of Biochemistry, Faculty of Medicine, MAHSA University, Bandar Saujana Putra 42610, Jenjarom, Selangor, Malaysia; wang.chee@mahsa.edu.my
[7] Institute of Basic Medical Sciences, Khyber Medical University, Peshawar 25100, Pakistan; shahzad.ibms@kmu.edu.pk
[8] Joint Unit of Research in Nutrition and Food Science, Ibn Tofail University, Kenitra 14000, Morocco; aglagoelom@gmail.com
[9] Division of Medicine, School of Life and Medical Sciences, University College London, London WC1E6BT, UK; yihe.zhang.16@ucl.ac.uk
[10] Department of Clinical Nutrition, The First People's Hospital of Wujiang District, Suzhou 215200, China; panbinyu19881102@126.com
* Correspondence: Zhengfeei.Ma@xjtlu.edu.cn; Tel.: +86-512-8188-4938
† These authors contributed equally to this work.

Received: 21 April 2018; Accepted: 18 May 2018; Published: 23 May 2018

Abstract: Whole grain foods have been promoted to be included as one of the important components of a healthy diet because of the relationship between the regular consumption of whole-grain foods and reduced risk of chronic diseases. Rice is a staple food, which has been widely consumed for centuries by many Asian countries. Studies have suggested that brown rice is associated with a wide spectrum of nutrigenomic implications such as anti-diabetic, anti-cholesterol, cardioprotective and antioxidant. This is because of the presence of various phytochemicals that are mainly located in bran layers of brown rice. Therefore, this paper is a review of publications that focuses on the bioactive compounds and nutrigenomic implications of brown rice. Although current evidence supports the fact that the consumption of brown rice is beneficial for health, these studies are heterogeneous in terms of their brown rice samples used and population groups, which cause the evaluation to be difficult. Future clinical studies should focus on the screening of individual bioactive compounds in brown rice with reference to their nutrigenomic implications.

Keywords: brown rice; nutrigenomics; phenolics; rice

1. Introduction

For centuries, rice (*Oryza sativa* L.), one of the most well-known cereal foods, has been a primary food for many people around the world and is known to feed half of the population [1]. Therefore, the role of rice as a staple food in providing nutrition to populations has been acknowledged. In 2015, the global rice paddy production was 739.1 million tonnes, yielding 490.5 million tonnes of white rice after milling. The rice paddy production in Asia was 668.4 million tonnes, accounting for 90% of the global production, indicating that rice consumption occurs mostly in Asian countries. The environmental flexibility of culturing rice paddies at various temperatures, humidities and soil conditions allows rice to become a globally-viable crop [2]. However, the health benefits of rice were never considered because rice is considered as a staple food based on the palatability and availability. The major producers of rice are China, India and Indonesia [3].

There are more than 8000 varieties of rice, which have different types of quality and nutritional content. After the post-harvest process, all the varieties of rice can be categorised as either white or brown rice [4]. The aromatic rice varieties, known collectively as "Basmati rice", have been sourced by people from Asian and European countries, because aroma has been considered as the highest preferred characteristic of cereal grain. Basmati rice possesses unique cereal quality features, such as long, supreme grains, characteristic aroma, swelling on cooking and tenderness of cooked rice. Basmati rice with a high amylose to amylopectin ratio and a medium glycaemic index is suitable for staple diets of diabetics [5].

Rough rice can be separated into husk and brown rice through a threshing process. The components in brown rice that was hulled from rough rice are bran layers (6–7%), an embryo (2–3%) and an endosperm (about 90%) [6]. Brown rice can be further separated into polished rice, commonly called white rice, which is obtained by removing the bran. Minor differences may exist in the degree of milling. Brown rice has a nutty flavour, chewier than white rice, but more easily goes rancid, as well [7]. The difference between brown rice and white rice can be obtained through milling [7]. White rice contains mainly the starchy endosperm. The removal of rice bran leads to a loss of nutrients. During milling, about 85% of the fat, 15% of protein, 75% of phosphorus, 90% of calcium and 70% of B vitamins (including B_1, B_2 and B_3) are removed [7].

As the degree of milling increases, the loss of phytochemical compounds beneficial to health occurs, and cellular antioxidant activity decreases. Furthermore, the contents of phenolic compounds have also been shown to decrease by increasing the degree of milling. Thus, by carefully controlling the degree of milling during rice processing, both the sensory quality and nutritional composition could be optimized. Thus, brown rice with a low degree of milling (<2.7%) exhibits a more ideal balance between sensory quality and retention of beneficial phytochemicals [8]. Brown rice is a rich source of various bioactive compounds, such as γ-oryzanol, tocopherol, tocotrienol, amino acids, dietary fibres and minerals. It is less consumed than white rice because its cooking is more difficult than white rice due to its slow water absorption, and the palatability quality of brown rice is inferior to white rice [9].

There are two types of brown rice, which are germinated and non-germinated. Germinated brown rice is obtained by immersing the brown rice grain in water to initiate germination [10]. The benefits of germinated brown rice are that the nutrients found in brown rice are more easily digested and the texture of brown rice is better [10]. Germination has been employed to improve the texture of cooked brown rice. It also initiates numerous changes in the composition and chemical structure of the bioactive components. Germination could induce the formation of new bioactive compounds, such as gamma-aminobutyric acid (GABA). The consumption of germinated brown rice is increasing in many Asian countries because of its improved palatability quality and potential health-promoting functions [11].

Advances in the human genome era have shown that diet plays an important factor in the health and the causation chronic diseases such as type 2 diabetes. This is because the diet-genome interactions can result in changes especially in the proteome, transcriptome and metabolome. For example, current healthcare practitioners recommend brown rice to be consumed rather than white rice. This is due to the fact that brown rice is more nutritious. One common trait between white rice and brown rice is that they are both gluten free and contain no trans fat or cholesterol [7]. Encouraging people to eat brown rice more is a difficult challenge due to its taste, which is less likeable compared to the taste of white rice [7]. In the United States, more than 70% of rice consumed is white rice, and rice consumption has reached 9.3 kg per capita since the 1930s [1]. In addition, the consumption of brown rice is beneficial for postprandial blood glucose control because brown rice has a lower glycaemic index than white rice (55 vs. 64) [12].

Rice is the main staple food for more than half of the world's population. The cereal was also utilised as a popular remedy since ancient times for several therapeutic purposes. Rice or rice-based products were also well documented in the traditional medicines of different Asian countries. The well-known popular uses are anti-diabetic, anti-inflammatory for the airway, ailment of gastrointestinal disorders and diarrhoea, diuretic, source of vitamins and skin preparations [13,14]. One of the rice varieties, red rice Rakthashali, is a staple food in India and has been described by Ayurveda practitioners as a functional food for a number of medications [15]. The medicinal rice Kullakar has high thiamine content, while the Karikalaveya variety is high in riboflavin and niacin [16].

Therefore, the aim of our work is to review the phytochemical constituents and nutrigenomic implications of brown rice in relation to animal and human studies. In addition, our work has also contributed significantly to the current understandings of brown rice with reference to the nutrigenomic implications of brown rice shown in human intervention studies. Therefore, this mini-review will provide a valuable reference resource for future studies in such areas.

Search Strategy

An electronic literature search was conducted using PubMed, Medline (OvidSP) Cochrane CENTRAL and Web of Science until December 2017. Additional articles were identified from references in the retrieved articles. Search terms included combinations of the following: rice, brown rice, phytochemicals, nutrigenomics and bioactives. The search was restricted to articles in English that addressed the phytochemical constituents and nutrigenomic implications of brown rice.

2. Phytochemical Compounds in Brown Rice

The advantages for health with the consumption of brown rice mainly come from the phytochemicals found in its bran layers [17]. Figure 1 shows the various parts of the rice grain. The phytochemical composition of brown rice cannot be dissociated from the scientific work of the Dutch Nobel prize scientist Christiaan Eijkman who initially reported the potential of brown rice and the story behind beriberi in humans in the previous centuries. Table 1 show the major phytochemical composition of brown rice. In addition to B vitamins, phytochemicals found in brown rice include dietary fibre, functional lipids, essential amino-acids, phytosterols, phenolic acids, flavonoids, anthocyanins, proanthocyanins, tocopherols, tocotrienols, minerals, gamma aminobutyric acid (GABA) and γ-oryzanol [11,17]. Brown rice also contains high levels of phytic acid [18].

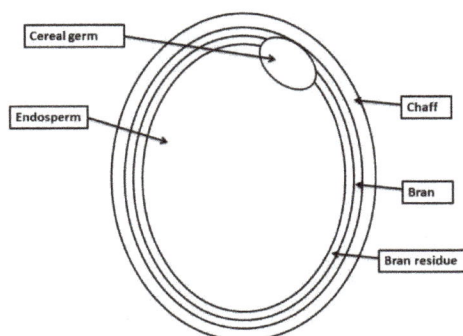

Figure 1. The various parts of the rice grain.

Table 1. Summary of the major phytochemical composition of brown rice.

Family	Compounds
Phenolics	Gallic acid, protocatechuic acid, *p*-hydroxybenzoic acid, vanillic acid, syringic acid, chlorogenic acid, caffeic acid, *p*-coumaric acid, sinapic acid, ferulic acid, cinnamic acid, ellagic acid
Flavonoids	Luteolin, apigenin, tricin, quercetin, kaempferol, isorhamnetin, myricetin
Anthocyanins and proanthocyanins	Peonidin-3-*O*-glucoside, cyanidin-3-*O*-glucoside, cyanidin-3-*O*-galactoside, cyanidin-3-*O*-rutinoside, catechin, epicatechin
Vitamins	Tocopherols, tocotrienols, B vitamins (B1, B3, B6)
Amino acids	Alanine, arginine, aspartic acid, cystine, glutamic acid, glycine, histidine, isoleucine, leucine, lysine, methionine, phenylalanine, proline, serine, threonine, tryptophan, tyrosine, valine
Phytosterols	Stigmasterol, stigmastanol, β-sitosterol, campesterol, δ5-avenasterol, δ7-avenasterol
γ-Oryzanol	Cycloartanyl ferulate, 24-methylene cycloartanyl ferulate, campesteryl ferulate, β-sitosteryl ferulate
Others	Dietary fibre, phytic acid, minerals

Amongst these nutritional factors, phenolic acids are the most common substances found in brown rice [19]. Phenolics are classified under phytochemicals having one or more aromatic rings with one or more hydroxyl groups [20]. Phenolic compounds are associated with diverse human health benefits including anti-inflammatory, hypoglycaemic, anticarcinogenic, antiallergenic and antiatherosclerotic properties [20]. Examples of phenolics are phenolic acids, flavonoids, tannins, coumarins and stilbenes [21]. In rice, phenolics are found in three distinct forms, which are free, soluble-conjugated and bound forms, and the bound form is the main form among the three [22]. High levels of phenolics exist in the germ and bran layers [23]. Since brown rice does not undergo any polishing or milling, these phenolics found in the germ and bran layers are easily preserved. Free phenolics are the most readily available for absorption in the small intestine, while bound phenolics tend to be preserved throughout the intestinal tract and released in the bowel, where they interact with the microbiome to favour the Firmicutes/Bacteroidetes ratio [24]. The two main groups of phenolic acids are p-hydroxybenzoic acid and p-hydroxycinnamic acid derivatives [21]. The primary phenolic compound found in brown rice is *trans*-ferulic acid (range: 161.42–374.81 µg/g), a hydroxycinnamic acid that exists in the bound form [19]. The second major phenolic acid found in brown rice is *trans-p*-coumaric acid (range: 35.49–81.52 µg/g), which is a hydroxycinnamic acid derivative with its bound form making up about 98% [19]. Another component that is widely found is *cis*-ferulic acid (range: 20.76–83.02 µg/g), an isomer of *trans*-ferulic acid, which is found abundantly in the bound form [19].

Soluble phenolic compounds consist of free phenolic acids and hydroxycinnamate sucrose esters [25]. The main soluble phenolic compounds in brown rice are feruloylsucrose, sinapoyl sucrose and ferulic acid [25]. The components that are notable in brown rice in bound forms are 8-*O*-4' diferulic acid (DFA) (range: 13.88–22.61 µg/g), 8-5' benzofuran DFA (range: 9.28–14.79 µg/g), 5-5' DFA (range: 7.29–13.86 µg/g) and 8-5' DFA (range: 3.26–8.79 µg/g) [19]. The three other hydroxycinnamic acid derivatives that were found in brown rice in small amounts are caffeic acid (range: 0.00–1.44 µg/g), sinapic acid (range: 1.19–1.25 µg/g) and chlorogenic acid (0.63 µg/g) [19]. Two examples of hydroxybenzoic acid derivatives that were found are vanillic acid (range: 2.65–4.74 µg/g) and syringic acid (range: 0.47–2.52 µg/g) [19]. In brown rice, catechin (range: 4.06–8.92 µg/g), quercetin (range: 3.27–6.53 µg/g) and kaempferol (range: 1.30–3.04 µg/g) are the three main flavonoids that are usually found in the free form [19].

In contrast to white rice, brown rice is still constituted by the germ and the bran layers, which contain diverse nutritional compounds, including anti-oxidants [26]. Therefore, despite its high nutritional value, brown rice is consumed less than white rice mainly due to its appearance, longer cooking time, cost, limited availability and bioavailability and poor appreciation of its nutritional value [27]. Apart from cooking, several approaches, including germination, have been emphasized to improve the palatability and the bioavailability of the nutrients present in brown rice. Germination improves the texture and the bioavailability of the nutrients and the phytochemicals [25,28].

In germinated (sprouted) brown rice, about a 70% drop is found in feruloylsucrose (from 1.09–0.27 mg/100 g of flour) and sinapoyl sucrose (from 0.41–0.13 mg/100 g of flour), whereas free ferulic acid content increased (0.48 mg/100 g of flour) when compared to brown rice [25]. However, in general germinated brown rice contains less soluble phenolic compounds when compared to brown rice (1.45 vs. 2.17 mg/100 g of flour) [25]. Apart from that, the sinapinic acid level also increases ten-fold in germinated brown rice (0.21 mg/100 g of flour) compared to brown rice (0.02 mg/100 g of flour) [25]. Germination also increases the levels of the GABA in brown rice [28]. Inositol hexaphosphate is a naturally-occurring molecule found in brown rice [18]. This compound has demonstrated anti-cancer properties [18]. Selenium is a trace mineral, which is found abundantly in brown rice [29]. The function of selenium is to induce DNA repair and combine in damaged cells to promote apoptosis, which is the self-destruction of the cells in the body to remove damaged and worn out cells [29]. Selenium also functions as a cofactor of glutathione peroxidase, which is an enzyme used in the liver to detoxify many possible harmful molecules [29]. Plant lignans are one type of phytonutrient that is found widely in brown rice, which are then converted to mammalian lignan, called enterolactone [30]. Brown rice also serves as a rich source of magnesium. Magnesium plays an important role in our body, as it works as a cofactor of more than 300 enzymes [31]. About 21% of the daily value of magnesium can be obtained by consuming a cup of brown rice [31].

Brown rice contains a high amount of dietary fibre, which has been shown to protect against colorectal cancer [32] and breast cancer [33]. In an animal study, rice bran from brown rice was shown to be beneficial against the development of polyps in the bowel [34]. Due to its high content in fibre, brown rice has a lower glycaemic index, compared to white rice [12]. Consumption of brown rice compared to white rice results in improved endothelial function, without changes in HbA1c levels, possibly through reducing glucose excursions [35]. Vitamin E is also found in brown rice, mainly in two types of structure, which are tocopherols (α, β, γ and δ forms) and tocotrienols (α, β, γ and δ forms) [17]. The function of vitamin E is antioxidant activity, maintenance of membrane integrity, DNA repair, immune support and metabolic processes [21].

Regarding insoluble phenolic compounds, germinated brown rice has at least twice the total content value of insoluble phenolic compounds than brown rice [25]. Ferulic acid and p-coumaric acid are found in the highest quantity in white rice, brown rice and germinated brown rice [25]. Generally, in germinated brown rice (24.78 mg/100 g of flour), insoluble phenolic compounds are 1–2-times more when compared to brown rice (18.47 mg/100 g of flour) [25]. The high levels of phenolic compounds in germinated brown rice are due to the increase in the free forms with alkaline hydrolysis, and this is because of the dismantling of the cell wall during germination [25]. The high levels of insoluble phenolic compounds can enhance the availability of hydrolyzable insoluble phenolic compounds during the germination of brown rice [25].

Amongst other antioxidants that constitute brown rice are the flavonoids. The chemical structure of flavonoids is constituted of a 15-carbon skeleton, which itself is constituted of two aromatic rings interlinked by a heterocyclic ring. The antioxidant activity of the flavonoids stems from the phenolic hydroxyls. Flavones are the most common flavonoids found in brown rice, and tricin is the major flavonoid, accounting for more than 75% of the flavonoids in brown rice [36]. Other flavonoids such as luteolin, apigenin, quercetin, isorhamnetin, kaempferol and myricetin are at relatively low concentrations, as well as isovitexin, naringenin, hesperidin, rutin, luteolin-7-*O*-glucoside, apigenin-7-*O*-glucoside and quercetin-3-*O*-glucoside, amongst others that have been reported [36].

Brown rice also contains sterols present in the bran. The most common sterol is γ-oryzanol, a ferulic acid ester of major phytosterols: campesterol, stigmasterol and β-sitosterol or triterpene alcohols [36]. The γ-oryzanol exhibits several physiological properties including effects on the anthropometry and muscles, cholesterol levels and potential anti-cancer properties [37]. Several analytical methods have been used for the determination of the phytochemical compounds in rice [38–47] (Table 2).

Table 2. Summary of analytical methods used to identify the phytochemical compounds in rice.

Phytochemical Compounds in Rice	Analytical Methods	References
Phenolic acids	Microwave-assisted extraction (MAE) Ultrasound-assisted extraction (UAE)	Sato et al. (2004) [38]
Antioxidants	Microwave-assisted extraction (MAE) Ultrasound-assisted extraction (UAE)	Sato et al. (2004) [38]
Anthocyanins and proanthocyanins	UV-visible spectroscopy	Sato et al. (2004) [38]
Dietary fibre	Enzymatic-gravimetric method	Tiansawang et al. (2016) [39]
Functional lipid	Gravimetric method	Zhou et al. (2003) [40]
Essential amino acid	HPLC method	Naomi et al. (2014) [41]
Phytosterols	Gas chromatography	Zubair et al. (2012). [42]
Flavonoids	Fluorescent DCF	Srisawat et al. (2011) [43]
Tocopherols and tocotrienols	Fluorescent DCF	Srisawat et al. (2011) [43]
Minerals	Ashing method	Horwitz (2000) [44]
Gamma aminobutyric acid (GABA)	Amino acid auto analyser	Cao et al. (2015) [45]
γ-oryzanol	Reversed-phase HPLC method	Xu and Godber (1999) [47]
Phytic acid	UV-Vis spectroscopy	Perera et al. (2018) [46]

Microbial Profiling in Brown Rice

Since the bran and embryo of brown rice are rich in vitamins and fibre, brown rice has the capacity to harbour more microbial association than white rice [48]. During germination, the quality of rice will be improved because the high molecular weight polymers undergo hydrolysis to produce GABA, amino acids, fibres and other bioactive compounds [9]. The germination usually takes place in a warm and humid condition, which favours the growth of microorganisms [49]. The germination of brown rice is initiated when soaking of brown rice occurs. This process involves fermentation because the microbial flora of the environment act upon it after the brown rice is soaked in the water for a certain period of time [50]. Some of these microorganisms can be either harmful or beneficial for consumers [51–55]. For example, some lactic acid bacteria including *Lactobacillus fermentum*, *Pediococcus pentosaceus* and *Weissella confuse* are detected in germinated brown rice [56]. Table 3 shows the types of microbial association in rice.

Table 3. Summary of major microbial association in rice.

Group	Microbes	Microbial Association	References
Gram positive bacteria	*Brevibacillus laterosporus*, *Brevibacillus brevis*, *Brevibacterium* spp.	Production of amino acids	Cottyn et al. [51]
	Cellulomonas flavigena	Degradation of cellulose	Cottyn et al. [51]
	Bacillus thuringiensis and *Bacillus cereus*	Production of enterotoxin	Kim et al. [52]
	Staphylococcus saprophyticus	Food-borne pathogen	Cottyn et al. [51]
Fungi	*Monascus purpureus*	Production of red pigment	Pengnoi et al. [53]
	Fusarium fujikuroi, *Aspergillus flavus* and *Candida*	Production of toxin	Tanaka et al. [54]
Yeast	*Torulopsis etchellsii*, *Hansenula anomala*, *Trichosporon pullulans*, *Geotrichum candidum*, *Saccharomyces* sp.	An increase in the essential amino acids; a decrease in phytic acid and enzyme inhibitors	Panneerselvam et al. [2]; Shortt [55]

3. Nutrigenomic Implications of Brown Rice

Similar to other plants [57–60], although the review of the literature has reported on the health benefits of brown rice, these studies often cannot provide the direct causal relationship between a bioactive compound of brown rice and the observed health benefits. Therefore, it is important to note that these studies should not be over-interpreted because this might be the simplification of the complicated mechanisms in the body that lead to such observed health benefits related to the consumption of brown rice. For example, the review of the literature has shown that brown rice is associated with a wide range of pharmacological properties such as anti-diabetic, anti-cholesterol, anti-hyperlipidemic, cardioprotective and antioxidant [61–67]. Table 4 shows the summary of some important nutrigenomic mechanisms involved in brown rice

Table 4. Summary of some important nutrigenomic mechanisms involved in brown rice.

Property	Potential Underlying Nutrigenomic Mechanism	References
Antioxidative	An increase in antioxidant status and a reduction in oxidative stress via v-akt murine thyromoma viral oncogene (AKT), nuclear factor beta (NF-Kβ), mitogen activated protein kinase (p38 MAPK), c-Jun N-terminal kinase (JNK), extracellular signal-regulated kinase (ERK1/2), p53 tumour suppressor genes, catalase, insulin-like growth factor 2 (IGF2) and superoxide dismutase (SOD)	Azmi et al. (2013) [61]; Imam et al. (2013) [62]; Imam et al. (2012a) [63]; Imam et al. (2012b) [64]
Anti-hyperglycemia	A decrease in the level of blood glucose via the suppression of fbp and pck genes, which are gluconeogenic	Imam and Ismail [62]
Anti-hypocholesterolaemia	A decrease in low density lipoprotein (LDL) and total cholesterol, as well as an increase in high density lipoprotein (HDL) via the transcriptional regulation of hepatic LDL receptor, lipoprotein lipase (LPL), adiponectin, peroxisome proliferator-activator receptor (PPAR) γ, ATP binding cassette (ABCA) 1, AKT and apolipoprotein genes	Imam et al. (2013) [66]; Imam et al. [67]

3.1. Anti-Diabetic Effect

Type 2 diabetes is a worldwide epidemic affecting millions of people across the world and associated with significant morbidity and mortality. Diet and life style factors play an important role in the pathogenicity of type 2 diabetes. Therapeutic management of the disease is only partially effective, costly and associated with adverse side effects. Therefore, scientists and healthcare professionals are looking for alternative management approaches that are safe, affordable and easily accessible to people, especially those residing in the low and middle income countries. In recent years, a considerable increase in scientific research has been observed regarding the use of brown rice for effective management of diabetes mellitus since it is the main staple food in many parts of the world, especially developing countries of Asia and Africa.

Several population-based studies have shown increased risk of type 2 diabetes associated with the intake of white rice, while higher dietary intake or substitution of white rice with brown rice in the diet may decrease the risk [1,27]. In the same context, results of clinical studies are also encouraging. Recently, a research group in Japan has reported a significant decrease in postprandial glucose level in diabetic patients following consumption of glutinous brown rice for one day [68]. The same group has also reported improved glycaemic control in diabetic patients even after eight weeks of ingestion of glutinous brown rice [69]. Using an open-labelled, randomized cross-over study design, they observed a significant decrease in postprandial plasma glucose, haemoglobin A1c (HbA1c) and glycoalbumin levels in patients who ate glutinous brown rice twice a day compared to those on white rice. Another study of similar duration and dietary intervention on Japanese diabetic patients has also reported decreased levels of postprandial plasma glucose levels and improved endothelial function. However, no significant changes were observed in the HbA1c level [35]. Similarly, a randomized controlled trial on Korean type 2 diabetic patients who followed a brown rice-based vegan diet for 12 weeks have also shown improved glycaemic control (larger reductions in HbA1c level) compared to those who followed the conventional diabetic diet [70].

Several other clinical studies have also reported a decreased glycaemic index and better glycaemic and insulin responses in healthy, diabetic and overweight subjects following consumption of brown rice [71–74]. All these beneficial effects are mainly attributed to several bioactive compounds present in brown rice. Brown rice has been shown to prevent type 2 diabetes in several studies [68,69,75]. Brown rice has a crucial role in lowering postprandial blood glucose levels in humans [75]. Apart from that, it also helped in weight management and ameliorated glucose and lipid dysmetabolism in individuals with metabolic syndrome. Brown rice contains high amounts of dietary fibre and other polysaccharides such as arabinoxylan and β-glucan. These fibres and polysaccharides help in regulating glucose absorption in the intestine, thus lowering the glycaemic index [76,77]. It also acts as growth substrates for these components to help in the growth of beneficial bacteria in the gut such as *Lactobacillus* and *Bifidobacterium* [78], thus modulating the gut microbial composition and helping in the prevention of diabetes and obesity [75,79]. It is found that in a recent study, there was a significant relationship between the composition of the intestinal microbiota, obesity and type 2 diabetes [75]. Brown rice is proven to have an important effect on the gut microbial composition in humans. This further supports that there is a relationship between the profile and activity of the intestinal microbiota with the anti-obesity and anti-diabetic effects of brown rice [75].

The rice bran in brown rice is rich in γ-oryzanol, which is responsible for many pharmacological properties, such as cholesterol lowering, anti-inflammatory, anti-cancer, anti-diabetic and antioxidant activities. Brown rice ameliorated glucose tolerance and insulin resistance. A lower glycaemic index was observed in healthy (12.1% lower) and diabetic subjects (35.6% lower) due to consumption of brown rice, and this could help to avoid type 2 diabetes and control glycemia, respectively [74]. These effects are due to the rich bioactive content found in brown rice [63]. A study by Sun et al. [1] has proven that consumption of white rice increased the chances of type 2 diabetes, while substituting one third of the daily serving with brown rice lowered the risk of type 2 diabetes in 200,000 subjects. Brown rice that undergoes germination also has differences in its properties, whereby it is less chewy and richer in bioactive compounds [63]. In germinated brown rice, the high content of dietary fibre helps to decrease the glycaemic index by regulating the absorption of glucose in the intestines [76]. Hypoadiponectinemia, which is implicated in reduced insulin sensitivity in diabetes, can be stopped by γ-oryzanol that is found in brown rice [80]. It also acts on pancreatic islets and increases glucose-stimulated insulin secretion [75]. In addition, GABA, another important bioactive present in brown rice, also has shown a similar effect against hypoadiponectinemia [81].

Similarly, acylated steryl glycoside (ASG), a component found in brown rice, regenerates sodium potassium adenosine triphosphatase and homocysteine thiolactonase enzymes with potential to reverse diabetic neuropathy and oxidative changes on biomolecules [82]. It also enhanced the overall metabolic condition in diabetes as a result of the induction of insulin-like growth factor-1 and reduced oxidative stress, which is a problem in type 2 diabetes [83]. The molecular targets of all these bioactive compounds discussed above are not known. However, it is believed that dysregulation of peroxisome proliferator-activated receptors gamma (PPARγ) is linked to the development of metabolic conditions including type 2 diabetes. Thus, PPARs can be a potential target for bioactive compounds, including those present in germinated brown rice. Indeed, a study by Imam et al. has already reported upregulation of PPARγ following treatment of HEP-G2 cells with germinated brown rice bioactive compounds. Upregulation of PPARγ has therapeutic potential in the management of diabetes.

3.2. Anti-Dyslipoproteinemia

Dyslipoproteinemia is a group of heterogeneous disorders characterized by elevated plasma cholesterol, triglycerides and lipoproteins level. Dyslipoproteinemia is an important risk factor for an array of clinical conditions including atherosclerosis, cardiovascular diseases and acute pancreatitis [84]. Diet plays an important role in inducing dyslipoproteinemia as evident by the rise in the incidence of the disease due to the intake of modern diets high in fats, sugars and refined grain products. Many studies have demonstrated that brown rice also has anti-dyslipoproteinemia and cholesterol lowering effects in animal models. A study by Shen et al. [85] reported an improved lipid profile (significantly decreased level of triglycerides, total cholesterol, high density lipoprotein and non-high density lipoprotein) in mice fed with pre-germinated brown rice-containing high fat diet for 16 weeks. The authors reported that feeding mice with a high fat diet induced dyslipidemia, which can be successfully averted when the mice were fed a high fat diet supplemented with pre-germinated brown rice [85]. The exact mechanisms are not known; however, this might be achieved by decreasing lipid absorption and synthesis and increasing lipid metabolism. The germinated brown rice extract administration in high fat diet-induced obese mice resulted in a significant reduction in serum triglycerides and total cholesterol levels by suppressing lipogenesis via downregulation of genes involved in lipid synthesis [86].

Another pre-clinical study by Miura et al. [87] reported that feeding hepatoma-bearing rats with white rice resulted in hypercholesteremia, which could be successfully suppressed when the rats were fed with a diet containing germinated brown rice. They probably do so by upregulating cholesterol catabolism. Other studies have also reported anti-dyslipoproteinemia and cholesterol lowering effects of germinated brown rice [88,89]. Human clinical studies evaluating the effects of germinated brown rice on dyslipoproteinemia are limited. In a clinical study involving sixty Vietnamese women (aged 45–65 years) with impaired glucose tolerance, the impact of germinated brown rice and white rice intake on blood glucose and lipid profile was evaluated. Following four months of intervention, Bui et al. [90] observed an improvement in blood glucose and lipid level in the pre-germinated brown rice diet group compared to the white rice group.

Similarly, a randomized control trial on 11 diabetic patients also reported a significant reduction in serum total cholesterol and triglyceride level following consumption of pre-germinated brown rice for 14 weeks compared to white rice group [91]. However, no such improvement in serum lipid profile and other metabolic parameters was observed in healthy volunteers who followed either a white rice diet or a white rice plus germinated brown rice diet (1:1, *w*/*w*) for 11–13 months [92]. However, this may be due to the presence of white rice in the diet, which diminishes the beneficial effects of germinated brown rice [72]. Hypercholesterolemia induced by hepatoma growth can be suppressed by means of upregulating cholesterol metabolism. Germinated brown rice also has a greater effect on the restorative effects on cholesterol levels compared to brown rice. This proves that germinated brown rice has a greater impact on high blood cholesterol [88]. All these beneficial activities of brown rice are mainly attributed to the presence of the high concentration of various biologically-active components such as GABA, dietary fibre, γ-oryzanol and other antioxidants in brown rice that help in preventing hyperlipidaemia. The risk of atherogenesis and coronary artery disease through its protection against LDL oxidation can be reduced by the antioxidant contents in brown rice and germinated brown rice [10,93].

3.3. Anti-Cancer Effect

Recent studies have reported the chemo-preventive and anticancer potential of some biologically-active molecules present in germinated brown rice. These molecules can prevent or suppress cancer development. Chemopreventive activities of germinated rough rice have been demonstrated in a recent study [94]. Using azoxymethane, colon cancer was induced in six-week-old male Sprague-Dawley rats followed by oral administration of either control diet or different doses of germinated rough rice crude extract (2000, 1000 and 5000 mg/kg body weight) once daily for eight weeks. The study showed a dose-dependent reduction in the size and number of aberrant crypt foci formation and β-catenin expression in rats fed with germinated rough rice crude extract.

Similarly, a study by Latifah et al. [95] also showed a significant decrease in aberrant crypt foci formation and β-catenin and cox-2 expression when azoxymethane-induced colon cancer rats were fed with different doses of germinated brown rice (2.5, 5 and 10 g/kg body weight). From germinated brown rice, GABA-enhanced parts were extracted, and they portrayed inhibitory action on the reproduction of some cancer cells and a stimulatory action on immune responses. GABA-enriched extracts from germinated brown rice had also been shown to inhibit effects on leukemic cells' proliferation and to stimulate cancer cells in terms of apoptosis [96]. GABA may also play a role in protecting smokers from pulmonary adenocarcinoma due to the reported tumour suppression activity in small airway epithelia [97]. Besides GABA, other bioactive compounds such as tocopherols and tocotrienols present in brown rice may also exhibit anticancer potential [98]. All these studies suggest the potential role of germinated brown rice in cancer prevention. However, further epidemiological and clinical studies are required in order to utilize germinated brown rice as a staple food in cancer prevention activity and for its inhibitory effect. The leukaemia cells that were treated with germinated brown rice extract showed greater DNA fragmentation compared to leukaemia cells treated with brown rice [11]. Apart from that, immunoregulatory activities found in germinated brown rice enhance the cell proliferation of mesenteric lymph node cells in vitro and also increase murine splenic B, T-helper cell subpopulations and nitric oxide c- interferon production [96].

3.4. Lowering Cholesterol

GABA found in brown rice also helps to nourish blood vessels, regulate insulin secretion, avoid increasing blood cholesterol, reduce emotional unrest, improvement from stroke, better the kidney and liver function and prevent chronic alcohol disease [11]. The rice bran oil (RBO) found in brown rice can help to reduce the atherogenic level and increase HDL cholesterol. The cholesterol-reducing activity induced by RBO was due to the decreased absorption-reabsorption of cholesterol and the interference of the plant sterols in cholesterol metabolism. When the unsaponifiable matters of rice bran were fed to hamsters in a study, the faecal fat and neutral sterol excretion was greater. This shows that there is a decrease in fat digestibility [11].

3.5. Cardio-Protective Effect

Cardiovascular disease (CVD) includes diseases of the heart and circulatory system including angina, hypertension, heart attack, congenital heart disease and stroke. On a global scale, CVD is the most common cause of death with an estimated 17.7 million people dying because of CVD in 2015 [99]. To reduce the risk and increase protection against CVD, effective nutritional intervention has always been a focus of public health strategies. In this regard, brown rice and its bioactive compounds have been reported to possess anti-hyperlipidemic, anti-hypercholesteraemic and antihypertensive potential and thus have a role in preventing CVD.

Recently, a randomized cross-over clinical trial was conducted to evaluate the effect of brown rice consumption on inflammatory markers and cardiovascular risk factors [100]. Forty non-menopausal overweight or obese women (BMI > 25) were recruited and divided into two groups. Participants in both groups were asked to consume 150 g cooked brown rice or white rice for six weeks, followed by a two-week wash out period and switching over to an alternate diet for six weeks. Results of the study revealed that consumption of brown rice can significantly reduce inflammatory markers (CRP) and other risk factors (weight, BMI, waist, diastolic blood pressure) associated with CVD.

Another clinical study on healthy female university students revealed that ingestion of brown rice as a staple food for 10 weeks improved general health and prevented hyperlipidaemia, thus protecting against CVD [101]. Several pre-clinical studies have also reported anti-hypertensive effects of germinated brown rice in spontaneously hypertensive rats [102,103]. Mechanistic studies revealed that the antihypertensive effects may be due to the presence of several bioactive compounds in brown rice such as GABA, dietary fibres and ferulic acids [104,105]. On this basis, germinated brown rice can be a good choice as a staple diet or functional food to prevent CVD.

3.6. Antioxidant Effect

Brown rice contains many types of phenolic acids, which are well known for their antioxidant activities and one of the most common antioxidants in our diet. They can protect cells against oxidative damage, thereby reducing the risk of diseases associated with oxidative damage. Prevention of diseases such as cardiovascular disease, type 2 diabetes, obesity and cancer is possible due to the high antioxidant levels found in brown rice. The phenolic acids from brown rice also are assumed to contain chemopreventive properties for breast and colon cancer [106]. Based on the sensory attributes, the whole rice grain is harder to chew and has less taste qualities. Thus, pre-germinated rice is favoured. Brown rice is first soaked in water to initiate germination and contains a higher nutritional value. It is also shown that pre-germinated brown rice increases mental health and immunity. It also helps to prevent diabetic decline [107]. Hepatic fibrosis is one of the most prevalent health problems, and it can be prevented by consuming brown rice. Ferulic acid, *p*-coumaric acid, γ-oryzanol, γ-tocotrienol, GABA and other components in pre-germinated brown rice can decrease liver inflammation and fibrosis and hence reduce the risk of liver cirrhosis and cancer [108].

4. Conclusions and Future Research

Our review has highlighted that brown rice contains certain bioactive phytochemical compounds that might be associated with some important nutrigenomic implications. Therefore, brown rice has received increasing attention from consumers who are health-conscious. In addition, our review also suggests that there are several opportunities for the food industry to develop a wide range of food products using brown rice as the main ingredient. Similar to other plants [59,60,109], future research should be designed to screen for the individual bioactive components that might be associated with the nutrigenomic implications of brown rice.

Author Contributions: The project idea was developed by Z.F.M., K.R. and Z.F.M. wrote the first draft of the manuscript. K.R., Z.F.M., H.Z., Y.C., C.W.W., S.M., E.K.A., Y.Z., Y.J. and B.P. conducted the literature review and revised the manuscript.

Funding: This research received no external funding.

Acknowledgments: Zheng Feei Ma would like to thank Siew Poh Tan, Peng Keong Ma and Zheng Xiong Ma for their active encouragement and support of this work. The authors received no specific funding for this work.

Conflicts of Interest: The authors declare no conflict of interest.

References

1. Sun, Q.; Spiegelman, D.; van Dam, R.M.; Holmes, M.D.; Malik, V.S.; Willett, W.C.; Hu, F.B. White rice, brown rice, and risk of type 2 diabetes in US men and women. *Arch. Intern. Med.* **2010**, *170*, 961–969. [CrossRef] [PubMed]
2. Panneerselvam, P.; Binodh, A.K.; Kumar, U.; Sugitha, T.; Anandan, A. Microbial association in brown rice and their influence on human health. In *Brown Rice*; Manickavasagan, A., Santhakumar, C., Venkatachalapathy, N., Eds.; Springer International Publishing: Cham, Switzerland, 2017; pp. 159–181.
3. Fairhurst, T.; Dobermann, A. Rice in the global food supply. *Better Crops Int.* **2002**, *16*, 3–6.
4. Zareiforoush, H.; Minaei, S.; Alizadeh, M.R.; Banakar, A. Qualitative classification of milled rice grains using computer vision and metaheuristic techniques. *J. Food Sci. Technol.* **2016**, *53*, 118–131. [CrossRef] [PubMed]
5. Kale, S.J.; Jha, S.K.; Jha, G.K.; Sinha, J.P.; Lal, S.B. Soaking induced changes in chemical composition, glycemic index and starch characteristics of basmati rice. *Rice Sci.* **2015**, *22*, 227–236. [CrossRef]
6. Chen, H.; Siebenmorgen, T.J.; Griffin, K. Quality characteristics of long-grain rice milled in two commercial systems. *Cereal Chem. J.* **1998**, *75*, 560–565. [CrossRef]
7. Vetha Varshini, P.; Azhagu Sundharam, K.; Vijay Praveen, P. Brown rice—Hidden nutrients. *J. Biosci. Tech.* **2013**, *4*, 503–507.
8. Liu, L.; Guo, J.; Zhang, R.; Wei, Z.; Deng, Y.; Guo, J.; Zhang, M. Effect of degree of milling on phenolic profiles and cellular antioxidant activity of whole brown rice. *Food Chem.* **2015**, *185*, 318–325. [CrossRef] [PubMed]

9. Ohtsubo, K.I.; Suzuki, K.; Yasui, Y.; Kasumi, T. Bio-functional components in the processed pre-germinated brown rice by a twin-screw extruder. *J. Food Compos. Anal.* **2005**, *18*, 303–316. [CrossRef]

10. Wu, F.; Yang, N.; Touré, A.; Jin, Z.; Xu, X. Germinated brown rice and its role in human health. *Crit. Rev. Food Sci. Nutr.* **2013**, *53*, 451–463. [CrossRef] [PubMed]

11. Cho, D.H.; Lim, S.T. Germinated brown rice and its bio-functional compounds. *Food Chem.* **2016**, *196*, 259–271. [CrossRef] [PubMed]

12. Foster-Powell, K.; Holt, S.H.; Brand-Miller, J.C. International table of glycemic index and glycemic load values: 2002. *Am. J. Clin. Nutr.* **2002**, *76*, 5–56. [CrossRef] [PubMed]

13. Umadevi, M.; Pushpa, R.; Sampathkumar, K.; Bhowmik, D. Rice—Traditional medicinal plant in India. *J. Pharmacogn. Phytochem.* **2012**, *1*, 6–12.

14. Burlando, B.; Cornara, L. Therapeutic properties of rice constituents and derivatives (*Oryza sativa* L.): A review update. *Trends Food Sci. Technol.* **2014**, *40*, 82–98. [CrossRef]

15. Hegde, S.; Yenagi, N.; Kasturiba, B. Indigenous knowledge of the traditional and qualified ayurveda practitioners on the nutritional significance and use of red rice in medications. *Indian J. Tradit. Knowl.* **2013**, *12*, 506–511.

16. Isaac, R.S.R.; Nair, A.S.; Varghese, E.; Chavali, M. Phytochemical, antioxidant and nutrient analysis of medicinal rice (*Oryza sativa* L.) varieties found in south India. *Adv. Sci. Lett.* **2012**, *11*, 86–90. [CrossRef]

17. Okarter, N.; Liu, R.H. Health benefits of whole grain phytochemicals. *Crit. Rev. Food Sci. Nutr.* **2010**, *50*, 193–208. [CrossRef] [PubMed]

18. Babu, P.D.; Subhasree, R.; Bhakyaraj, R.; Vidhyalakshmi, R. Brown rice-beyond the color reviving a lost health food—A review. *Am.-Eurasian J. Agron.* **2009**, *2*, 67–72.

19. Gong, E.S.; Luo, S.J.; Li, T.; Liu, C.M.; Zhang, G.W.; Chen, J.; Zeng, Z.C.; Liu, R.H. Phytochemical profiles and antioxidant activity of brown rice varieties. *Food Chem.* **2017**, *227*, 432–443. [CrossRef] [PubMed]

20. Tan, B.L.; Norhaizan, M.E. Scientific evidence of rice by-products for cancer prevention: Chemopreventive properties of waste products from rice milling on carcinogenesis in vitro and in vivo. *BioMed Res. Int.* **2017**, *2017*, 9017902. [CrossRef] [PubMed]

21. Liu, R.H. Whole grain phytochemicals and health. *J. Cereal Sci.* **2007**, *46*, 207–219. [CrossRef]

22. Adom, K.K.; Liu, R.H. Antioxidant activity of grains. *J. Agric. Food Chem.* **2002**, *50*, 6182–6187. [CrossRef] [PubMed]

23. Adom, K.K.; Sorrells, M.E.; Liu, R.H. Phytochemicals and antioxidant activity of milled fractions of different wheat varieties. *J. Agric. Food Chem.* **2005**, *53*, 2297–2306. [CrossRef] [PubMed]

24. Martínez, I.; Lattimer, J.M.; Hubach, K.L.; Case, J.A.; Yang, J.; Weber, C.G.; Louk, J.A.; Rose, D.J.; Kyureghian, G.; Peterson, D.A.; et al. Gut microbiome composition is linked to whole grain-induced immunological improvements. *ISME J.* **2013**, *7*, 269–280. [CrossRef] [PubMed]

25. Tian, S.; Nakamura, K.; Kayahara, H. Analysis of phenolic compounds in white rice, brown rice, and germinated brown rice. *J. Agric. Food Chem.* **2004**, *52*, 4808–4813. [CrossRef] [PubMed]

26. Cáceres, P.J.; Peñas, E.; Martinez-Villaluenga, C.; Amigo, L.; Frias, J. Enhancement of biologically active compounds in germinated brown rice and the effect of sun-drying. *J. Cereal Sci.* **2017**, *73*, 1–9. [CrossRef]

27. Adebamowo, S.N.; Eseyin, O.; Yilme, S.; Adeyemi, D.; Willett, W.C.; Hu, F.B.; Spiegelman, D.; Adebamowo, C.A. A mixed-methods study on acceptability, tolerability, and substitution of brown rice for white rice to lower blood glucose levels among Nigerian adults. *Front. Nutr.* **2017**, *4*, 33. [CrossRef] [PubMed]

28. Patil, S.B.; Khan, M.K. Germinated brown rice as a value added rice product: A review. *J. Food Sci. Technol.* **2011**, *48*, 661–667. [CrossRef] [PubMed]

29. Vogt, T.M.; Ziegler, R.G.; Graubard, B.I.; Swanson, C.A.; Greenberg, R.S.; Schoenberg, J.B.; Swanson, G.M.; Hayes, R.B.; Mayne, S.T. Serum selenium and risk of prostate cancer in U.S. blacks and whites. *Int. J. Cancer* **2003**, *103*, 664–670. [CrossRef] [PubMed]

30. Anderson, J.W.; Hanna, T.J. Whole grains and protection against coronary heart disease: What are the active components and mechanisms? *Am. J. Clin. Nutr.* **1999**, *70*, 307–308. [CrossRef] [PubMed]

31. Jiamyangyuen, S.; Ooraikul, B. The physico-chemical, eating and sensorial properties of germinated brown rice. *J. Food Agric. Environ.* **2008**, *6*, 119–124.

32. Aune, D.; Chan, D.S.M.; Lau, R.; Vieira, R.; Greenwood, D.C.; Kampman, E.; Norat, T. Dietary fibre, whole grains, and risk of colorectal cancer: Systematic review and dose-response meta-analysis of prospective studies. *Br. Med. J.* **2011**, *343*, d6617. [CrossRef] [PubMed]

33. Aune, D.; Chan, D.S.; Greenwood, D.C.; Vieira, A.R.; Rosenblatt, D.A.; Vieira, R.; Norat, T. Dietary fiber and breast cancer risk: A systematic review and meta-analysis of prospective studies. *Ann. Oncol.* **2012**, *23*, 1394–1402. [PubMed]

34. Verschoyle, R.D.; Greaves, P.; Cai, H.; Edwards, R.E.; Steward, W.P.; Gescher, A.J. Evaluation of the cancer chemopreventive efficacy of rice bran in genetic mouse models of breast, prostate and intestinal carcinogenesis. *Br. J. Cancer* **2007**, *96*, 248–254. [CrossRef] [PubMed]

35. Kondo, K.; Morino, K.; Nishio, Y.; Ishikado, A.; Arima, H.; Nakao, K.; Nakagawa, F.; Nikami, F.; Sekine, O.; Nemoto, K.I.; et al. Fiber-rich diet with brown rice improves endothelial function in type 2 diabetes mellitus: A randomized controlled trial. *PLoS ONE* **2017**, *12*, e0179869. [CrossRef] [PubMed]

36. Goufo, P.; Trindade, H. Rice antioxidants: Phenolic acids, flavonoids, anthocyanins, proanthocyanidins, tocopherols, tocotrienols, gamma-oryzanol, and phytic acid. *Food Sci. Nutr.* **2014**, *2*, 75–104. [CrossRef] [PubMed]

37. Cicero, A.F.; Gaddi, A. Rice bran oil and gamma-oryzanol in the treatment of hyperlipoproteinaemias and other conditions. *Phytother. Res.* **2001**, *15*, 277–289. [PubMed]

38. Sato, S.; Soga, T.; Nishioka, T.; Tomita, M. Simultaneous determination of the main metabolites in rice leaves using capillary electrophoresis mass spectrometry and capillary electrophoresis diode array detection. *Plant J.* **2004**, *40*, 151–163. [CrossRef] [PubMed]

39. Tiansawang, K.; Luangpituksa, P.; Varanyanond, W.; Hansawasdi, C. GABA production, antioxidant activity in some germinated dietary seeds and the effect of cooking on their GABA content. *Food Sci. Technol.* **2016**, *36*, 313–321. [CrossRef]

40. Zhou, Z.; Blanchard, C.; Helliwell, S.; Robards, K. Fatty acid composition of three rice varieties following storage. *J. Cereal Sci.* **2003**, *37*, 327–335. [CrossRef]

41. Eshun Naomi, A.; Emmanuel, A.; Barimah, J.; VM, D.; Van Twisk, C. Amino acid profiles of some varieties of rice, soybean and groundnut grown. *J. Food Process. Technol.* **2015**, *6*, 420.

42. Zubair, M.; Anwar, F.; Ashraf, M.; Uddin, M.K. Characterization of high-value bioactives in some selected varieties of Pakistani Rice (*Oryza sativa* L.). *Int. J. Mol. Sci.* **2012**, *13*, 4608–4622. [CrossRef] [PubMed]

43. Srisawat, U.; Panunto, W.; Kaendee, N.; Tanuchit, S.; Itharat, A.; Lerdvuthisopon, N.; Hansakul, P. Determination of phenolic compounds, flavonoids, and antioxidant activities in water extracts of Thai red and white rice cultivars. *J. Med. Assoc. Thail.* **2010**, *93*, S83–S91.

44. Horwitz, W. *Official Methods of Analysis of AOAC International*; AOAC International: Gaithersburg, MD, USA, 2000.

45. Cao, Y.; Jia, F.; Han, Y.; Liu, Y.; Zhang, Q. Study on the optimal moisture adding rate of brown rice during germination by using segmented moisture conditioning method. *J. Food Sci. Technol.* **2015**, *52*, 6599–6606. [CrossRef] [PubMed]

46. Perera, I.; Seneweera, S.; Hirotsu, N. Manipulating the phytic acid content of rice grain toward improving micronutrient bioavailability. *Rice* **2018**, *11*, 4. [CrossRef] [PubMed]

47. Xu, Z.; Godber, J.S. Purification and identification of components of γ-oryzanol in rice bran oil. *J. Agric. Food Chem.* **1999**, *47*, 2724–2728. [CrossRef] [PubMed]

48. Cui, L.; Pan, Z.; Yue, T.; Atungulu, G.G.; Berrios, J. Effect of ultrasonic treatment of brown rice at different temperatures on cooking properties and quality. *Cereal Chem. J.* **2010**, *87*, 403–408. [CrossRef]

49. Lu, Z.H.; Zhang, Y.; Li, L.T.; Curtis, R.B.; Kong, X.L.; Fulcher, R.G.; Zhang, G.; Cao, W. Inhibition of microbial growth and enrichment of gamma-aminobutyric acid during germination of brown rice by electrolyzed oxidizing water. *J. Food Prot.* **2010**, *73*, 483–487. [CrossRef] [PubMed]

50. Puri, S.; Dhillon, B.; Sodhi, N.S. Effect of degree of milling (Dom) on overall quality of rice—A review. *Int. J. Adv. Biotechnol. Res.* **2014**, *5*, 474–489.

51. Cottyn, B.; Regalado, E.; Lanoot, B.; De Cleene, M.; Mew, T.W.; Swings, J. Bacterial populations associated with rice seed in the tropical environment. *Phytopathology* **2001**, *91*, 282–292. [CrossRef] [PubMed]

52. Kim, B.; Bang, J.; Kim, H.; Kim, Y.; Kim, B.S.; Beuchat, L.R.; Ryu, J.H. *Bacillus cereus* and *Bacillus thuringiensis* spores in Korean rice: Prevalence and toxin production as affected by production area and degree of milling. *Food Microbiol.* **2014**, *42*, 89–94. [CrossRef] [PubMed]

53. Pengnoi, P.; Mahawan, R.; Khanongnuch, C.; Lumyong, S. Antioxidant properties and production of monacolin k, citrinin, and red pigments during solid state fermentation of purple rice (*Oryzae sativa*) varieties by *Monascus purpureus*. *Czech J. Food Sci.* **2017**, *35*, 32–39.

54. Tanaka, K.; Sago, Y.; Zheng, Y.; Nakagawa, H.; Kushiro, M. Mycotoxins in rice. *Int. J. Food Microbiol.* **2007**, *119*, 59–66. [CrossRef] [PubMed]

55. Shortt, C. Living it up for dinner. *Chem. Ind.* **1998**, *8*, 300–303.

56. Kim, K.-S.; Kim, B.-H.; Kim, M.-J.; Han, J.-K.; Kum, J.-S.; Lee, H.-Y. Quantitative microbiological profiles of brown rice and germinated brown rice. *Food Sci. Biotechnol.* **2012**, *21*, 1785–1788. [CrossRef]

57. Ma, Z.F.; Zhang, H. Phytochemical constituents, health benefits, and industrial applications of grape seeds: A mini-review. *Antioxidants* **2017**, *6*, 71. [CrossRef] [PubMed]

58. Ma, Z.F.; Lee, Y.Y. Virgin coconut oil and its cardiovascular health benefits. *Nat. Prod. Commun.* **2016**, *11*, 1151–1152.

59. Zhang, H.; Ma, Z.F. Phytochemical and pharmacological poperties of *Capparis spinosa* as a medicinal plant. *Nutrients* **2018**, *10*, 116. [CrossRef] [PubMed]

60. Cao, Y.; Ma, Z.; Zhang, H.; Jin, Y.; Zhang, Y.; Hayford, F. Phytochemical properties and nutrigenomic implications of yacon as a potential source of prebiotic: Current evidence and future directions. *Foods* **2018**, *7*, 59. [CrossRef] [PubMed]

61. Azmi, N.H.; Ismail, N.; Imam, M.U.; Ismail, M. Ethyl acetate extract of germinated brown rice attenuates hydrogen peroxide-induced oxidative stress in human SH-SY5Y neuroblastoma cells: Role of anti-apoptotic, pro-survival and antioxidant genes. *BMC Complement. Altern. Med.* **2013**, *13*, 177. [CrossRef] [PubMed]

62. Imam, M.U.; Ismail, M. Nutrigenomic effects of germinated brown rice and its bioactives on hepatic gluconeogenic genes in type 2 diabetic rats and HEPG2 cells. *Mol. Nutr. Food Res.* **2013**, *57*, 401–411. [CrossRef] [PubMed]

63. Imam, M.U.; Musa, S.N.A.; Azmi, N.H.; Ismail, M. Effects of white rice, brown rice and germinated brown rice on antioxidant status of type 2 diabetic rats. *Int. J. Mol. Sci.* **2012**, *13*, 12952–12969. [CrossRef] [PubMed]

64. Imam, M.U.; Ismail, M.; Omar, A.R. Nutrigenomic effects of germinated brown rice bioactives on antioxidant genes. *Free Radic. Biol. Med.* **2012**, *53*, S105–S106. [CrossRef]

65. Imam, M.U.; Azmi, N.H.; Ismail, M. Upregulation of superoxide dismutase gene is involved in antioxidant effects of germinated rown rice. *Free Radic. Biol. Med.* **2012**, *53*, S84. [CrossRef]

66. Imam, M.U.; Ismail, M.; Omar, A.R.; Ithnin, H. The hypocholesterolemic effect of germinated brown rice involves the upregulation of the apolipoprotein A1 and low-density lipoprotein receptor genes. *J. Diabetes Res.* **2013**, *2013*, 134694. [CrossRef] [PubMed]

67. Imam, M.U.; Ishaka, A.; Ooi, D.-J.; Zamri, N.D.M.; Sarega, N.; Ismail, M.; Esa, N.M. Germinated brown rice regulates hepatic cholesterol metabolism and cardiovascular disease risk in hypercholesterolaemic rats. *J. Funct. Foods* **2014**, *8*, 193–203. [CrossRef]

68. Terashima, Y.; Nagai, Y.; Kato, H.; Ohta, A.; Tanaka, Y. Eating glutinous brown rice for one day improves glycemic control in Japanese patients with type 2 diabetes assessed by continuous glucose monitoring. *Asia Pac. J. Clin. Nutr.* **2017**, *26*, 421–426. [PubMed]

69. Nakayama, T.; Nagai, Y.; Uehara, Y.; Nakamura, Y.; Ishii, S.; Kato, H.; Tanaka, Y. Eating glutinous brown rice twice a day for 8 weeks improves glycemic control in Japanese patients with diabetes mellitus. *Nutr. Diabetes* **2017**, *7*, e273. [CrossRef] [PubMed]

70. Lee, Y.M.; Kim, S.A.; Lee, I.K.; Kim, J.G.; Park, K.G.; Jeong, J.Y.; Jeon, J.H.; Shin, J.Y.; Lee, D.H. Effect of a brown rice based vegan diet and conventional diabetic diet on glycemic control of patients with type 2 diabetes: A 12-week randomized clinical trial. *PLoS ONE* **2016**, *11*, e0155918. [CrossRef] [PubMed]

71. Shobana, S.; Lakshmipriya, N.; Bai, M.R.; Gayathri, R.; Ruchi, V.; Sudha, V.; Malleshi, N.G.; Krishnaswamy, K.; Henry, C.K.; Anjana, R.M.; et al. Even minimal polishing of an Indian parboiled brown rice variety leads to increased glycemic responses. *Asia Pac. J. Clin. Nutr.* **2017**, *26*, 829–836. [PubMed]

72. Ito, Y.; Mizukuchi, A.; Kise, M.; Aoto, H.; Yamamoto, S.; Yoshihara, R.; Yokoyama, J. Postprandial blood glucose and insulin responses to pre-germinated brown rice in healthy subjects. *J. Med. Investig.* **2005**, *52*, 159–164. [CrossRef]

73. Mohan, V.; Spiegelman, D.; Sudha, V.; Gayathri, R.; Hong, B.; Praseena, K.; Anjana, R.M.; Wedick, N.M.; Arumugam, K.; Malik, V.; et al. Effect of brown rice, white rice, and brown rice with legumes on blood glucose and insulin responses in overweight Asian Indians: A randomized controlled trial. *Diabetes Technol. Ther.* **2014**, *16*, 317–325. [CrossRef] [PubMed]

74. Panlasigui, L.N.; Thompson, L.U. Blood glucose lowering effects of brown rice in normal and diabetic subjects. *Int. J. Food Sci. Nutr.* **2006**, *57*, 151–158. [CrossRef] [PubMed]

75. Kozuka, C.; Yabiku, K.; Takayama, C.; Matsushita, M.; Shimabukuro, M. Natural food science based novel approach toward prevention and treatment of obesity and type 2 diabetes: Recent studies on brown rice and gamma-oryzanol. *Obes. Res. Clin. Pract.* **2013**, *7*, e165–e172. [CrossRef] [PubMed]

76. Ou, S.; Kwok, K.; Li, Y.; Fu, L. In vitro study of possible role of dietary fiber in lowering postprandial serum glucose. *J. Agric. Food Chem.* **2001**, *49*, 1026–1029. [CrossRef] [PubMed]

77. Seki, T.; Nagase, R.; Torimitsu, M.; Yanagi, M.; Ito, Y.; Kise, M.; Mizukuchi, A.; Fujimura, N.; Hayamizu, K.; Ariga, T. Insoluble fiber is a major constituent responsible for lowering the post-prandial blood glucose concentration in the pre-germinated brown rice. *Biol. Pharm. Bull.* **2005**, *28*, 1539–1541. [CrossRef] [PubMed]

78. Jaskari, J.; Kontula, P.; Siitonen, A.; Jousimies-Somer, H.; Mattila-Sandholm, T.; Poutanen, K. Oat beta-glucan and xylan hydrolysates as selective substrates for *Bifidobacterium* and *Lactobacillus* strains. *Appl. Microbiol. Biotechnol.* **1998**, *49*, 175–181. [CrossRef] [PubMed]

79. Benno, Y.; Endo, K.; Miyoshi, H.; Okuda, T.; Koishi, H.; Mitsuoka, T. Effect of rice fiber on human fecal microflora. *Microbiol. Immunol.* **1989**, *33*, 435–440. [CrossRef] [PubMed]

80. Nagasaka, R.; Yamsaki, T.; Uchida, A.; Ohara, K.; Ushio, H. Gamma-oryzanol recovers mouse hypoadiponectinemia induced by animal fat ingestion. *Phytomedicine* **2011**, *18*, 669–671. [CrossRef] [PubMed]

81. Ohara, K.; Kiyotani, Y.; Uchida, A.; Nagasaka, R.; Maehara, H.; Kanemoto, S.; Hori, M.; Ushio, H. Oral administration of gamma-aminobutyric acid and gamma-oryzanol prevents stress-induced hypoadiponectinemia. *Phytomedicine* **2011**, *18*, 655–660. [CrossRef] [PubMed]

82. Usuki, S.; Ito, Y.; Morikawa, K.; Kise, M.; Ariga, T.; Rivner, M.; Yu, R.K. Effect of pre-germinated brown rice intake on diabetic neuropathy in streptozotocin-induced diabetic rats. *Nutr. Metab.* **2007**, *4*, 25. [CrossRef] [PubMed]

83. Usuki, S.; Tsai, Y.Y.; Morikawa, K.; Nonaka, S.; Okuhara, Y.; Kise, M.; Yu, R.K. IGF-1 induction by acylated steryl beta-glucosides found in a pre-germinated brown rice diet reduces oxidative stress in streptozotocin-induced diabetes. *PLoS ONE* **2011**, *6*, e28693. [CrossRef] [PubMed]

84. Nordestgaard, B.G.; Varbo, A. Triglycerides and cardiovascular disease. *Lancet* **2014**, *384*, 626–635. [CrossRef]

85. Shen, K.-P.; Hao, C.-L.; Yen, H.-W.; Chen, C.-Y.; Chen, J.-H.; Chen, F.-C.; Lin, H.-L. Pre-germinated brown rice prevented high fat diet induced hyperlipidemia through ameliorating lipid synthesis and metabolism in C57BL/6J mice. *J. Clin. Biochem. Nutr.* **2016**, *59*, 39–44. [CrossRef] [PubMed]

86. Ho, J.N.; Son, M.E.; Lim, W.C.; Lim, S.T.; Cho, H.Y. Anti-obesity effects of germinated brown rice extract through down-regulation of lipogenic genes in high fat diet-induced obese mice. *Biosci. Biotechnol. Biochem.* **2012**, *76*, 1068–1074. [CrossRef] [PubMed]

87. Miura, D.; Ito, Y.; Mizukuchi, A.; Kise, M.; Aoto, H.; Yagasaki, K. Hypocholesterolemic action of pre-germinated brown rice in hepatoma-bearing rats. *Life Sci.* **2006**, *79*, 259–264. [CrossRef] [PubMed]

88. Roohinejad, S.; Omidizadeh, A.; Mirhosseini, H.; Saari, N.; Mustafa, S.; Yusof, R.M.; Hussin, A.S.; Hamid, A.; Abd Manap, M.Y. Effect of pre-germination time of brown rice on serum cholesterol levels of hypercholesterolaemic rats. *J. Sci. Food Agric.* **2010**, *90*, 245–251. [CrossRef] [PubMed]

89. Albarracin, M.; Weisstaub, A.R.; Zuleta, A.; Drago, S.R. Extruded whole grain diets based on brown, soaked and germinated rice. Effects on cecum health, calcium absorption and bone parameters of growing Wistar rats. Part I. *Food Funct.* **2016**, *7*, 2722–2728. [CrossRef] [PubMed]

90. Bui, T.N.; Le, T.H.; Nguyen, D.H.; Tran, Q.B.; Nguyen, T.L.; Le, D.T.; Do, V.A.; Vu, A.L.; Aoto, H.; Okuhara, Y.; et al. Pre-germinated brown rice reduced both blood glucose concentration and body weight in Vietnamese women with impaired glucose tolerance. *J. Nutr. Sci. Vitaminol.* **2014**, *60*, 183–187. [CrossRef] [PubMed]

91. Hsu, T.F.; Kise, M.; Wang, M.F.; Ito, Y.; Yang, M.D.; Aoto, H.; Yoshihara, R.; Yokoyama, J.; Kunii, D.; Yamamoto, S. Effects of pre-germinated brown rice on blood glucose and lipid levels in free-living patients with impaired fasting glucose or type 2 diabetes. *J. Nutr. Sci. Vitaminol.* **2008**, *54*, 163–168. [CrossRef] [PubMed]

92. Morita, H.; Uno, Y.; Umemoto, T.; Sugiyama, C.; Matsumoto, M.; Wada, Y.; Ishizuka, T. Effect of gamma-aminobutyric acid-rich germinated brown rice on indexes of life-style related diseases. *Nihon Ronen Igakkai Zasshi* **2004**, *41*, 211–216. [CrossRef] [PubMed]

93. Wu, C.; Ye, Z.; Li, H.; Wu, S.; Deng, D.; Zhu, Y.; Wong, M. Do radial oxygen loss and external aeration affect iron plaque formation and arsenic accumulation and speciation in rice? *J. Exp. Bot.* **2012**, *63*, 2961–2970. [CrossRef] [PubMed]

94. Saki, E.; Saiful Yazan, L.; Mohd Ali, R.; Ahmad, Z. Chemopreventive effects of germinated rough rice crude extract in inhibiting azoxymethane-induced aberrant crypt foci formation in sprague-dawley rats. *BioMed Res. Int.* **2017**, *2017*, 9517287. [CrossRef] [PubMed]

95. Latifah, S.Y.; Armania, N.; Tze, T.H.; Azhar, Y.; Nordiana, A.H.; Norazalina, S.; Hairuszah, I.; Saidi, M.; Maznah, I. Germinated brown rice (GBR) reduces the incidence of aberrant crypt foci with the involvement of beta-catenin and COX-2 in azoxymethane-induced colon cancer in rats. *Nutr. J.* **2010**, *9*, 16. [CrossRef] [PubMed]

96. Oh, C.H.; Oh, S.H. Effects of germinated brown rice extracts with enhanced levels of GABA on cancer cell proliferation and apoptosis. *J. Med. Food* **2004**, *7*, 19–23. [CrossRef] [PubMed]

97. Schuller, H.M.; Al-Wadei, H.A.; Majidi, M. Gamma-aminobutyric acid, a potential tumor suppressor for small airway-derived lung adenocarcinoma. *Carcinogenesis* **2008**, *29*, 1979–1985. [CrossRef] [PubMed]

98. Har, C.H.; Keong, C.K. Effects of tocotrienols on cell viability and apoptosis in normal murine liver cells (BNL CL.2) and liver cancer cells (BNL 1ME A.7R.1), in vitro. *Asia Pac. J. Clin. Nutr.* **2005**, *14*, 374–380. [PubMed]

99. Go, A.S.; Mozaffarian, D.; Roger, V.L.; Benjamin, E.J.; Berry, J.D.; Blaha, M.J.; Dai, S.; Ford, E.S.; Fox, C.S.; Franco, S.; et al. Heart Disease and Stroke Statistics—2014 Update: A Report From the American Heart Association. *Circulation* **2014**, *129*, e28–e292. [CrossRef] [PubMed]

100. Kazemzadeh, M.; Safavi, S.M.; Nematollahi, S.; Nourieh, Z. Effect of brown rice consumption on inflammatory marker and cardiovascular risk factors among overweight and obese non-menopausal female adults. *Int. J. Prev. Med.* **2014**, *5*, 478–488. [PubMed]

101. Ebizuka, H.; Sasaki, C.; Kise, M.; Arita, M. Effects of retort pouched rice containing pre-germinated brown rice on daily nutrition and physical status in healthy subjects. *J. Integr. Study Diet. Habits* **2007**, *18*, 216–222. [CrossRef]

102. Ebizuka, H.; Ihara, M.; Arita, M. Antihypertensive effect of pre-germinated brown rice in spontaneously hypertensive rats. *Food Sci. Technol. Res.* **2009**, *15*, 625–630. [CrossRef]

103. Choi, H.-D.; Kim, Y.-S.; Choi, I.-W.; Park, Y.-K.; Park, Y.-D. Hypotensive effect of germinated brown rice on spontaneously hypertensive rats. *Korean J. Food Sci. Technol.* **2006**, *38*, 448–451.

104. Suzuki, A.; Kagawa, D.; Fujii, A.; Ochiai, R.; Tokimitsu, I.; Saito, I. Short- and long-term effects of ferulic acid on blood pressure in spontaneously hypertensive rats. *Am. J. Hypertens.* **2002**, *15*, 351–357. [CrossRef]

105. Ohsaki, Y.; Shirakawa, H.; Koseki, T.; Komai, M. Novel effects of a single administration of ferulic acid on the regulation of blood pressure and the hepatic lipid metabolic profile in stroke-prone spontaneously hypertensive rats. *J. Agric. Food Chem.* **2008**, *56*, 2825–2830.

106. Shao, Y.; Bao, J. Polyphenols in whole rice grain: Genetic diversity and health benefits. *Food Chem.* **2015**, *180*, 86–97. [CrossRef] [PubMed]

107. Sakamoto, S.; Hayashi, T.; Hayashi, K.; Murai, F.; Hori, M.; Kimoto, K.; Murakami, K. Pre-germinated brown rice could enhance maternal mental health and immunity during lactation. *Eur. J. Nutr.* **2007**, *46*, 391–396. [CrossRef] [PubMed]

108. Wunjuntuk, K.; Kettawan, A.; Rungruang, T.; Charoenkiatkul, S. Anti-fibrotic and anti-inflammatory effects of parboiled germinated brown rice (*Oryza sativa* 'KDML 105') in rats with induced liver fibrosis. *J. Funct. Foods* **2016**, *26*, 363–372. [CrossRef]

109. Zhang, H.; Ma, Z.F.; Luo, X.; Li, X. Effect of mulberry fruit (*Morus alba* L.) consumption on health outcomes: A mini-review. *Antioxidants* **2018**, *7*, 69. [CrossRef]

![antioxidants logo] *antioxidants*

MDPI

Review

Effects of Cocoa Antioxidants in Type 2 Diabetes Mellitus

Sonia Ramos [1], María Angeles Martín [1,2] and Luis Goya [1,*]

[1]   Department of Metabolism and Nutrition. Institute of Food Science and Technology and Nutrition (ICTAN-CSIC). José Antonio Novais 10. Ciudad Universitaria, 28040 Madrid, Spain; s.ramos@ictan.csic.es (S.R.); amartina@ictan.csic.es (M.A.M.)

[2]   Spanish Biomedical Research Centre in Diabetes and Associated Metabolic Disorders (CIBERDEM), Instituto de Salud Carlos III (ISCIII), 28040 Madrid, Spain

*   Correspondence: luisgoya@ictan.csic.es; Tel.: +34-915-445-607

Received: 28 September 2017; Accepted: 27 October 2017; Published: 31 October 2017

Abstract: Type 2 Diabetes mellitus (T2D) is the most common form of diabetes and one of the most common chronic diseases. Control of hyperglycaemia by hypoglycaemic drugs is insufficient in for patients and nutritional approaches are currently being explored. Natural dietary compounds such as flavonoids, abundant in fruits and vegetables, have received broad attention because of their potential capacity to act as anti-diabetic agents. Especially cocoa flavonoids have been proved to ameliorate important hallmarks of T2D. In this review, an update of the most relevant reports published during the last decade in cell culture, animal models and human studies is presented. Most results support an anti-diabetic effect of cocoa flavonoids by enhancing insulin secretion, improving insulin sensitivity in peripheral tissues, exerting a lipid-lowering effect and preventing the oxidative and inflammatory damages associated to the disease. While it could be suggested that daily consumption of flavanols from cocoa or dark chocolate would constitute a potential preventive tool useful for the nutritional management of T2D, this recommendation should be cautious since most of commercially available soluble cocoa products or chocolates contain low amount of flavanols and are rich in sugar and calories that may aggravate glycaemic control in T2D patients.

Keywords: beta cell; cocoa flavonoids; flavanols; hyperglycaemia; insulin resistance; procyanidins

1. Introduction

Type 2 Diabetes mellitus (T2D) is the most common form of diabetes and one of the most common chronic diseases, and its prevalence is raising worldwide [1]. T2D is characterized by a sustained hyperglycaemia due to the persistent damage in insulin secretion by pancreatic β-cell dysfunction and by insulin resistance at the peripheral tissues [1]. However, administration of glucose-lowering medications is insufficient to maintain glycaemic control in many patients and changes in life-style, such as physical exercise and nutrition, both with lowest adverse side effects, are presumed to be the most promising approaches to prevent or delay the onset of T2D. Accordingly identification of dietary components as potential antidiabetic agents has become an essential subject in the current research. Flavonoids, natural dietary compounds abundant in fruits and vegetables, have raised attention because of their prospective capacity to act as anti-diabetic agents with very few adverse side effects [1]. In particular, cocoa and its flavanols, a subfamily of flavonoids, have been proved to ameliorate important hallmarks of T2D [1]. In this review, an update of the most relevant reports published on this topic during the last decade in cell culture, animal models and human studies is presented.

2. Studies in Cultured Cells

Cultured cells constitute a useful tool to study the poorly understood anti-diabetic mechanism by which cocoa flavanols act in vivo. By using several cultured cell lines, very appealing and encouraging results have been reported within the last five years (Table 1).

Cocoa flavanols may improve glucose homeostasis by slowing carbohydrate digestion and absorption in the gut. Indeed, Gu et al. [2] have shown that cocoa extracts and procyanidins dose-dependently inhibited pancreatic α-amylase (PA), pancreatic lipase (PL), and secreted phospholipase A2 (PLA2). These inhibitory activities were related to the polyphenolic content in cocoa extracts and the degree of polymerization of cocoa procyanidins, showing an inverse correlation between the inhibition and degree of polymerization (Table 1). Similarly, Yamashita et al. [3] found that a cocoa liquor procyanidin extract (CLPr) (1–10 µg/mL) inhibited α-glucosidase activity in the muscle L6 myotube cells. In addition, cocoa polyphenols may also inhibit certain glucose transporters. Thus, the same CLPr increased glucose uptake in a dose-dependent manner and stimulated GLUT-4 translocation in skeletal muscle L6 cells, whereas levels of GLUT-1 and -4 remained unchanged in the plasma membrane [3] (Table 1).

Deterioration of functional β-cell mass is observed during T2D, and this critically affects to the maintenance of normoglycemia [1]. There are various studies that suggest that cocoa polyphenols may protect β-cells against death-inducing damaging factors, enhance glucose stimulated insulin secretion and induce β-cell replication. In this regard, a monomeric catechin-rich cocoa flavanol fraction enhanced glucose-stimulated insulin secretion, while cells cultured with total cocoa extract or with oligomeric or polymeric procyanidin-rich fractions did not show any effect demonstrated no improvement (all substances were used at concentrations ranged 0.75–25 µg/mL) [4] (Table 1). The increased glucose-stimulated insulin secretion in the presence of the monomeric catechin-rich fraction was associated with enhanced mitochondrial respiration (upregulation of mitochondrial complex III, IV and V components and increased cellular ATP production), suggesting improvements in β-cell fuel utilization. Additionally, the monomer-rich fraction improved cellular redox state, as increased reduced glutathione (GSH) levels and nuclear factor, erythroid 2 like 2 (Nrf2) in the nucleus and its target genes that are essential for increasing mitochondrial function [nrf1 and GA binding protein transcription factor alpha subunit (GABPA)] [4] (Table 1). Similarly, the main cocoa flavanol epicatechin (EC) (100 nM) reverted mitochondrial function-associated and biogenesis-associated indicators such as mitochondrial transcription factor A (TFAM), sirtuin (SIRT)-1, mitofilin and peroxisome proliferator-activated receptor γ co-activator 1α (PGC-1α) in high-glucose-treated endothelial cells [5] (Table 1). Further, Martín et al. [6,7] showed that a cocoa phenolic extract (CPE) and EC protected against oxidative stress in rat insulin-secreting INS-1 cells. Pre-treatment of cells with EC (5–20 µM) prevented the prooxidant *tert*-butylhydroperoxide (t-BOOH) induced reactive oxygen species (ROS), carbonyl groups, phosphorylated jun N-terminal kinase (p-JNK) expression and cell death, and recovered insulin secretion [6]. Similarly, pre-treatment of cells with CPE significantly prevented the t-BOOH-induced ROS and carbonyl groups and returned antioxidant defences [GSH levels and glutathione peroxidase (GPx) and reductase (GR) activities] to control values [7] (Table 1). Moreover, the microbial-derived flavonoid metabolites 3,4-dihydroxyphenylacetic acid (DHPAA, 5 µM), and 3-hydroxyphenylpropionic acid (HPPA, 1 µM) derived from flavanols metabolized by gut microflora potentiated glucose-stimulated insulin secretion in INS-1E cells and in rat pancreatic islets [8]. Pre-treatment of cells with both compounds protected against beta cell dysfunction and death induced by t-BOOH through the activation of protein kinase C (PKC) and extracellular-regulated kinases (ERK) pathways [8] (Table 1). In addition, procyanidin A2 (3–300 µM) prevented the damage induced on insulin secretion by bisphenol A (100 µg/L) in mice isolated islets [9] (Table 1).

Cocoa flavanols may also modulate insulin signaling, which is impaired in T2D [1]. In this line, Cordero-Herrera et al. [10,11] have reported that a CPE and EC at physiologically relevant doses (1–10 µg/mL and 1–10 µM, respectively) enhanced the tyrosine phosphorylated and total levels of insulin receptor (IR), insulin receptor substrate (IRS)-1, and IRS-2, and activated the phosphatidylinositol-3-kinase/protein kinase B (PI3K/AKT) pathway. Both natural substances also modulated hepatic gluconeogenic and phosphoenolpyruvate carboxykinase (PEPCK) expression through AKT and AMP-activated protein kinase (AMPK), and CPE increased GLUT-2 levels [10]. Indeed, EC and CPE alleviated the hepatic insulin resistance induced on HepG2 by treating the cells with a high dose of glucose [11] (Table 1). Thus, EC and CPE decreased IRS-1 Ser636/639 phosphorylation, enhanced tyrosine phosphorylated and total levels of IR, IRS-1 and IRS-2 and activated the PI3K/AKT pathway and AMPK. In addition, EC and CPE preserved HepG2 cell functionality by restoring the levels of GLUT-2, increasing glucose uptake, maintaining glycogen synthesis and decreasing glucose production [11]. Accordingly, EC and CPE pre-treatment prevented high-glucose-induced ROS generation and protein carboxylation, as well as avoided the diminution of GSH, normalized the activity of enzymatic antioxidant defences GPx, GR, catalase (CAT) and glutathione-S-transferase (GST), and phosphorylated levels of mitogen-activated protein kinases (p-MAPK)s, and maintained Nrf2 stimulation [12]. In this regard, the presence of selective MAPK inhibitors induced changes in redox status, glucose uptake, p-(Ser)- and total IRS-1 levels that were observed in CPE-mediated protection [12] (Table 1).

Early this year, cocoa procyanidin extracts have also demonstrated an insulinomimetic effect in human primary skeletal muscle cells [13]. Indeed, a procyanidin-rich cocoa extract elicited an antidiabetic effect by stimulating glycogen synthesis and glucose uptake, independent of insulin, being more pronounced this effect with larger procyanidins. In this case, cocoa procyanidins did not appear to act via stimulation of AMPK or Ca^{2+}/calmodulin-dependent protein kinase II activities [13] (Table 1). Similarly, previous studies suggested that cocoa also have insulinomimetic effects in adipose tissue. Thus, a CPE (100–200 µg/mL) did not affect the levels of IR, but inhibited the IR kinase activity by direct binding, without altering total tyrosine phosphorylation of IR or inhibiting its auto-phosphorylation in 3T3-L1 adipocytes [14]. This inhibitory effect, which resulted in reduced lipid accumulation and differentiation in preadipocytes in vitro, has been related to a suppression of ERK and AKT-mediated signaling cascades to facilitate an anti-adipogenic effect; indeed, it is thought that this is one mechanism by which cocoa flavanols may inhibit the onset of obesity [14] (Table 1). In this regard, cocoa flavanols may improve blood glucose control indirectly, by modulating lipid digestion and thus, reducing hyperlipidemia and its subsequent deleterious effects on glucose homeostasis. Accordingly, Cordero-Herrera et al. [15] have reported that EC (1–10 µM) alleviated the altered lipid values induced liver cells submitted to high glucose. The lipid-lowering effect was related to diminished fatty acid synthesis (sterol-regulatory-element-binding protein-1-c and fatty-acid synthase down-regulation), and increased fatty-acid oxidation [proliferator-activated receptor (PPAR)-α up-regulation]. These effects depended on AMPK, AKT and protein kinase C ζ (PKCζ), which phosphorylated levels returned to control values upon EC treatment, playing PKCζ a role on AKT and AMPK regulation [15] (Table 1).

Cocoa flavanols may also contribute to prevent the chronic, low-grade, inflammation of T2D [16]. EC (0.5–10 µM) attenuated the TNFα-mediated down-regulation of peroxisome PPAR-γ expression and decreased nuclear DNA binding in 3T3-L1 adipocytes [17]. EC also inhibited tumor necrosis factor (TNF)α-mediated altered transcription of protein tyrosine phosphatase 1B (PTP1B), leading both effects to an attenuation of the TNFα-mediated triggering of signaling cascades involved in insulin resistance [17] (Table 1).

Table 1. Anti-diabetic effects of cocoa flavanols in cultured cells [a].

Effects Related to an Anti-Diabetic Action	Cell	Cocoa Flavanol	Treatment	Reference
Glucose uptake				
↑ glucose uptake, ↑ GLUT-4 translocation; =GLUT-2, =GLUT-1	L6 (skeletal muscle)	Cocoa liquor procyandin extract	0.05–10 μg/mL, 15 min	[3]
Insulin signaling				
↑ insulin secretion, ↑ mitochondrial complex III-V, ↑ ATP, ↑ GSH, ↑ Nrf2, ↑ Nrf1, ↑ GABPA	INS-1E (pancreas)	Cocoa extract or oligomeric or polymeric-rich fraction	0.75–25 μg/mL, 24 h	[4]
↑ TFAM, ↑ SIRT-1, ↑ mitofilin, ↑ PGC-1α	HCAEC (endothelia)	EC	100 nM, 10 min or 48 h	[5]
↓ ROS, ↓ carbonyls, ↓ p-JNK, ↓ cell death, ↑ insulin secretion	INS-1E (pancreas)	EC	5–20 μM, 20 h	[6]
↓ ROS, ↓ carbonyls, ↑ GSH, ↑ GPx, ↑ GR	INS-1E (pancreas)	Cocoa phenolic extract	5–20 μg/mL, 20 h	[7]
↑ Insulin secretion, ↑ β-cell survival	INS-1E (pancreas)	3,4-dihydroxyphenylacetic acid (DHPAA), and 3-hydroxyphenylpropionic acid (HPPA)	5 μM DHPAA, 20 h; 1 μM HPPA, 20 h	[8]
↑ Insulin secretion	Mice isolated islets (pancreas)	Procyanindin A2	3–300 μM, 48 h	[9]
↑ p(Tyr)-IR, ↑ IR, ↑ p(Tyr)IRS-1, ↑ IRS-1, ↑ p(Tyr)IRS-2, ↑ IRS-2, ↑ p-AKT, ↑ p-GSK-3, ↑ p-AMPK, ↑ GLUT-2; ↓ p-GS, ↓ PEPCK, ↓ glucose production	HepG2	Cocoa phenolic extract	1–10 μg/mL, 24 h	[10]
↑ p(Tyr)-IR, ↑ IR, ↑ p(Tyr)IRS-1, ↑ IRS-1, ↑ p(Tyr)IRS-2, ↑ IRS-2, ↑ p-AKT, ↑ p-GSK-3, ↑ p-AMPK, ↓ p-GS, ↓ PEPCK, ↓ glucose production; =GLUT-2	HepG2	Epicatechin	1–10 μM, 24 h	[10]
↑ p(Tyr)-IR, ↑ IR, ↑ p(Tyr)IRS-1, ↑ IRS-1, ↑ p(Tyr)IRS-2, ↑ IRS-2, ↑ p-AKT, ↑ p-GSK-3, ↑ p-AMPK, ↑ glucose uptake; ↓ p(Ser)-IRS-1; ↓ p-GS, ↓ PEPCK, ↓ glucose production; =GLUT-2, =glycogen content	HepG2 (insulin-resistant cells)	Cocoa phenolic extract	1–10 μg/mL, 24 h	[11]
↑ p(Tyr)-IR, ↑ IR, ↑ p(Tyr)IRS-1, ↑ IRS-1, ↑ p(Tyr)IRS-2, ↑ IRS-2, ↑ p-AKT, ↑ p-GSK-3, ↑ p-AMPK, ↑ glucose uptake; ↓ p(Ser)-IRS-1; ↓ p-GS, ↓ PEPCK, ↓ glucose production; =GLUT-2, =glycogen content	HepG2 (insulin-resistant cells)	EC	1–10 μM, 24 h	[11]
↓ ROS, ↓ carbonyls, ↑ GSH, ↑ GPx, ↑ GR, ↑ catalase, ↑ GST, ↓ p-ERK, ↓ p-JNK, ↓ p-p38, ↑ Nrf2	HepG2 (insulin-resistant cells)	Cocoa phenolic extract	1–10 μg/mL, 24 h	[12]
↓ ROS, ↓ carbonyls, ↑ GSH, ↑ GPx, ↑ GR, ↑ catalase, ↑ GST, ↓ p-ERK, ↓ p-JNK, ↓ p-p38, ↑ Nrf2	HepG2 (insulin-resistant cells)	EC	1–10 μM, 24 h	[12]
↑ glycogen synthesis, ↑ glucose uptake	Human primary skeletal muscle cells	Procyanidin-rich cocoa extract	10 and 25 μM, 2 h	[13]
↑ p-ERK, ↓ p-AKT; =IR	3T3-L1 (adipocyte)	Cocoa polyphenols	100–200 μg/mL, 4 h	[14]
↓ SREBP-1c, ↓ FAS, ↑ PPAR-α, ↓ PKCζ	HepG2 (insulin-resistant cells)	EC	1–10 μM, 24 h	[15]
↓ PPARγ, ↓ PTP1B	3T3-L1 (adipocyte)	EC	0.5–10 μM, 4 h	[17]

[a] The arrow indicates an increase (↑) or decrease (↓) in the levels or activity of the different parameters analyzed. "=" symbol indicates no changes in the parameter.

3. Animal Studies

Animal studies offer an outstanding opportunity to assess the contribution of the physiological effects of consumption of cocoa and cocoa components in different models of diabetes. Interestingly, supplementation of experimental diets with cocoa has regularly shown high acceptation by the animals and no toxicity even in chronic studies for more than 100 weeks [18]. Within the last decade different models for experimental T2D in rats and mice have been successfully used in order to study the nutritional prevention and treatment of the disease, and supplementation of the diet with cocoa has proved one of the most effective nutritional approaches (Table 2).

Administration of a diet enriched with 10% cocoa for 9 weeks to Zucker diabetic fatty (ZDF) rats reduced hyperglycaemia, enhanced insulin sensitivity and increased β-cell mass and function [19]. In particular, cocoa intake prevented β-cell apoptosis and enhanced antioxidant defences to avoid the oxidative damage and reduce lipid and protein oxidative stress (Table 2). The same treatment ameliorated circulating and hepatic lipid profiles [15], improved insulin resistance [20] and reverted hepatic oxidative stress by enhancing the antioxidant capacity of hepatocytes in ZDF rats [21] (Table 2). The lipid-lowering effect was associated to diminished fatty acid synthesis and increased fatty-acid oxidation [15]. The decreased levels of hepatic PEPCK and increased values of GK and GLUT-2 strongly collaborate to the hypoglycaemic effect of cocoa. Moreover, increased JNK and p38 induced by insulin resistance [20] and oxidative stress [21] was also reverted by cocoa (Table 2). Likewise, administration of EC to obese diabetic mice for 15 weeks, prevented fat deposition and degeneration in hepatocytes [22] (Table 2). These findings demonstrate that a diet enriched in cocoa or its main flavanol relieves hepatic insulin resistance and oxidative stress, and improves lipid metabolism, which are critical landmarks in the development and progression of T2D.

In the same line, feeding of ZDF rats for 7 weeks with a diet supplemented with a 5% of a soluble cocoa product enriched in cocoa fiber decreased blood glucose and insulin and, thus, insulin resistance measured as homeostasis model assessment of insulin resistance (HOMA-IR) index [23] (Table 2). Likewise, administration of 8% cocoa powder for 10 to male C57BL/6J high-fat-fed obese mice resulted in an improved HOMA-IR, indicating a reduced insulin resistance [24]. In the same study, cocoa feeding also reduced plasma concentration of interleukin (IL)-6, monocyte chemoattractant protein-1 (MCP-1), and increased adiponectin, changes related to a reduced inflammation characteristic of obesity and amelioration of fatty liver disease [24]. All these effects were partly mediated through the regulation of dietary fat absorption and inhibition of macrophage infiltration in white adipose tissue (Table 2). Likewise, decreased values of blood glucose without changes in body weight and food consumption were observed in diabetic obese mice fed with 0.5% and 1% cocoa liquor procyanidins for 3 weeks [25] (Table 2). Furthermore, in the same mice model, administration of a cocoa liquor procyanidin extract for 13 weeks evoked a stimulation of AMPK-α, GLUT-4 translocation and uncoupling protein (UCP)-1 and -3 expression in skeletal muscle and adipose tissues, resulting in a reduced obesity, hyperglycaemia, and insulin resistance [26] (Table 2). On the contrary, no changes in glycaemia, insulinemia levels and insulin sensitivity were found in obese–diabetic rats fed a cocoa extract enriched with polyphenol and methylxanthines [27] (Table 2).

Table 2. Anti-diabetic effects of cocoa and cocoa flavanols in animal studies [a].

Effects Related to an Anti-Diabetic Action	Animal Model	Treatment	Duration	Reference
↓ Glucose, ↓ insulin, ↓ HOMA-IR, ↓ TG, ↓ LDL-Cho, ↑ HDL-Cho, ↓ NEFA	Zucker diabetic fatty (ZDF) rats	10% cocoa powder	9 weeks	[15]
↑ β-cell mass, ↑ Bcl-xL, ↓ Bax, ↓caspase-3 activity, ↑ GPx, ↑ GR, ↓ TBARS, ↓ carbonyl groups	Zucker diabetic fatty (ZDF) rats (Pancreas)	10% cocoa powder	9 weeks	[19]
↓ p-(Ser)-IRS-1, =IR, =IRS-2, ↑ p-GSK3, ↓ p-GS, ↓ PEPCK, ↑ GK, ↑ GLUT-2, =p-ERK, ↓ p-JNK, ↓ p-p38 ↓ ROS, ↓ carbonyl groups, =GSH, =GPx, =GR, =CAT, ↑ SOD, ↓ GST, ↓ HO-1, ↓ p-Nrf2, ↓ Nrf2, ↓ p65-NFkB	Zucker diabetic fatty (ZDF) rats (Liver)	10% cocoa powder	9 weeks	[20,21]
↓ fat deposition, ↑ p-AMPK	Obese-diabetic (db-db) mice (Liver)	0.25% EC	15 weeks	[22]
↓ Glucose, ↓ insulin, ↓ HOMA-IR	Obese Zucker fatty (ZF) rats	5% soluble cocoa fiber	7 weeks	[23]
= Glucose, ↓ insulin ↓ HOMA-IR, ↓ IL-6 ↑ adiponectin, ↓ MCP-1	High-fat-fed obese C57BL/6J mice	8% cocoa powder	10 weeks	[24]
↓ Glucose, ↓ fructosamine	High-fat-fed obese C57BL/6J mice (Adipose tissue and skeletal muscle)	0.5% and 1% cacao liquor proanthocyanidins	3 weeks	[25]
↑ p-AMPKα, ↑ GLUT-4, ↑ UCP-1,3	High-fat-fed obese C57BL/6J mice	0.5% and 0.2% cacao liquor procyanidin extract	13 weeks	[26]
=Glucose, =insulin, =HOMA-IR	Obese-diabetic (ob-db) rats	600 mg cocoa polyphenols/Kg body weight/day	4 weeks	[27]
↑ p-IR, ↑ p-IRS-1, ↑ ERK, ↑AKT, ↓ JNK, ↓ PKC, ↑ PTP1B, ↓ p-IKβ, ↓ IKK, ↓ p-p65-NFkB, ↓ TNFα, ↓ MCP1, ↓ p-IRE1α, ↓ p-PERK, ↓ p-IRE1α, =sXBP-1, =p-eIF2α, =ATF6, ↓ NADPH oxidase ↓ Glucose, ↓ insulin, ↓ HOMA-IR	High-fructose (HFr)-fed rats (Liver and adipose tissue)	20 mg EC/Kg body weight/day	8 weeks	[28]
↑ p-IR, ↑ p-IRS-1, ↑ ERK, ↑AKT ↓ JNK, ↓ PKC, ↓ IKK, ↑ PTP1B	High-fat-fed obese C57BL/6J mice (Liver and adipose tissue)	20 mg EC/Kg body weight/day	15 weeks	[29]
↓ Glucose, ↓ insulin, ↓ ITT	High-fat-fed obese C57BL/6J mice	25 mg oligomeric procyanidins/Kg body weight/day	12 weeks	[30]

[a] The arrow indicates an increase (↑) or decrease (↓) in the levels or activity of the different parameters analyzed. "=" symbol designates unchanged parameters.

An EC-supplemented diet for 8 weeks alleviates insulin resistance in high-fructose-fed rats [28]. High-fructose feeding deactivated key proteins of the insulin signaling pathway resulting in a compromised response to insulin in the liver and adipose tissue; administration of a diet supplemented with EC partial or totally avoided these alterations. In addition, EC administration also inhibited the activation of redox-sensitive signals, expression of pro-inflammatory cytokines and chemokines, and endoplasmic reticulum stress proteins, effects that help to attenuate insulin resistance [28] (Table 2). Very recently, Cremonini and colleagues [29] have reported that administration of a high-fat diet to male C57BL/6J obese mice for 15 weeks caused obesity and insulin resistance in C57BL/6J mice as evidenced by high fasted and fed plasma glucose and insulin levels, and impaired insulin tolerance test (ITT) and glucose tolerance test (GTT) tests. This was associated with alterations in the activation of components of the insulin-triggered signaling cascade in adipose and liver tissues. In this scenario, EC supplementation ameliorated all these parameters in the high-fat fed mice, specially improving insulin sensitivity through a downregulation of the inhibitory molecules JNK, PKC and protein tyrosine phosphatase 1B (PTP1B) (Table 2). Finally, feeding of C57BL/6J mice submitted to a high-fat-diet, with a cocoa flavanol extract or three flavanol fractions enriched with monomeric, oligomeric or polymeric procyanidins for 12 weeks exerted different effects depending on the degree of flavanol polymerization [30]. Although insulin levels were lowered by all flavanol fractions, impaired glucose tolerance, insulin resistance, weight gain, and fat mass were most effectively avoided by the oligomer-rich fraction [30] (Table 2).

Therefore, most studies carried out in experimental animal models endorse the favorable effect of cocoa and its flavanols on T2D; this effect appears to be associated both to their proved beneficial effects on vascular function and on glycaemic control through the regulation of key proteins of the insulin signaling pathway and critical biomarkers of inflammation and stress in adipose tissue and skeletal muscle [1].

4. Human Studies

Although most cell culture and animal studies have shown an anti-diabetic activity of cocoa and its main flavonoids, translation of the results to humans is challenging and confirmation of cocoa preventive efficacy against diabetes requires large and long-lasting controlled clinical trials. Within the last few years, a significant number of systematic reviews and meta-analyses [1,31–40] point to a positive effect of cocoa and dark chocolate on improving insulin resistance, endothelial function, blood pressure (BP) and lipid profile. A number of prospective observations in longitudinal studies and intervention clinical assays have yielded analogous conclusions and a few potential molecular and biochemical mechanisms have been proposed to explain the observed benefits of cocoa and dark chocolate for diabetics [1,33,41–43]. All this information has been pulled out from both epidemiological and intervention studies.

Epidemiological evidence: Epidemiological evidence for an association between dietary intake of flavanols (whatever the source) and the risk of type 2 diabetes is large and somehow inconsistent. Three major studies could be reported as representative examples: (1) the European Prospective Investigation into Cancer and Nutrition-InterAct (EPIC-InterAct) case-cohort study including 340,234 participants with 3.99 million person-years of follow-up in eight European countries showed that a higher intake of flavanols was associated with a significantly reduced hazard of diabetes [44]; (2) the Health, Alcohol and Psychosocial factors In Eastern Europe (HAPIEE) study with 5806 participants during 4 years also concluded that intake of flavanols was significantly associated with decreased risk of T2D [45]; however, (3) in a major epidemiological study in US men and women recruited from the Nurses' Health Studies (NHS and NHS II) and Health Professionals Follow-Up Study (HPFS) and during a 3,645,585 person-years of follow-up, only a higher consumption of anthocyanins and anthocyanin-rich fruit was associated with a lower risk of T2D, whereas no association was observed with any of the other flavonoid subfamilies [46].

Regarding studies focusing specifically on intake of chocolate and cocoa, a recent epidemiological study suggest that long-term intake of any type of chocolate may induce an inverse relation with incident T2D in young and normal–body weight men [47]. In the same line, results from the Atherosclerosis Risk in Communities (ARIC) Cohort show that the higher the frequency of chocolate intake the lower the risk of developing diabetes; a statement valid for up to 2–6 servings (1 oz) per week of chocolate. Consuming more than 1 serving per day did not yield significantly lower relative risk [48]. Very recently, two large epidemiological studies have revealed contradictory data regarding the intake of chocolate in T2D. In a prospective study longer than 30 year, a reduced incidence of T2D was observed after a moderate chocolate intake of several times per week [49]. However, in a long-term prospective cohort study in American postmenopausal women, Greenberg et al. concluded that long-term consumption any type of chocolate was unable to decrease the risk of T2D [50]. Nevertheless, the beneficial effect of cocoa and chocolate in T2D has lately received unequivocal support from meta-analysis. Hence, cocoa flavanol ingestion significantly enhanced insulin sensitivity and ameliorated lipid profile in a recent meta-analysis of a large number of randomized controlled trials [51]. Furthermore, the latest meta-analysis of prospective studies has concluded that a moderate consumption of chocolate (up to 6 servings per week) is associated with a reduced risk of coronary heart disease, stroke, and T2D [52].

Intervention studies with chocolate: Very promising results regarding chocolate intake and T2D have been reported during the last decade (Table 3). Thus, BP was reduced and insulin sensitivity increased in glucose-intolerant, hypertensive subjects after 15 days of consuming dark chocolate containing 1080 mg of total polyphenols/day [53]. In this study, intake of dark chocolate enriched with cocoa polyphenols diminished BP and HOMA-IR, augmented insulin sensitivity, and improved β-cell function as compared to white chocolate (Table 3). In a longer intervention during 8 weeks in diabetic patients, Mellor and colleagues [54] showed that the consumption of a high-polyphenol chocolate containing 50 mg of EC did not evoke any change in body weight, insulin resistance, BP or glycaemic control but significantly reduced the atherosclerotic cholesterol (Table 3). A later clinical trial by Almoosawi and colleagues [55] demonstrated that a 4 weeks consumption of dark chocolate containing 500 mg polyphenols by lean and overweight females evoked a significant reduction of BP and improved blood glucose control as shown by the reduced fasting glucose and HOMA-IR (Table 3). Furthermore, a relevant improvement in lipoprotein status and a significant decrease of insulin resistance have been reported after administration of chocolate enriched in flavanols and isoflavones to statin-treated diabetic women in a large study carried out in the United Kingdom (the FLAVO study) [56,57] (Table 3). Finally and more recently, Rostami and colleagues [58] reported that administration of chocolate enriched with cocoa polyphenols decreased fasting glucose and ameliorated BP in patients with diabetes and hypertension (Table 3). But not all intervention studies with chocolate have reported positive results, since fasting plasma glucose was slightly but significantly increased in overweight men after a 4 weeks intervention with dark chocolate [59] (Table 3).

Table 3. Anti-diabetic effects of cocoa and chocolate intake in humans [a].

Effects Related to an Anti-Diabetic Action	Design	Population	Size	Duration (Days)	Dose (Day)	Reference
↓ HOMA-IR, ↑ QUICKI, ↑ ISI, ↑FMD, ↓BP, ↓LDL-Cho, =HDL-Cho	Randomized crossover	Hypertensive, glucose intolerant	38	15	1080 mg polyphenols	[53]
= HOMA-IR, =BP; =LDL-Cho, ↑HDL-Cho, =Glucose, =Insulin, =HbA1c	Randomized crossover	Diabetic	24	56	50 mg epicatechin	[54]
↓ HOMA-IR, ↓ BP; =Insulin, ↓ Glucose	Randomized crossover	Overweight/obese females	42	28	500 mg polyphenols	[55]
↓ HOMA-IR, =BP; ↓ LDL-Cho, =HDL-Cho, =Glucose, ↓ Insulin, =HbA1c	Randomized, placebo controlled	Diabetic	93	365	850 mg flavanols	[56,57]
↓ HbA1c, ↓ Glucose, =BP	Randomized, placebo controlled	Diabetic	60	56	450 mg flavonoids	[58]
↑ Glucose	Randomized crossover	Overweight men	44	28	1078 mg flavanols	[59]
↓ IR, ↓ BP	Randomized, controlled	Overweight/obese Volunteers	49	84	902 mg flavanols	[60]
↑ GSH, ↑ SOD, ↑ Catalase, ↓ nitrotyrosilation and carbonylation of proteins	Open label protocol	Diabetic	5	90	100 mg epicatechin	[61]
↓ Glycaemia, ↓ BP, ↓ MDA, ↑ HDL-Cho	Randomized, controlled, crossover, free-living	Moderately hypercholesterolaemic	21	60	283 mg polyphenols	[62]
↓ Glycaemia, ↓ IL-1b, IL-10, =VCAM1	Randomized, controlled, crossover, free-living	Moderately hypercholesterolaemic	44	28	416 mg flavanols	[63]
↓ Glycaemia, ↓ IL-1b, ↑ HDL-Cho	Randomized, controlled, crossover, free-living	Moderately hypercholesterolaemic	44	28	43.8 mg flavanols	[64]
↓ LDL-Cho, ↓ HDL-Cho, ↓ inflammatory markers	Randomized	Diabetic	100	42	10 g cocoa powder	[65]
↑ HDL-Cho, ↑ Ins, =LDL, Cho, =TG, =Glucose, =IR, =BP	Randomized, crossover trial	Diabetic	18	Acute, 6 h	960 mg polyphenols (480 flavanols)	[66]
=BP, =glycaemic parameters	Randomized, placebo-controlled, double-blind, crossover trial	Hypertensive	20	14	Cocoa beverage (900 mg flavanols/day)	[67]
=glycaemic parameters, =BP	Randomized, double-masked fashion	Diabetic	41	30	Flavanol-rich cocoa (963 mg flavanols/day)	[68]
=Glycaemic parameters, =IL-6, =CRP	Randomized crossover design	Obese adults	20	5	Cocoa beverage (900 mg flavanols/day)	[69]
↓ IR (HOMA-IR), =Glucose, =BP	Randomized, double-blind, placebo-controlled, crossover trial	Healthy	37	28	100 mg epicatechin	[70]

[a] The arrow indicates an increase (↑) or decrease (↓) in the levels or activity of the different parameters analyzed. "=" symbol designates unchanged parameters.

Intervention studies with cocoa and pure flavanols: Cocoa supplementation of the diet has proved both effective and innocuous for glycaemic control in humans (Table 3). Thus, consumption of a cocoa diet enriched with flavanols (902 mg flavanols/day) for 12 weeks by overweight and obese adults considerably improved endothelial function, decreased insulin resistance and reduced diastolic and mean arterial BP as compared to those that received a low-flavanol cocoa diet [60] (Table 3). Treatment for three months with cocoa rich in EC evoked a positive regulation of oxidative stress biomarkers in skeletal muscle of patients with heart failure and T2D [61] (Table 3). A slight but significant hypoglycaemic effect has been reported in moderately hypercholesterolemic humans after administration for 2 months of a fiber-rich cocoa product providing a daily dose of 12 g of dietary fiber and 283 mg of soluble polyphenols [62] or after administration for 4 weeks of a cocoa product providing a daily dose of 416 mg of flavanols [63] (Table 3). In agreement with these results, administration for 4 weeks of a commercialized soluble cocoa product rich in dietary fiber providing 43.8 mg flavanols daily to healthy and moderately hypercholesterolemic subjects induced a slight decrease of postprandial blood glucose [64] (Table 3). In patients with T2D, cocoa powder intake for six weeks reduced total blood cholesterol, LDL-cholesterol and biomarkers of inflammation [65], and, in an acute assay, cocoa supplementation of a high-fat breakfast raised postprandial serum HDL-cholesterol and insulin [66] (Table 3). However, some studies failed to show an effect of a cocoa diet on T2D biomarkers. In a study by Muniyappa and co-workers [67], an improved endothelial function without changes in BP or insulin sensitivity was reported after ingestion of a cocoa drink rich in flavanols (nearly 900 mg of flavanols in 150 mL twice a day) for 2 weeks in patients with hypertension (Table 3). Moreover, in a study by Balzer and colleagues [68], a substantial increase in fasting flow-mediated vascular dilation along with no changes in glycaemic control, BP and heart rate were observed in T2D patients that received a diet supplemented with cocoa with a high daily dose of 963 mg of flavanols for 30 days (Table 3). Furthermore, a short-term intake of a cocoa beverage rich in flavanols by obese adults at risk for insulin resistance reduced critical markers of oxidative stress and inflammation but did not improve glucose metabolism [69] (Table 3).

Regarding administration of pure flavanols, EC supplementation to healthy adults decreased fasting blood insulin and insulin resistance (HOMA-IR) but had no effect on fasting blood glucose, BP and arterial stiffness, nitric oxide and endothelin 1 concentration, or blood lipid profile [70] (Table 3). Overall, most of the above studies support the notion that regular intake of foods rich in cocoa or cocoa flavanols could endorse a dietary strategy to appease insulin resistance. Accordingly, EC has been very recently suggested as adjuvant of metformin in the therapy for T2D patients [71].

Cocoa and body weight: A very exciting outcome in recent human intervention studies with cocoa is that cocoa supplementation of diets did not evoke any increase in body weight or other anthropometric changes [72–74]. Thus, administration of diets supplemented with up to 12.5% of cocoa powder has unequivocally shown anti-obesity effects in rats [19–21,24,75–77]. In addition, despite the fact that cocoa products commercially available are frequently high-caloric foodstuffs, they have been reported to have a similar effect in humans [72–74,78]. However, this anti-obesity effect of cocoa and its derivatives in humans has lately been challenged; i.e., in a prospective cohort study, Greenberg and co-workers [79] have reported a dose-response greater prospective weight gain over time after a regular chocolate consumption. In fact, the highest weight gain was reported in volunteers with the largest frequency of chocolate ingestion, which could be partially related to diminished satiety prompted by the habitual intake of chocolate. It is worth remarking that no differences among different types of chocolate (white, milk and plain) were considered in the previous work [79]. Another study by the same group in the Women's Health Initiative cohort, reported that a greater ingestion of chocolate-candy, usually milk chocolate, was associated to a higher weight gain during a 3-year study period with postmenopausal women [80]. It ought to be stated that habitual consumption of dark chocolate, rather than milk chocolate, will prompt long-term cardiovascular benefits with a minor risk of weight gain [76]. In addition, a stronger sense of satiety and reduced need for energy intake can be more easily achieved after consumption of dark chocolate rather than milk chocolate [81], but a recent study by Esser and colleagues [59] has shown that increased flavanol content in chocolate does affect taste and has a negative effect on the motivation to eat chocolate.

In summary, most of the data above suggest that the beneficial effect of cocoa and its flavanols on T2D seems to be associated to their substantiated favorable effects on vascular function and on glycaemic control mediated through the regulation of main proteins involved in the insulin signalling pathway, as well as in the processes of inflammation and oxidative stress. The precise biochemical and molecular mechanisms have recently been reviewed [1,43]. Additionally, the European Food Safety Authority (EFSA) sanctions that cocoa flavanols help maintain normal blood pressure [82] and endothelium-dependent vasodilation [83], and, in the context of a healthy diet, the claimed effect can be obtained with a daily consumption of 200 mg of cocoa flavanols. This amount of flavanols can be acquired with the intake of 100 g of most cocoa soluble products or 40 g of any >70% cocoa chocolate in the market.

Finally, some gaps in the research to delineate the anti-diabetic effect of a cocoa intake have been detected in the literature; thus, the role of three compounds should be promptly investigated, should be investigated: (i) host and microbiota-derived flavanol metabolites; (ii) cocoa fiber and (iii) theobromine. All three have shown promising beneficial effects on cardiovascular health and body weight control and they should be considered in future controlled clinical trials.

5. Conclusions

Most studies within the last decade support a substantial role for cocoa and its flavanols in the nutritional prevention of T2D. Cocoa flavanols act by (a) regulating carbohydrate absorption in the gut; (b) protecting β-pancreatic cells function and enhancing insulin secretion; (c) improving insulin sensitivity in peripheral tissues such as liver, adipose tissue and skeletal muscle through regulation of glucose transporters and main proteins of the insulin signalling pathway; (d) exerting a lipid-lowering effect and; (e) preventing the exacerbated oxidative stress and inflammation characteristics of the disease. All these effects contribute to improve the insulin sensitivity and to maintain normoglycaemia, and thus, to avert and/or significantly delay the onset of T2D and development of its complications. Consequently, a moderate daily consumption of flavanols from cocoa or dark chocolate, along with an everyday intake of other dietary flavonoids, could be a valuable recommendation for the nutritional management of this disease. However, it is worth remembering that most of commercially available soluble cocoa products or chocolates contain low amount of flavanols and are rich in sugar and calories that may aggravate glycaemic control in T2D patients. Hence, recommendation of consumption of chocolate or other cocoa derivatives to this population still requires further research, especially extensive well-designed human epidemiological and intervention studies, to delineate the amount of cocoa and variety of its products that might be beneficial to prevent, delay or contribute to the treatment of T2D.

Acknowledgments: This work was supported by the grant AGL2015-67087 (MINECO/FEDER, UE) from the Spanish Ministry of Science and Innovation (MINECO).

Conflicts of Interest: The authors declare that there are no conflicts of interest.

References

1. Martín, M.A.; Goya, L.; Ramos, S. Antidiabetic actions of cocoa flavanols. *Mol. Nutr. Food Res.* **2016**, *60*, 1756–1769. [CrossRef] [PubMed]
2. Gu, Y.; Hurst, W.J.; Stuart, D.A.; Lambert, J.D. Inhibition of key digestive enzymes by cocoa extracts and procyanidins. *J. Agric. Food Chem.* **2011**, *59*, 5305–5311. [CrossRef] [PubMed]
3. Yamashita, Y.; Okabe, M.; Natsume, M.; Ashida, H. Cacao liquor procyanidin extract improves glucose tolerance by enhancing GLUT4 translocation and glucose uptake in skeletal muscle. *J. Nutr. Sci.* **2012**, *1*, e2. [CrossRef] [PubMed]

4. Rowley, T.J.; Bitner, B.F.; Ray, J.D.; Lathen, D.R.; Smithson, A.T.; Dallon, B.W.; Plowman, C.J.; Bikman, B.T.; Hansen, J.M.; Dorenkott, M.R.; et al. Monomeric cocoa catechins enhance β-cell function by increasing mitochondrial respiration. *J. Nutr. Biochem.* **2017**, *49*, 30–41. [CrossRef] [PubMed]

5. Ramírez-Sánchez, I.; Rodríguez, A.; Moreno-Ulloa, A.; Ceballos, G.; Villarreal, F. (−)-Epicatechin-induced recovery of mitochondria from simulated diabetes: Potential role of endothelial nitric oxide synthase. *Diabetes Vasc. Res.* **2016**, *13*, 201–210. [CrossRef] [PubMed]

6. Martín, M.A.; Fernandez-Millan, E.; Ramos, S.; Bravo, L.; Goya, L. Cocoa flavonoid epicatechin protects pancreatic beta cell viability and function against oxidative stress. *Mol. Nutr. Food Res.* **2013**, *58*, 447–456. [CrossRef] [PubMed]

7. Martin, M.A.; Ramos, S.; Cordero-Herrero, I.; Bravo, L.; Goya, L. Cocoa phenolic extract protects pancreatic beta cells against oxidative stress. *Nutrients* **2013**, *5*, 2955–2968. [CrossRef] [PubMed]

8. Fernández-Millán, E.; Ramos, S.; Alvarez, C.; Bravo, L.; Goya, L.; Martín, M.A. Microbial phenolic metabolites improve glucose-stimulated insulin secretion and protect pancreatic beta cells against tert-butylhydroperoxide-induced toxicity via ERKs and PKC pathways. *Food Chem. Toxicol.* **2014**, *66*, 245–253. [CrossRef] [PubMed]

9. Ahangarpour, A.; Afshari, G.; Mard, S.A.; Khodadadi, A.; Hashemitabar, M. Preventive effects of procyanidin A2 on glucose homeostasis, pancreatic and duodenal homebox 1, and glucose transporter 2 gene expression disturbance induced by bisphenol A in male mice. *J. Physiol. Pharmacol.* **2016**, *67*, 243–252. [PubMed]

10. Cordero-Herrera, I.; Martín, M.A.; Bravo, L.; Goya, L.; Ramos, S. Cocoa flavonoids improve insulin signalling and modulate glucose production via AKT and AMPK in HepG2 cells. *Mol. Nutr. Food Res.* **2013**, *57*, 974–985. [CrossRef] [PubMed]

11. Cordero-Herrera, I.; Martín, M.A.; Goya, L.; Ramos, S. Cocoa flavonoids attenuate high glucose-induced insulin signalling blockade and modulate glucose uptake and production in human HepG2 cells. *Food Chem. Toxicol.* **2014**, *64*, 10–19. [CrossRef] [PubMed]

12. Cordero-Herrera, I.; Martín, M.A.; Goya, L.; Ramos, S. Cocoa flavonoids protect hepatic cells against high glucose-induced oxidative stress. Relevance of MAPKs. *Mol. Nutr. Food Res.* **2015**, *59*, 597–609. [CrossRef] [PubMed]

13. Bowser, S.M.; Moore, W.T.; McMillan, R.P.; Dorenkottc, M.R.; Goodrich, K.M.; Ye, L.; O'Keefe, S.F.; Hulver, M.W.; Neilson, A.P. High-molecular-weight cocoa procyanidins possess enhanced insulin-enhancing and insulin mimetic activities in human primary skeletal muscle cells compared to smaller procyanidins. *J. Nutr. Biochem.* **2017**, *39*, 48–58. [CrossRef] [PubMed]

14. Min, S.Y.; Yang, H.; Seo, S.G.; Shin, S.H.; Chung, M.-Y.; Kim, J.; Lee, S.J.; Lee, H.J.; Lee, K.W. Cocoa polyphenols suppress adipogenesis in vitro and obesity in vivo by targeting insulin receptor. *Int. J. Obes. (Lond.)* **2013**, *37*, 584–592. [CrossRef] [PubMed]

15. Cordero-Herrera, I.; Martín, M.A.; Fernández-Millán, E.; Álvarez, C.; Goya, L.; Ramos, S. Cocoa and cocoa flavanol epicatechin improve hepatic lipid metabolism in in vivo and in vitro models. Role of PKCζ. *J. Funct. Food.* **2015**, *17*, 761–773. [CrossRef]

16. Gu, Y.; Lambert, J.D. Modulation of metabolic syndrome-related inflammation by cocoa. *Mol. Nutr. Food Res.* **2013**, *57*, 948–961. [CrossRef] [PubMed]

17. Vazquez-Prieto, M.A.; Bettaie, A.; Haj, F.G.; Fraga, C.G.; Oteiza, P.I. (−)-Epicatechin prevents TNFa-induced activation of signaling cascades involved in inflammation and insulin sensitivity in 3T3-L1 adipocytes. *Arch. Biochem. Biophys.* **2012**, *527*, 113–118. [CrossRef] [PubMed]

18. Tarka, S.M.; Morrissey, R.B.; Apgar, J.L.; Hostetler, K.A.; Shively, C.A. Chronic toxicity/carcinogenicity studies of cocoa powder in rats. *Food Chem. Toxicol.* **1991**, *29*, 7–19. [CrossRef]

19. Fernández-Millán, E.; Cordero-Herrera, I.; Ramos, S.; Escrivá, F.; Álvarez, C.; Goya, L.; Martín, M.A. Cocoa-rich diet attenuates beta cell mass loss and function in young Zucker diabetic fatty rats by preventing oxidative stress and beta cell apoptosis. *Mol. Nutr. Food Res.* **2015**, *59*, 820–824. [CrossRef] [PubMed]

20. Cordero-Herrera, I.; Martín, M.A.; Escrivá, F.; Álvarez, C.; Goya, L.; Ramos, S. Cocoa-rich diet ameliorates hepatic insulin resistance by modulating insulin signaling and glucose homeostasis in Zucker diabetic fatty rats. *J. Nutr. Biochem.* **2015**, *26*, 704–712. [CrossRef] [PubMed]

21. Cordero-Herrera, I.; Martín, M.A.; Goya, L.; Ramos, S. Cocoa intake ameliorates hepatic oxidative stress in young Zucker diabetic fatty rats. *Food Res. Int.* **2015**, *69*, 194–201. [CrossRef]

22. Si, H.; Fu, Z.; Babu, P.V.A.; Zhen, W.; Leroith, T.; Meaney, M.P.; Voelker, K.A.; Jia, Z.; Grange, R.W.; Liu, D. Dietary epicatechin promotes survival of obese diabetic mice and Drosophila melanogaster. *J. Nutr.* **2011**, *141*, 1095–1100. [CrossRef] [PubMed]

23. Sánchez, D.; Moulay, L.; Muguerza, B.; Quiñones, M.; Miguel, M.; Aleixandre, A. Effect of a soluble cocoa fiber-enriched diet in Zucker fatty rats. *J. Med. Food* **2010**, *13*, 621–628. [CrossRef] [PubMed]

24. Gu, Y.; Yu, S.; Lambert, J.D. Dietary cocoa ameliorates obesity-related inflammation in high fat-fed mice. *Eur. J. Nutr.* **2014**, *53*, 149–158. [CrossRef] [PubMed]

25. Tomaru, M.; Takano, H.; Osakabe, N.; Yasuda, A.; Inoue, K.; Yanahisawa, R.; Ohwatari, T.; Uematsu, H. Dietary supplementation with cacao liquor proanthocyanidins prevents elevation of blood glucose levels in diabetic obese mice. *Nutrition* **2007**, *23*, 351–355. [CrossRef] [PubMed]

26. Yamashita, Y.; Okabe, M.; Natsume, M.; Ashida, H. Prevention mechanisms of glucose intolerance and obesity by cacao liquor procyanidin extract in high-fat diet-fed C57BL/6 mice. *Arch. Biochem. Biophys.* **2012**, *527*, 95–104. [CrossRef] [PubMed]

27. Jalil, A.M.M.; Ismail, A.; Chong, P.P.; Hamid, M.; Kamaruddin, S.H.S. Effects of cocoa extract containing polyphenols and methylxanthines on biochemical parameters of obese-diabetic rats. *J. Sci. Food Agric.* **2009**, *89*, 130–137. [CrossRef]

28. Bettaieb, A.; Vazquez-Prieto, M.A.; Rodriguez-Lanzi, C.; Miatello, R.; Haj, F.G.; Fraga, C.G.; Oteiza, P.I. (−)-Epicatechin mitigates high-fructose-associated insulin resistance by modulating redox signalling and endoplasmic reticulum stress. *Free Radic. Biol. Med.* **2014**, *72*, 247–256. [CrossRef] [PubMed]

29. Cremonini, E.; Bettaieb, A.; Fawaz, G.H.; Fraga, C.G.; Oteiza, P.I. (−)-Epicatechin improves insulin sensitivity in high fat diet-fed mice. *Arch. Biochem. Biophys.* **2016**, *599*, 13e21. [CrossRef] [PubMed]

30. Dorenkott, M.R.; Griffin, L.E.; Goodrich, K.M.; Thompson-Witrick, K.A.; Fundaro, G.; Ye, L.; Stevens, J.R.; Ali, M.; O'Keefe, S.F.; Hulver, M.W.; et al. Oligomeric cocoa procyanidins possess enhanced bioactivity compared to monomeric and polymeric cocoa procyanidins for preventing the development of obesity, insulin resistance, and impaired glucose tolerance during high-fat feeding. *J. Agric. Food Chem.* **2014**, *62*, 2216–2227. [CrossRef] [PubMed]

31. Buitrago-Lopez, A.; Sanderson, J.; Johnson, L.; Warnakula, S.; Wood, A.; Di Angelantonio, E.; Franco, O.H. Chocolate consumption and cardiometabolic disorders: Systematic review and meta-analysis. *Br. Med. J.* **2011**, *343*, 4488–4495. [CrossRef] [PubMed]

32. Ellam, S.; Williamson, G. Cocoa and human health. *Annu. Rev. Nutr.* **2013**, *33*, 105–128. [CrossRef] [PubMed]

33. Grassi, D.; Desideri, G.; Mai, F.; Martella, L.; De Feo, M.; Soddu, D.; Fellini, E.; Veneri, M.; Stamerra, C.A.; Ferri, C. Cocoa, Glucose Tolerance, and Insulin Signaling: Cardiometabolic Protection. *J. Agric. Food Chem.* **2015**, *63*, 9919–9926. [CrossRef] [PubMed]

34. Hooper, L.; Kay, C.; Abdelhamid, A.; Kroon, P.A.; Kohn, J.S.; Rimm, E.B.; Cassidy, A. Effects of chocolate, cocoa, and flavan-3-ols on cardiovascular health: A systematic review and meta-analysis of randomized trials. *Am. J. Clin. Nutr.* **2012**, *95*, 740–751. [CrossRef] [PubMed]

35. Kim, Y.; Keogh, J.B.; Clifton, P.M. Polyphenols and Glycemic Control. *Nutrients* **2016**, *8*, 17. [CrossRef] [PubMed]

36. Ludovici, V.; Barthelmes, J.; Nägele, M.P.; Enseleit, F.; Ferri, C.; Flammer, A.J.; Ruschitzka, F.; Sudano, I. Cocoa, Blood Pressure, and Vascular Function. *Front. Nutr.* **2017**, *4*, 36. [CrossRef] [PubMed]

37. Ríos, J.L.; Francini, F.; Schinella, G.R. Natural Products for the Treatment of Type 2 Diabetes Mellitus. *Planta Med.* **2015**, *81*, 975–994. [CrossRef] [PubMed]

38. Shrime, M.G.; Bauer, S.R.; McDonald, A.C.; Chowdhury, N.H.; Coltart, C.E.; Ding, E.L. Flavonoid-rich cocoa consumption affects multiple cardiovascular risk factors in a meta-analysis of short-term studies. *J. Nutr.* **2011**, *141*, 1982–1988. [CrossRef] [PubMed]

39. Vitale, M.; Masulli, M.; Rivellese, A.A.; Bonora, E.; Cappellini, F.; Nicolucci, A.; Squatrito, S.; Antenucci, D.; Barrea, A.; Bianchi, C.; et al. Dietary intake and major food sources of polyphenols in people with type 2 diabetes: The TOSCA.IT Study. *Eur. J. Nutr.* **2016**. [CrossRef] [PubMed]

40. Zamora-Ros, R.; Forouhi, N.G.; Sharp, S.J.; González, C.A.; Buijsse, B.; Guevara, M.; Van der Schouw, Y.T.; Amiano, P.; Boeing, H.; Bredsdorff, L.; et al. Dietary Intakes of Individual Flavanols and Flavonols Are Inversely Associated with Incident Type 2 Diabetes in European Populations. *J. Nutr.* **2014**, *144*, 335–343. [CrossRef] [PubMed]

41. Grassi, D.; Desideri, G.; Ferri, C. Protective effects of dark chocolate on endothelial function and diabetes. *Curr. Opin. Clin. Nutr. Metab. Care* **2013**, *16*, 662–668. [CrossRef] [PubMed]

42. Mellor, D.D.; Sathyapalan, T.; Kilpatrick, E.S.; Atkin, S.L. Diabetes and chocolate: Friend or foe? *J. Agric. Food Chem.* **2015**, *63*, 9910–9918. [CrossRef] [PubMed]

43. Strat, K.M.; Rowley, T.J., 4th; Smithson, A.T.; Tessem, J.S.; Hulver, M.W.; Liu, D.; Davy, B.M.; Davy, K.P.; Neilson, A.P. Mechanisms by which cocoa flavanols improve metabolic syndrome and related disorders. *J. Nutr. Biochem.* **2016**, *35*, 1–21. [CrossRef] [PubMed]

44. Zamora-Ros, R.; Forouhi, N.G.; Sharp, S.J.; González, C.A.; Buijsse, B.; Guevara, M.; Van der Schouw, Y.T.; Amiano, P.; Boeing, H.; Bredsdorff, L.; et al. The association between dietary flavonoid and lignan intakes and incident type 2 diabetes in European populations: The EPIC-InterAct study. *Diabetes Care* **2013**, *36*, 3961–3970. [CrossRef] [PubMed]

45. Grosso, G.; Stepaniak, U.; Micek, A.; Kozela, M.; Stefler, D.; Bobak, M.; Pajak, A. Dietary polyphenol intake and risk of type 2 diabetes in the Polish arm of the Health, Alcohol and Psychosocial factors in Eastern Europe (HAPIEE) study. *Br. J. Nutr.* **2017**, *118*, 60–68. [CrossRef] [PubMed]

46. Wedick, N.M.; Pan, A.; Cassidy, A.; Rimm, E.B.; Sampson, L.; Rosner, B.; Willett, W.; Hu, F.B.; Sun, Q.; van Dam, R.M. Dietary flavonoid intakes and risk of type 2 diabetes in US men and women. *Am. J. Clin. Nutr.* **2012**, *95*, 925–933. [CrossRef] [PubMed]

47. Matsumoto, C.; Petrone, A.B.; Sesso, H.D.; Gaziano, J.M.; Djouss, L. Chocolate consumption and risk of diabetes mellitus in the Physicians' Health Study. *Am. J. Clin. Nutr.* **2015**, *101*, 362–367. [CrossRef] [PubMed]

48. Greenberg, J.A. Chocolate intake and diabetes risk. *Clin. Nutr.* **2015**, *34*, 129–133. [CrossRef] [PubMed]

49. Crichton, G.E.; Elias, M.F.; Dearborn, P.; Robbins, M. Habitual chocolate intake and type 2 diabetes mellitus in the Maine-Syracuse Longitudinal Study: (1975–2010): Prospective observations. *Appetite* **2017**, *108*, 263–269. [CrossRef] [PubMed]

50. Greenberg, J.A.; Manson, J.E.; Tinker, L.; Neuhouser, M.L.; Garcia, L.; Vitolins, M.Z.; Phillips, L.S. Chocolate intake and diabetes risk in postmenopausal American women. *Eur. J. Clin. Nutr.* **2017**. [CrossRef] [PubMed]

51. Lin, X.; Zhang, I.; Li, A.; Manson, J.E.; Sesso, H.D.; Wang, L.; Liu, S. Cocoa flavanol intake and biomarkers for cardiometabolic health: A systematic review and meta-analysis of randomized controlled trials. *J. Nutr.* **2016**, *146*, 2325–2333. [CrossRef] [PubMed]

52. Yuan, S.; Li, X.; Jin, Y.; Lu, J. Chocolate consumption and risk of coronary heart disease, stroke, and diabetes: A meta-analysis of prospective studies. *Nutrients* **2017**, *9*, 688. [CrossRef]

53. Grassi, D.; Desideri, G.; Necozione, S.; Lippi, C.; Casale, R.; Properzi, G.; Blumberg, J.B.; Ferri, C. Blood pressure is reduced and insulin sensitivity increased in glucose-intolerant, hypertensive subjects after 15 days of consuming high-polyphenol dark chocolate. *J. Nutr.* **2008**, *138*, 1671–1676. [PubMed]

54. Mellor, D.D.; Sathyapalan, T.; Kilpatrick, E.S.; Beckett, S.; Atkin, S.L. High cocoa polyphenol-rich chocolate improves HDL cholesterol in type 2 diabetes patients. *Diabet. Med.* **2010**, *27*, 1318–1321. [CrossRef] [PubMed]

55. Almoosawi, S.; Tsang, C.; Ostertag, L.M.; Fyfed, L.; Al-Dujaili, E.A.S. Differential effect of polyphenol-rich dark chocolate on biomarkers of glucose metabolism and cardiovascular risk factors in healthy, overweight and obese subjects: A randomized clinical trial. *Food Funct.* **2012**, *3*, 1035–1043. [CrossRef] [PubMed]

56. Curtis, P.J.; Sampson, M.; Potter, J.; Dhatariya, K.; Kroon, P.A.; Cassidy, A. Chronic ingestion of flavan-3-ols and isoflavones improves insulin sensitivity and lipoprotein status and attenuates estimated 10-year CVD risk in medicated postmenopausal women with type 2 diabetes: A 1-year, double-blind, randomized, controlled trial. *Diabetes Care* **2012**, *35*, 226–232. [CrossRef] [PubMed]

57. Curtis, P.J.; Potter, J.; Kroon, P.A.; Wilson, P.; Dhatariya, K.; Sampson, M.; Cassidy, A. Vascular function and atherosclerosis progression after 1 year of flavonoid intake in statin-treated postmenopausal women with type 2 diabetes: A double-blind randomized controlled trial. *Am. J. Clin. Nutr.* **2013**, *97*, 936–942. [CrossRef] [PubMed]

58. Rostami, A.; Khalili, M.; Haghighat, N.; Eghtesad, S.; Shidfar, F.; Heidari, I.; Ebrahimpour-Koujan, S.; Eghtesadi, M. High-cocoa polyphenol-rich chocolate improves blood pressure in patients with diabetes and hypertension. *ARYA Atheroscler.* **2015**, *11*, 21–29. [PubMed]

59. Esser, D.; Mars, M.; Oosterink, E.; Stalmach, A.; Müller, M.; Afman, L.A. Dark chocolate consumption improves leukocyte adhesion factors and vascular function in overweight men. *FASEB J.* **2014**, *28*, 1464–1473. [CrossRef] [PubMed]

60. Davison, K.; Coates, A.M.; Buckley, J.D.; Howe, P.R.C. Effect of cocoa flavanols and exercise on cardiometabolic risk factors in overweight and obese subjects. *Int. J. Obes. (Lond.)* **2008**, *32*, 1289–1296. [CrossRef] [PubMed]

61. Ramírez-Sánchez, I.; Taub, P.R.; Ciaraldi, T.P.; Nogueira, L.; Coe, T.; Perkins, G.; Hogan, M.; Maisel, A.S.; Henry, R.R.; Ceballos, G.; et al. (−)-Epicatechin rich cocoa mediated modulation of oxidative stress regulators in skeletal muscle of heart failure and type 2 diabetes patients. *Int. J. Cardiol.* **2013**, *168*, 3982–3990. [CrossRef] [PubMed]

62. Sarriá, B.; Mateos, R.; Sierra-Cinos, J.L.; Goya, L.; García-Diz, L.; Bravo, L. Hypotensive, hypoglycaemic and antioxidant effects of consuming a cocoa product in moderately hypercholesterolemic humans. *Food Funct.* **2012**, *3*, 867–874. [CrossRef] [PubMed]

63. Sarriá, B.; Martínez-López, S.; Sierra-Cinos, J.L.; García-Diz, L.; Mateos, R.; Bravo-Clemente, L. Regular consumption of a cocoa product improves the cardiometabolic profile in healthy and moderately hypercholesterolaemic adults. *Br. J. Nutr.* **2014**, *111*, 122–134. [CrossRef] [PubMed]

64. Sarriá, B.; Martínez-López, S.; Sierra-Cinos, J.L.; García-Diz, L.; Goya, L.; Mateos, R.; Bravo, L. Effects of bioactive constituents in functional cocoa products on cardiovascular health in humans. *Food Chem.* **2015**, *174*, 214–218. [CrossRef] [PubMed]

65. Parsaeyan, N.; Mozaffari-Khosravi, H.; Absalan, A.; Mozayan, M.R. Beneficial effects of cocoa on lipid peroxidation and inflammatory markers in type 2 diabetic patients and investigation of probable interactions of cocoa active ingredients with prostaglandin synthase-2 (PTGS-2/COX-2) using virtual analysis. *J. Diabetes Metab. Disord.* **2014**, *13*, 30–38. [CrossRef] [PubMed]

66. Basu, A.; Betts, N.M.; Leyva, M.J.; Fu, D.; Aston, C.E.; Lyons, T.J. Acute Cocoa Supplementation Increases Postprandial HDL Cholesterol and Insulin in Obese Adults with Type 2 Diabetes after Consumption of a High-Fat Breakfast. *J. Nutr.* **2015**, *145*, 2325–2332. [CrossRef] [PubMed]

67. Muniyappa, R.; Hall, G.; Kolodziej, T.L.; Karne, R.J.; Crandon, S.K.; Quon, M.J. Cocoa consumption for 2 wk enhances insulin-mediated vasodilatation without improving blood pressure or insulin resistance in essential hypertension. *Am. J. Clin. Nutr.* **2008**, *88*, 1685–1696. [CrossRef] [PubMed]

68. Balzer, J.; Rassaf, T.; Heiss, C.; Kleinbongard, P.; Lauer, T.; Merx, M.; Heussen, N.; Gross, H.B.; Keen, C.L.; Schroeter, H.; et al. Sustained benefits in vascular function through flavanol-containing cocoa in medicated diabetic patients a double-masked, randomized, controlled trial. *J. Am. Coll. Cardiol.* **2008**, *51*, 2141–2149. [CrossRef] [PubMed]

69. Stote, K.S.; Clevidence, B.A.; Novotny, J.A.; Henderson, T.; Radecki, S.V.; Baer, D.J. Effect of cocoa and green tea on biomarkers of glucose regulation, oxidative stress, inflammation and hemostasis in obese adults at risk for insulin resistance. *Eur. J. Clin. Nutr.* **2012**, *66*, 1153–1159. [CrossRef] [PubMed]

70. Dower, J.I.; Geleijnse, J.M.; Gijsbers, L.; Zock, P.L.; Kromhout, D.; Hollman, P.C.H. Effects of the pure flavonoids epicatechin and quercetin on vascular function and cardiometabolic health: A randomized, double-blind, placebo-controlled, crossover trial. *Am. J. Clin. Nutr.* **2015**, *101*, 914–921. [CrossRef] [PubMed]

71. Moreno-Ulloa, A.; Moreno-Ulloa, J. Mortality reduction among persons with type 2 diabetes: Epicatechin as add-on therapy to metformin? *Med. Hypotheses* **2016**, *91*, 86–89. [CrossRef] [PubMed]

72. Bohannon, J.; Koch, D.; Homm, P.; Driehaus, A. Chocolate with high Cocoa content as a weight-loss accelerator. *Int. Arch. Med.* **2015**, *8*. [CrossRef]

73. Golomb, B.A.; Koperski, S.; White, H.L. Association between More Frequent Chocolate Consumption and Lower Body Mass Index. *Arch. Intern. Med.* **2012**, *172*, 519–521. [CrossRef] [PubMed]

74. Martínez-López, S.; Sarriá, B.; Sierra-Cinos, J.L.; Goya, L.; Mateos, R.; Bravo, L. Realistic intake of a flavanol-rich soluble cocoa product increases HDL-cholesterol without inducing anthropometric changes in healthy and moderately hypercholesterolemic subjects. *Food Funct.* **2014**, *5*, 364–374. [CrossRef] [PubMed]

75. Camps-Bossacoma, M.; Pérez-Cano, F.J.; Franch, À.; Untersmayr, E.; Castell, M. Effect of a cocoa diet on the small intestine and gut-associated lymphoid tissue composition in an oral sensitization model in rats. *J. Nutr. Biochem.* **2017**, *42*, 182–193. [CrossRef] [PubMed]

76. Farhat, G.; Drummond, S.; Fyfe, L.; Al-Dujaili, E.A.S. Dark Chocolate: An Obesity Paradox or a Culprit for Weight Gain? *Phytother. Res.* **2014**, *28*, 791–797. [CrossRef] [PubMed]

77. Matsui, N.; Itoa, R.; Nishimura, E.; Kato, M.; Kamei, M.; Shibata, H.; Kamei, M.; Shibata, H.; Matsumoto, I.; Abe, K.; et al. Ingested cocoa can prevent high-fat diet–induced obesity by regulating the expression of genes for fatty acid metabolism. *Nutrition* **2005**, *21*, 594–601. [CrossRef] [PubMed]

78. Visioli, F.; Bernaert, H.; Corti, R.; Ferri, C.; Heptinstall, S.; Molinari, E.; Poli, A.; Serafini, M.; Smit, H.J.; Vinson, J.A.; et al. Chocolate, lifestyle, and health. *Crit. Rev. Food Sci. Nutr.* **2009**, *49*, 299–312. [CrossRef] [PubMed]

79. Greenberg, J.A.; Buijsse, B. Habitual chocolate consumption may increase body weight in a dose-response manner. *PLoS ONE* **2013**, *8*, e70271. [CrossRef] [PubMed]

80. Greenberg, J.A.; Manson, J.E.; Buijsse, B.; Wang, L.; Allison, M.A.; Neuhouser, M.L.; Tinker, L.; Waring, M.E.; Isasi, C.R.; Martin, L.W.; et al. Chocolate-candy comsumption and three-year weight gain among postmenopausal American women. *Obesity* **2015**, *23*, 677–683. [CrossRef] [PubMed]

81. Sørensen, L.B.; Astrup, A. Eating dark and milk chocolate: A randomized crossover study of effects on appetite and energy intake. *Nutr. Diabetes* **2011**, *1*, e21. [CrossRef] [PubMed]

82. European Food Safety Authority. Scientific Opinion on the substantiation of health claims related to cocoa flavanols and protection of lipids from oxidative damage and maintenance of normal blood pressure. *EFSA J.* **2010**, *8*, 1792.

83. European Food Safety Authority. Scientific opinion on the substantiation of a health claim related to cocoa flavanols and maintenance of normal endothelium-dependent vasodilation. *EFSA J.* **2012**, *10*, 2809.

antioxidants

MDPI

Review

Bioactive Components in *Moringa Oleifera* Leaves Protect against Chronic Disease

Marcela Vergara-Jimenez [1], Manal Mused Almatrafi [2] and Maria Luz Fernandez [2,*

[1]   Department of Nutrition, Universidad Autonoma de Sinaloa, Culiacan 80019, Mexico; marveji@hotmail.com
[2]   Department of Nutritional Sciences, University of Connecticut, Storrs, CT 06269, USA; manal.almatrafi@uconn.edu
*   Correspondence maria-luz.fernandez@uconn.edu; Tel.: +1-860-486-5547

Received: 20 October 2017; Accepted: 15 November 2017; Published: 16 November 2017

Abstract: *Moringa Oleifera* (MO), a plant from the family Moringacea is a major crop in Asia and Africa. MO has been studied for its health properties, attributed to the numerous bioactive components, including vitamins, phenolic acids, flavonoids, isothiocyanates, tannins and saponins, which are present in significant amounts in various components of the plant. *Moringa Oleifera* leaves are the most widely studied and they have shown to be beneficial in several chronic conditions, including hypercholesterolemia, high blood pressure, diabetes, insulin resistance, non-alcoholic liver disease, cancer and overall inflammation. In this review, we present information on the beneficial results that have been reported on the prevention and alleviation of these chronic conditions in various animal models and in cell studies. The existing limited information on human studies and *Moringa Oleifera* leaves is also presented. Overall, it has been well documented that *Moringa Oleifera* leaves are a good strategic for various conditions associated with heart disease, diabetes, cancer and fatty liver.

Keywords: *Moringa Oleifera*; bioactive components; hepatic steatosis; heart disease; diabetes; cancer

1. Introduction

Moringa, a native plant from Africa and Asia, and the most widely cultivated species in Northwestern India, is the sole genus in the family Moringaceae [1]. It comprises 13 species from tropical and subtropical climates, ranging in size from tiny herbs to massive trees. The most widely cultivated species is *Moringa Oleifera (MO)* [1]. MO is grown for its nutritious pods, edible leaves and flowers and can be utilized as food, medicine, cosmetic oil or forage for livestock. Its height ranges from 5 to 10 m [1].

Several studies have demonstrated the beneficial effects in humans [2]. MO has been recognized as containing a great number of bioactive compounds [3,4] The most used parts of the plant are the leaves, which are rich in vitamins, carotenoids, polyphenols, phenolic acids, flavonoids, alkaloids, glucosinolates, isothiocyanates, tannins and saponins [5]. The high number of bioactive compounds might explain the pharmacological properties of *MO* leaves. Many studies, in vitro and in vivo, have confirmed these pharmacological properties [5].

The leaves of *MO* are mostly used for medicinal purposes as well as for human nutrition, since they are rich in antioxidants and other nutrients, which are commonly deficient in people living in undeveloped countries [6]. *MO* leaves have been used for the treatment of various diseases from malaria and typhoid fever to hypertension and diabetes [7].

The roots, bark, gum, leaf, fruit (pods), flowers, seed, and seed oil of *MO* are reported to have various biological activities, including protection against gastric ulcers [8], antidiabetic [9], hypotensive [10] and anti-inflammatory effects [11]. It has also been shown to improve hepatic and renal functions [12] and the regulation of thyroid hormone status [13]. *MO* leaves also protect against oxidative stress [14], inflammation [15], hepatic fibrosis [16], liver damage [17], hypercholesterolemia [18,19], bacterial activity [20], cancer [14] and liver injury [21].

2. Bioactive Components in *Moringa Oleifera*

2.1. Vitamins

Fresh leaves from *MO* are a good source of vitamin A [22]. It is well established that vitamin A has important functions in vision, reproduction, embryonic growth and development, immune competence and cell differentiation [23]. *MO* leaves are a good source of carotenoids with pro-vitamin A potential [24].

MO leaves also contain 200 mg/100 g of vitamin C, a concentration greater than what is found in oranges [22,25]. *MO* leaves also protect the body from various deleterious effects of free radicals, pollutants and toxins and act as antioxidants [26]. *MO* fresh leaves are a good source of vitamin E, with concentrations similar to those found in nuts [21]. This is important because vitamin E not only acts as an antioxidant, but it has been shown to inhibit cell proliferation [27].

2.2. Polyphenols

The dried leaves of *MO* are a great source of polyphenol compounds, such as flavonoids and phenolic acids.

Flavonoids, which are synthesized in the plant as a response to microbial infections, have a benzo-γ-pyrone ring as a common structure [28,29]. Intake of flavonoids has been shown to protect against chronic diseases associated with oxidative stress, including cardiovascular disease and cancer. *MO* leaves are a good source of flavonoids [30].

The main flavonoids found in *MO* leaves are myrecytin, quercetin and kaempferol, in concentrations of 5.8, 0.207 and 7.57 mg/g, respectively [31,32].

Quercetin is found in dried *MO* leaves, at concentrations of 100 mg/100 g, as quercetin-3-*O*-β-d-glucoside (iso-quercetin or isotrifolin) [33,34]. Quercetin is a strong antioxidant, with multiple therapeutic properties [35]. It has hypolipidemic, hypotensive, and anti-diabetic properties in obese Zucker rats with metabolic syndrome [36]. It can reduce hyperlipidemia and atherosclerosis in high cholesterol or high-fat fed rabbits [37,38]. It can protect insulin-producing pancreatic β cells from *Streptozotocin (STZ)* induced oxidative stress and apoptosis in rats [39].

Phenolic acids are a sub-group of phenolic compounds, derived from hydroxybenzoic acid and hydroxycinnamic acid, naturally present in plants, and these compounds have antioxidant, anti-inflammatory, antimutagenic and anticancer properties [40,41]. In dried leaves, Gallic acid is the most abundant, with a concentration of 1.034 mg/g of dry weight. The concentration of chlorogenic and caffeic acids range from 0.018 to 0.489 mg/g of dry weight and 0.409 mg/g of dry weight, respectively [42,43].

Chlorogenic acid (CGA) is an ester of dihydrocinnamic acid and a major phenolic acid in *MO* [44]. CGA has a role in glucose metabolism. It inhibits glucose-6-phosphate translocase in rat liver, reducing hepatic gluconeogenesis and glycogenolysis [45]. CGA has also been found to lower post-prandial blood glucose in obese Zucker rats [46] and to reduce the glycemic response in rodents [47]. CGA has anti-dyslipidemic properties, as it reduces plasma total cholesterol and triglycerides (TG) in obese Zucker rats or mice fed a high fat diet [48] and reverses STZ-induced dyslipidemia in diabetic rats [41].

2.3. Alkaloids, Glucosinolates and Isothiocyonates

Alkaloids are a group of chemical compounds, which contain mostly basic nitrogen atoms. Several of these compounds, including N,α-L-rhamnopyranosyl vincosamide, phenylacetonitrile pyrrolemarumine,4′-hydroxyphenylethanamide-α-L-rhamnopyranoside and its glucopyranosyl derivative, have been isolated from *Moringa Oleifera* leaves [49,50].

Glucosinolates are a group of secondary metabolites in plants [51]. Both glucosinolates and isothiocyanates have been found to have important health-promoting properties [52].

2.4. Tannins

Tannins are water-soluble phenolic compounds that precipitate alkaloids, gelatin and other proteins. Their concentrations in dried leaves range between 13.2 and 20.6 g tannin/kg [53] being a little higher in freeze-dried leaves [54]. Tannins have been reported to have anti-cancer, antiatherosclerotic, anti-inflammatory and anti-hepatoxic properties [55].

2.5. Saponins

MO leaves are also a good source of saponins, natural compounds made of an isoprenoidal-derived aglycone, covalently linked to one or more sugar moieties [56]. The concentrations of saponins in *MO* freeze-dried leaves range between 64 and 81 g/kg of dry weight [57]. Saponins have anti-cancer properties [58].

3. Effects of *Moringa Oleifera* on the Prevention of Chronic Disease

3.1. Hypolipidemic Effects

Many bioactive compounds found in *MO* leaves may influence lipid homeostasis. Phenolic compounds, as well as flavonoids, have important roles in lipid regulation [59]. They are involved in the inhibition of pancreatic cholesterol esterase activity, thereby reducing and delaying cholesterol absorption, and binding bile acids, by forming insoluble complexes and increasing their fecal excretion, thereby decreasing plasma cholesterol concentrations [60]. The extracts of *MO* have shown hypolipidemic activity, due to inhibition of both lipase and cholesterol esterase, thus showing its potential for the prevention and treatment of hyperlipidemia [61].

MO has a strong effect on lipid profile through cholesterol reducing effects. Cholesterol homeostasis is maintained by two processes: cholesterol biosynthesis, in which 3-hydroxymethyl glutaryl CoA (HMG-Co-A) reductase catalyzes the rate limiting process and cholesterol absorption of both dietary cholesterol and cholesterol cleared from the liver through biliary secretion. The activity of HMG-CoA reductase was depressed by the ethanolic extract of *MO*, further supporting its hypolipidemic action [62]. *Moringa Oleifera (MO)* leaves also contain the bioactive β-sitosterol, with documented cholesterol lowering effects, which might have been responsible for the cholesterol lowering action in plasma of high fat fed rats [18].

Saponins, found in *MO* leaves, prevented the absorption of cholesterol, by binding to this molecule and to bile acids, causing a reduction in the enterohepatic circulation of bile acids and increasing their fecal excretion [9]. The increased bile acid excretion is offset by enhanced bile acid synthesis from cholesterol in the liver, leading to the lowering of plasma cholesterol [9].

3.2. Antioxidant Effects

Due to the high concentrations of antioxidants present in *MO* leaves [14,63,64], they can be used in patients with inflammatory conditions, including cancer, hypertension, and cardiovascular diseases [17,65]. The β carotene found in *MO* leaves has been shown to act as an antioxidant. The antioxidants have the maximum effect on the damage caused by free radicals only when they are ingested in combination. A combination of antioxidants found in *MO* leaves was proven to be more effective than a single antioxidant, possibly due to synergistic mechanisms and increased antioxidant cascade mechanisms [22,66,67]. A recent study in children demonstrated that *MO* leaves could be an important source of vitamin A [68].

The extract of *MO* leaves also contains tannins, saponins, flavonoids, terpenoids and glycosides, which have medicinal properties. These compounds have been shown to be effective antioxidants, antimicrobial and anti-carcinogenic agents [69,70]. Phenolic compounds are known to act as primary antioxidants [71], due to their properties for the inactivation of lipid free radicals or prevention of the decomposition of hydroperoxides into free radicals, due to their redox properties. These properties play a key role in neutralizing free radicals, quenching singlet or triplet oxygen, or decomposing peroxides [72,73].

The radical scavenging and antioxidant activities of the aqueous and aqueous ethanol extracts of freeze-dried leaves of *MO*, from different agro-climatic regions, were investigated by Siddhuraju and Becker [74]. They found that different leaf extracts inhibited 89.7–92.0% of peroxidation of linoleic acid and had scavenging activities on superoxide radicals in a dose-dependent manner in the β-carotene-linoleic acid system. Iqbal and Bhanger [75] showed that the environmental temperature and soil properties have significant effects on antioxidant activity of *MO* leaves.

3.3. Anti-Inflammatory and Immunomodulatory Effect

The extract of *MO* leaves inhibited human macrophage cytokine production (tumor necrosis factor alpha (TNF-α), interleukin-6 (IL-6) and IL-8), which were induced by cigarette smoke and by lipopolysaccharide (LPS) [76]. Further, Waterman et al. [77] reported that both *MO* concentrate and isothiocyanates decreased the gene expression and production of inflammatory markers in RAW macrophages.

The extracts of *MO* leaves stimulated both cellular and humoral immune responses in cyclophosphamide-induced immunodeficient mice, through increases in white blood cells, percent of neutrophils and serum immunoglobulins [78,79]. In addition, quercetin may have been involved in the reduction of the inflammatory process by inhibiting the action of neutral factor kappa-beta (NF-kβ) and subsequent NF-kB-dependent downstream events and inflammation [80]. Further, fermentation of *MO* appears to enhance the anti-inflammatory properties of *MO* [81]. C57BL/6 mice, fed for 10 weeks with distilled water, fermented and non-fermented *MO* [81]. Investigators reported decreases in the mRNA levels of inflammatory cytokines and reductions in endoplasmic reticulum stress in those animals fed the fermented product.

3.4. Hepato-Protective Effects

The methanol extract of *MO* leaves has a hepatoprotective effect, which might be due to the presence of quercetin [14,67]. *MO* leaves had substantial effects on the levels of aspartate amino transferase (AST), alanine amino transferase (ALT) and alkaline phosphatase (ALP), in addition to reductions in lipids and lipid peroxidation levels in the liver of rats [18].

MO leaves have been shown to reduce plasma ALT, AST, ALP and creatinine [82,83] and to ameliorate hepatic and kidney damage induced by drugs. In rats, co-treated with *MO* leaves and NiSO4, in order to induce nephrotoxicity, similar findings were observed [84]. Also, Das et al. [75] observed the same reductions in hepatic enzymes in rats fed a high fat diet, in combination with *MO* leaves. Also, the administration of the extract of *MO* leaves in mice was followed by decreases in serum ALT, AST, ALP, and creatinine [85,86]. In guinea pigs, treatment of *MO* leaves prevented non-alcoholic fatty liver disease (NAFLD) in a model of hepatic steatosis, as measured by lower concentrations of hepatic cholesterol and triglycerides in animals treated with *MO* compared to controls [87]. This lowering of hepatic lipids was associated with lower inflammation and expression of genes involved in lipid uptake and inflammation [87]. Further, the *MO* treated guinea pigs had lower concentrations of plasma ASP. In contrast, *MO* leaves did not reduce the inflammation of lipid accumulation in the adipose tissue of guinea pigs [88].

3.5. Anti-Hyperglycemic (Antidiabetic) Effect

Many compounds found in *MO* leaves might be involved in glucose homeostasis. For example, isothiocyanates have been reported to reduce insulin resistance as well as hepatic gluconeogenesis [89,90]. Phenolic acids and flavonoids affect glucose homeostasis, influencing β-cell mass and function, and increasing insulin sensitivity in peripheral tissues [91,92]. Phenolic compounds, flavonoids and tannins also inhibit intestinal sucrase and to a certain extent, pancreatic α-amylase activities [56].

The beneficial activities of MO leaves on carbohydrate metabolism have been shown by different mechanisms, including preventing and restoring the integrity and function of β-cells, increasing insulin activity, improving glucose uptake and utilization [57]. Hypoglycemic and antihyperglycemic activity of the leaves of MO might be due to the presence of terpenoids, which are involved in the stimulation of β-cells and the subsequent secretion of insulin. Also, flavonoids have been shown to play an important role in the hypoglycemic action [93]. In another study, where diabetes was induced peritoneally by injection with streptozotocin, rats were fed the equivalent of 250 mg/kg of MO for 6 weeks, using control and diabetic animals [94]. The groups consuming MO extract had significant decreases in malonaldehyde and improvements in the inflammatory cytokines—TNF-α and IL-6—when compared to control animals [94].

3.6. Hypotensive Effects

MO leaves contain several bioactive compounds, which have been used for stabilizing blood pressure, including nitrile, mustard oil glycosides and thiocarbamate glycosides. The isolated four pure compounds, niazinin A, niazinin B, niazimicin and niazinin A + B—from ethanol extract of *MO* leaves showed a blood pressure lowering effect in rats, mediated possibly through a calcium antagonist effect [14,95]. A recent study reported that *MO* reduced vascular oxidation in spontaneously hypertensive rats [96].

3.7. Effects on Ocular Diseases

The major cause of blindness, which ranges from impaired dark adaptation to night blindness, is vitamin A deficiency. *MO* leaves, pods and leaf powder contain high concentrations of vitamin A, which can help to prevent night blindness and eye problems. Also, consumption of leaves with oils improved vitamin A nutrition and delayed the development of cataracts [14].

3.8. Anticancer Effects

MO has been studied for its chemopreventive properties and has been shown to inhibit the growth of several human cancer cells [97]. The capacity of *MO* leaves to protect organisms and cells from oxidative DNA damage, associated with cancer and degenerative diseases, has been reported in several studies [98]. Khalafalla et al. [99] found that the extract of *MO* leaves inhibited the viability of acute myeloid leukemia, acute lymphoblastic leukemia and hepatocellular carcinoma cells. Several bioactive compounds, including 4-(α-L-rhamnosyloxy) benzyl isothiocyanate, niazimicin and β-sitosterol-3-*O*-β-D-glucopyranoside present in *MO*, may be responsible for its anti-cancer properties [100]. *MO* leaf extract has also been proven to be efficient in pancreatic and breast cancer cells [98,99].

In pancreatic cells, *MO* was shown to contain the growth of pancreatic cancer cells, by inhibiting NF-kB signaling as well as increasing the efficacy of chemotherapy, by enhancing the effect of the drug in these cells [101]. In breast cancer cells, the antiproliferative effects of *MO* were also demonstrated [102]. A recent study by Abd-Rabou et al. [103] evaluated the effects of various extracts from *Moringa Oleifera*, including leaves and roots, and preparations of nanocomposites of these compounds against HepG, breast MCF7 and colorectal HCT116/Caco2 cells. All these preparations were effective on their cytotoxic impact, as measured by apoptosis [103]. Several animal studies have also confirmed the efficacy of *Moringa Oleifera* leaves in preventing cancer in rats with hepatic carcinomas induced by diethyl nitrosamine [104] and in suppressing azoxymethane-induced colon carcinogenesis in mice [105]. A list of some bioactive components present in MO leaves, their postulated actions in the animal model used, their protection against a specific disease and the corresponding reference are presented in Table 1.

Table 1. Bioactive Components in *Moringa Oleifera* and their Positive Effects on Chronic Disease.

Compounds	Postulated Function	Model Used	Disease Protection	References
Flavonoids: Quercitin	Hypolipidemic and anti-diabetic properties	Zucker rat	Diabetes	[36]
	Lower hyperlipidemia	Rabbits	Atherosclerosis	[37,38]
	Decrease expression of DGAT	Guinea Pigs	NAFLD	[80]
	Inhibition of cholesterol esterase and α-glucosidase	In vitro study	Cardiovascular disease and Diabetes	[60]
	Inhibits activation of NF-kB	High fat fed Mice	Cardiovascular disease	[74]
Chlorogenic Acid	Glucose lowering effect	Diabetic rats	Diabetes	[45]
	Cholesterol lowering in plasma and liver	Zucker rat	Cardiovascular disease	[46]
	Decrease expression of CD68, SERBP1c	Guinea pigs	NAFLD	[87]
	Anti-obesity properties	High-fat induced obesity rats	Obesity	[49]
	Inhibit enzymes linked to T2D		Diabetes	[90]
Alkaloids	Cardioprotection	Cardiotoxic-induced rats	Cardiovascular disease	[49]
Tannins	Anti-inflammatory	Rats	Cardiovascular/Cancer	[54]
Isothiocyanates	Decreased expression of inflammatory markers	RAW Macrophages	Cardiovascular disease	[76]
	Reduction in insulin resistance	Mice	Diabetes	[88]
	Inhibition of NF-kB signaling	Cancer breast cells	Cancer	[99]
B-Sitosterol	Decrease cholesterol absorption	High-fat fed rats	Cardiovascular disease	[18]

Abbreviations used: CD68: cluster of differentiation 68; DGAT: diacyl glycerol transferase; NF-kB: nuclear factor-kB; SRBP1c: sterol regulatory binding protein 1c; T2D: type 2 diabetes; NAFLD: non-alcoholic fatty liver disease.

3.9. Protection Against Alzheimer's Disease (AD) and Parkinson's Disease (PD)

It is recognized that the monoaminegistic system has a modulatory role in memory processing and that this system is disturbed by AD [106]. Some plants, including *MO*, have been demonstrated to enhance memory by nootropics activity and protect against the oxidative stress present in AD [107] Ganguly et al. [108] have an established model for AD involving the infusion of colchicine into the brain of rats and they demonstrated that *MO* led to the alteration of brain monoamines and electrical patterns. A recent study was conducted to evaluate the effects of an isothiocynate isolated from *MO*, both in a mouse model of PD and in RAW 264.7 macrophages stimulated with LPS [109]. Results demonstrated great efficacy of a bioactive compound in *MO*, which results from myrosinase hydrolysis in favorably modulating the inflammatory and apoptotic pathways as well as oxidative stress [109].

4. Conclusions

In summary, there are a number of animal studies documenting the effects of *MO* leaves in protecting against cardiovascular disease, diabetes, NAFLD, Alzheimer's, hypertension and others, due the actions of the bioactive components in preventing lipid accumulation, reducing insulin resistance and inflammation. Additional studies in humans, including clinical trials are needed to confirm these effects of *MO* on chronic diseases. In addition, some studies have found that the compounds in *MO* may also protect against Alzheimer's disease and Parkinson's disease. A summary of the effects of the bioactive component of *MO* leaves in protecting against these conditions is shown in Figure 1.

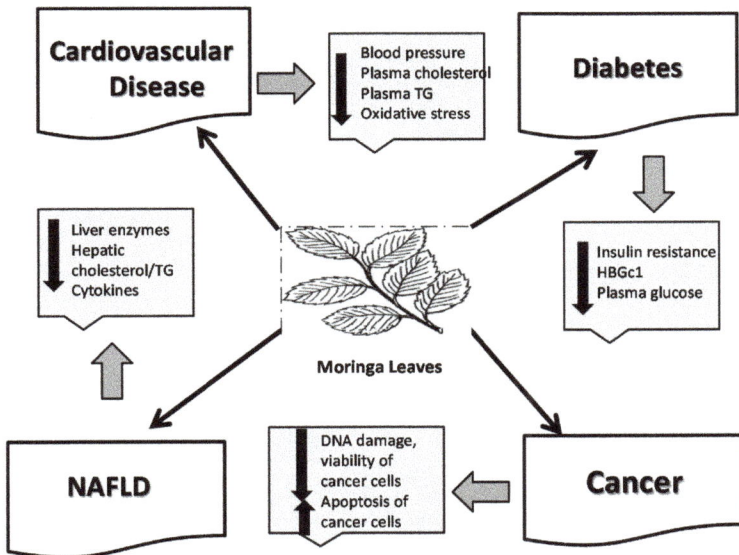

Figure 1. Protective effects of *MO* leaves against chronic diseases: **cardiovascular disease**, by lowering plasma lipids including triglycerides (TG) [45,60] decreasing blood pressure [92] and reducing oxidative stress [73]; **diabetes**, by lowering plasma glucose [61], reducing insulin resistance [89] and increasing β cell function [90]; **NAFLD**, by reducing hepatic lipids [82,87], reducing liver enzymes [82,83,88] and decreasing hepatic inflammation [88] and **cancer**, by reducing DNA damage [97], viability of cancer cells [99,100] and increasing apoptosis [104,105].

Author Contributions: Marcela Vergara-Jimenez searched the literature, provided a number of references and the input in the final version of the paper; Manal Mused Almatrafi contributed substantially to the writing of the paper; Maria Luz Fernandez reviewed all articles, did a summary of the most relevant literature, put together all the information, produced the final version of this manuscript and created the Table and the Figure.

Conflicts of Interest: The authors declare no conflict of interest.

References

1. Padayachee, B.; Baijnath, H. An overview of the medicinal importance of *Moringaceae*. *J. Med. Plants Res.* **2012**, *6*, 5831–5839.
2. Stohs, S.; Hartman, M.J. Review of the Safety and Efficacy of *Moringa oleifera*. *Phytother. Res.* **2015**, *29*, 796–804. [CrossRef] [PubMed]
3. Saini, R.K.; Sivanesan, I.; Keum, Y.S. Phytochemicals of *Moringa oleifera*: A review of their nutritional, therapeutic and industrial significance. *3 Biotech* **2016**, *6*. [CrossRef] [PubMed]
4. Martin, C.; Martin, G.; Garcia, A.; Fernández, T.; Hernández, E.; Puls, L. Potential applications of *Moringa oleifera*. A critical review. *Pastosy Forrajes* **2013**, *36*, 150–158.
5. Leone, A.; Spada, A.; Battezzati, A.; Schiraldi, A.; Aristil, J.; Bertoli, S. Cultivation, genetic, ethnopharmacology, phytochemistry and pharmacology of *Moringa oleifera* Leaves: An overview. *Int. J. Mol. Sci.* **2015**, *16*, 12791–12835. [CrossRef] [PubMed]
6. Popoola, J.O.; Obembe, O.O. Local knowledge, use pattern and geographical distribution of *Moringa oleifera* Lam. (*Moringaceae*) in Nigeria. *J. Ethnopharmacol.* **2013**, *150*, 682–691. [CrossRef] [PubMed]
7. Sivasankari, B.; Anandharaj, M.; Gunasekaran, P. An ethnobotanical study of indigenous knowledge on medicinal plants used by the village peoples of Thoppampatti, Dindigul district, Tamilnadu, India. *J. Ethnopharmacol.* **2014**, *153*, 408–423. [CrossRef] [PubMed]
8. Pal, S.K.; Mukherjee, P.K.; Saha, B.P. Studies on the antiulcer activity of *Moringa oleifera* leaf extract on gastric ulcer models in rats. *Phytother. Res.* **1995**, *9*, 463–465. [CrossRef]
9. Oyedepo, T.A.; Babarinde, S.O.; Ajayeoba, T.A. Evaluation of the antihyperlipidemic effect of aqueous leaves extract of *Moringa oleifera* in alloxan induced diabetic rats. *Int. J. Biochem. Res. Rev.* **2013**, *3*, 162–170. [CrossRef]
10. Faizi, S.; Siddiqui, B.; Saleem, R.; Aftab, K.; Shaheen, F.; Gilani, A. Hypotensive constituents from the pods of *Moringa oleifera*. *Planta Med.* **1998**, *64*, 225–228. [CrossRef] [PubMed]
11. Rao, K.S.; Mishra, S.H. Anti-inflammatory and antihepatoxic activities of the roots of Moringa pterygosperma Gaertn. *Indian J. Pharm. Sci.* **1998**, *60*, 12–16.
12. Bennett, R.N.; Mellon, F.A.; Foidl, N.; Pratt, J.H.; Dupont, M.S.; Perkins, L.; Kroon, P.A. Profiling glucosinolates and phenolics in vegetative and reproductive tissues of the multi-purpose trees *Moringa oleifera* L. (horseradish tree) and *Moringa stenopetala* L. *J. Agric. Food Chem.* **2003**, *51*, 3546–3553. [CrossRef] [PubMed]
13. Tahiliani, P.; Kar, A. Role of *Moringa oleifera* leaf extract in the regulation of thyroid hormone status in adult male and female rats. *Pharmacol. Res.* **2000**, *41*, 319–323. [CrossRef] [PubMed]
14. Anwar, F.; Latif, S.; Ashraf, M.; Gilani, A.H. *Moringa oleifera*: A food plant with multiple medicinal uses. *Phytother. Res.* **2007**, *21*, 17–25. [CrossRef] [PubMed]
15. Mahajan, S.; Banerjee, A.; Chauhan, B.; Padh, H.; Nivsarkar, M.; Mehta, A. Inhibitory effect of *N*-butanol fraction of *Moringa oleifera* Lam seeds on ovalbumin-induced airway inflammation in a guinea pig model of asthma. *Int. J. Toxicol.* **2009**, *28*, 519–527. [CrossRef] [PubMed]
16. Hamza, A.A. Ameliorative effects of *Moringa oleifera* Lam seed extract on liver fibrosis in rats. *Food Chem. Toxicol.* **2010**, *48*, 345–355. [CrossRef] [PubMed]
17. Pari, L.; Kumar, N.A. Hepatoprotective activity of *Moringa oleifera* on antitubercular drug-induced liver damage in rats. *J. Med. Food* **2002**, *5*, 171–177. [CrossRef] [PubMed]
18. Halaby, M.S.; Metwally, E.M.; Omar, A.A. Effect of *Moringa oleifera* on serum lipids and kidney function of hyperlipidemic rats. *J. Appl. Sci. Res.* **2013**, *9*, 5189–5198.
19. Okwari, O.; Dasofunjo, K.; Asuk, A.; Alagwu, E.; Mokwe, C. Anti-hypercholesterolemic and hepatoprotective effect of aqueous leaf extract of *Moringa oleifera* in rats fed with thermoxidized palm oil diet. *J. Pharm. Biol. Sci.* **2013**, *8*, 57–62.

20. Walter, A.; Samuel, W.; Peter, A.; Joseph, O. Antibacterial activity of *Moringa oleifera* and *Moringa stenopetala* methanol and *N*-hexane seed extracts on bacteria implicated in water borne diseases. *Afr. J. Microbiol. Res.* **2011**, *5*, 153–157.

21. Efiong, E.E.; Igile, G.O.; Mgbeje, B.I.A.; Out, E.A.; Ebong, P.E. Hepatoprotective and anti-diabetic effect of combined extracts of *Moringa oleifera* and *Vernoniaamygdalina* in streptozotocin-induced diabetic albino Wistar rats. *J. Diabetes Endocrinol.* **2013**, *4*, 45–50.

22. Ferreira, P.M.P.; Farias, D.F.; Oliveira, J.T.D.A.; Carvalho, A.D.F.U. *Moringa oleifera*: Bioactive compounds and nutritional potential. *Rev. Nutr.* **2008**, *21*, 431–437. [CrossRef]

23. Alvarez, R.; Vaz, B.; Gronemeyer, H.; de Lera, A.R. Functions, therapeutic applications, and synthesis of retinoids and carotenoids. *Chem. Rev.* **2014**, *114*, 1–125. [CrossRef] [PubMed]

24. Slimani, N.; Deharveng, G.; Unwin, I.; Southgate, D.A.; Vignat, J.; Skeie, G.; Salvini, S.; Parpinel, M.; Møller, A.; Ireland, J.; et al. The EPIC nutrient database project (ENDB): A first attempt to standardize nutrient databases across the 10 European countries participating in the EPIC study. *Eur. J. Clin. Nutr.* **2007**, *61*, 1037–1056. [CrossRef] [PubMed]

25. Ramachandran, C.; Peter, K.V.; Gopalakrishnan, P.K. Drumstick (*Moringa oleifera*): A multipurpose Indian vegetable. *Econ. Bot.* **1980**, *34*, 276–283. [CrossRef]

26. Chambial, S.; Dwivedi, S.; Shukla, K.K.; John, P.J.; Sharma, P. Vitamin C in disease prevention and cure: An overview. *Indian J. Clin. Biochem.* **2013**, *28*, 314–328. [CrossRef] [PubMed]

27. Borel, P.; Preveraud, D.; Desmarchelier, C. Bioavailability of vitamin E in humans: An update. *Nutr. Rev.* **2013**, *71*, 319–331. [CrossRef] [PubMed]

28. Kumar, S.; Pandey, A.K. Chemistry and biological activities of flavonoids: An overview. *Sci. World J.* **2013**, *2013*, 162750. [CrossRef] [PubMed]

29. Bovicelli, P.; Bernini, R.; Antonioletti, R.; Mincione, E. Selective halogenation of flavanones. *Tetrahedron Lett.* **2002**, *43*, 5563–5567. [CrossRef]

30. Pandey, K.B.; Rizvi, S.I. Plant polyphenols as dietary antioxidants in human health and disease. *Oxid. Med. Cell Longev.* **2009**, *2*, 270–278. [CrossRef] [PubMed]

31. Sultana, B.; Anwar, F. Flavonols (kaempeferol, quercetin, myricetin) contents of selected fruits, vegetables and medicinal plants. *Food Chem.* **2008**, *108*, 879–884. [CrossRef] [PubMed]

32. Coppin, J.P.; Xu, Y.; Chen, H.; Pan, M.H.; Ho, C.T.; Juliani, R.; Simon, J.E.; Wu, Q. Determination of flavonoids by LC/MS and anti-inflammatory activity in *Moringa oleifera*. *J. Funct. Foods* **2013**, *5*, 1892–1899. [CrossRef]

33. Lako, J.; Trenerry, V.C.; Wahlqvist, M.; Wattanapenpaiboon, N.; Sotheeswaran, S.; Premier, R. Phytochemical flavonols, carotenoids and the antioxidant properties of a wide selection of Fijian fruit, vegetables and other readily available foods. *Food Chem.* **2007**, *101*, 1727–1741. [CrossRef]

34. Atawodi, S.E.; Atawodi, J.C.; Idakwo, G.A.; Pfundstein, B.; Haubner, R.; Wurtele, G.; Bartsch, H.; Owen, R.W. Evaluation of the polyphenol content and antioxidant properties of methanol extracts of the leaves, stem, and root barks of *Moringa oleifera* Lam. *J. Med. Food.* **2010**, *13*, 710–716. [CrossRef] [PubMed]

35. Bischoff, S.C. Quercetin: Potentials in the prevention and therapy of disease. *Curr. Opin. Clin. Nutr. Metab. Care* **2008**, *11*, 733–740. [CrossRef] [PubMed]

36. Rivera, L.; Moron, R.; Sanchez, M.; Zarzuelo, A.; Galisteo, M. Quercetin ameliorates metabolic syndrome and improves the inflammatory status in obese Zucker rats. *Obesity (Silver Spring)* **2008**, *16*, 2081–2087. [CrossRef] [PubMed]

37. Juzwiak, S.; Wojcicki, J.; Mokrzycki, K.; Marchlewicz, M.; Bialecka, M.; Wenda-Rozewicka, L.; Gawrońska-Szklarz, B.; Droździk, M. Effect of quercetin on experimental hyperlipidemia and atherosclerosis in rabbits. *Pharmacol. Rep.* **2005**, *57*, 604–609. [PubMed]

38. Kamada, C.; da Silva, E.L.; Ohnishi-Kameyama, M.; Moon, J.H.; Terao, J. Attenuation of lipid peroxidation and hyperlipidemia by quercetin glucoside in the aorta of high cholesterol-fed rabbit. *Free Radic. Res.* **2005**, *39*, 185–194. [CrossRef] [PubMed]

39. Coskun, O.; Kanter, M.; Korkmaz, A.; Oter, S. Quercetin, a flavonoid antioxidant, prevents and protects streptozotocin-induced oxidative stress and beta-cell damage in rat pancreas. *Pharmacol. Res.* **2005**, *51*, 117–123. [CrossRef] [PubMed]

40. El-Seedi, H.R.; El-Said, A.M.; Khalifa, S.A.; Göransson, U.; Bohlin, L.; Borg-Karlson, A.K.; Verpoorte, R. Biosynthesis, natural sources, dietary intake, pharmacokinetic properties, and biological activities of hydroxycinnamic acids. *J. Agric. Food Chem.* **2012**, *60*, 10877–10895. [CrossRef] [PubMed]

41. Verma, S.; Singh, A.; Mishra, A. Gallic acid: Molecular rival of cancer. *Environ. Toxicol. Pharmacol.* **2013**, *35*, 473–485. [CrossRef] [PubMed]
42. Prakash, D.; Suri, S.; Upadhyay, G.; Singh, B.N. Total phenol, antioxidant and free radical scavenging activities of some medicinal plants. *Int. J. Food Sci. Nutr.* **2007**, *58*, 18–28. [CrossRef] [PubMed]
43. Singh, B.N.; Singh, B.R.; Singh, R.L.; Prakash, D.; Dhakarey, R.; Upadhyay, G.; Singh, H.B. Oxidative DNA damage protective activity, antioxidant and anti-quorum sensing potentials of *Moringa oleifera*. *Food Chem. Toxicol.* **2009**, *47*, 1109–1116. [CrossRef] [PubMed]
44. Amaglo, N.K.; Bennett, R.N.; LoCurto, R.B.; Rosa, E.A.S.; LoTurco, V.; Giuffrid, A.; LoCurto, A.; Crea, F.; Timpo, G.M. Profiling selected phytochemicals and nutrients in different tissues of the multipurpose tree *Moringa oleifera* L., grown in Ghana. *Food Chem.* **2010**, *122*, 1047–1054. [CrossRef]
45. Karthikesan, K.; Pari, L.; Menon, V.P. Combined treatment of tetrahydrocurcumin and chlorogenic acid exerts potential antihyperglycemic effect on streptozotocin-nicotinamide-induced diabetic rats. *Gen. Physiol. Biophys.* **2010**, *29*, 23–30. [CrossRef] [PubMed]
46. De Sotillo Rodriguez, D.V.; Hadley, M. Chlorogenic acid modifies plasma and liver concentrations of: Cholesterol, triacylglycerol, and minerals in (fa/fa) Zucker rats. *J. Nutr. Biochem.* **2002**, *13*, 717–726. [CrossRef]
47. Tunnicliffe, J.M.; Eller, L.K.; Reimer, R.A.; Hittel, D.S.; Shearer, J. Chlorogenic acid differentially affects postprandial glucose and glucose-dependent insulin otropic polypeptide response in rats. *Appl. Physiol. Nutr. Metab.* **2011**, *36*, 650–659. [CrossRef] [PubMed]
48. Cho, A.S.; Jeon, S.M.; Kim, M.J.; Yeo, J.; Seo, K.I.; Choi, M.S.; Lee, M.K. Chlorogenic acid exhibits anti-obesity property and improves lipid metabolism in high-fat diet-induced-obese mice. *Food Chem. Toxicol.* **2010**, *48*, 937–943. [CrossRef] [PubMed]
49. Panda, S.; Kar, A.; Sharma, P.; Sharma, A. Cardioprotective potential of N, α-L-rhamnopyranosyl vincosamide, an indole alkaloid, isolated from the leaves of *Moringa oleifera* in isoproterenol induced cardiotoxic rats: In vivo and in vitro studies. *Bioorg. Med. Chem. Lett.* **2013**, *23*, 959–962. [CrossRef] [PubMed]
50. Sahakitpichan, P.; Mahidol, C.; Disadee, W.; Ruchirawat, S.; Kanchanapoom, T. Unusual glycosides of pyrrole alkaloid and 4′-hydroxyphenylethanamide from leaves of *Moringa oleifera*. *Phytochemistry* **2011**, *72*, 791–795. [CrossRef] [PubMed]
51. Forster, N.; Ulrichs, C.; Schreiner, M.; Muller, C.T.; Mewis, I. Development of a reliable extraction and quantification method for glucosinolates in *Moringa oleifera*. *Food Chem.* **2015**, *166*, 456–464. [CrossRef] [PubMed]
52. Dinkova-Kostova, A.T.; Kostov, R.V. Glucosinolates and isothiocyanates in health and disease. *Trends Mol. Med.* **2012**, *18*, 337–347. [CrossRef] [PubMed]
53. Teixeira, E.M.B.; Carvalho, M.R.B.; Neves, V.A.; Silva, M.A.; Arantes-Pereira, L. Chemical characteristics and fractionation of proteins from *Moringa oleifera* Lam. leaves. *Food Chem.* **2014**, *147*, 51–54. [CrossRef] [PubMed]
54. Richter, N.; Siddhuraju, P.; Becker, K. Evaluation of nutritional quality of moringa (*Moringa oleifera* Lam.) leaves as an alternative protein source for Nile tilapia (*Oreochromis niloticus* L.). *Aquaculture* **2003**, *217*, 599–611. [CrossRef]
55. Adedapo, A.A.; Falayi, O.O.; Oyagvemi, A.A.; Kancheva, V.D.; Kasaikina, O.T. Evaluation of the analgesic, anti-inflammatory, anti-oxidant, phytochemical and toxicological properties of the methanolic leaf extract of commercially processed *Moringa oleifera* in some laboratory animals. *J. Basic Clin. Physiol. Pharmacol.* **2015**, *26*, 491–499. [CrossRef] [PubMed]
56. Augustin, J.M.; Kuzina, V.; Andersen, S.B.; Bak, S. Molecular activities, biosynthesis and evolution of triterpenoid saponins. *Phytochemistry* **2011**, *72*, 435–457. [CrossRef] [PubMed]
57. Makkar, H.P.S.; Becker, K. Nutritional value and anti-nutritional components of whole and ethanol extracted *Moringa oleifera* Leaves. *Anim. Feed Sci. Technol.* **1996**, *63*, 211–228. [CrossRef]
58. Tian, X.; Tang, H.; Lin, H.; Cheng, G.; Wang, S.; Zhang, X. Saponins: The potential chemotherapeutic agents in pursuing new anti-glioblastoma drugs. *Mini Rev. Med. Chem.* **2013**, *13*, 1709–1724. [CrossRef] [PubMed]
59. Siasos, G.; Tousoulis, D.; Tsigkou, V.; Kokkou, E.; Oikonomou, E.; Vavuranakis, M.; Basdra, E.K.; Papavassiliou, A.G.; Stefanadis, C. Flavonoids in atherosclerosis: An overview of their mechanisms of action. *Curr. Med. Chem.* **2013**, *20*, 2641–2660. [CrossRef] [PubMed]

60. Adisakwattana, S.; Chanathong, B. Alpha-glucosidase inhibitory activity and lipid-lowering mechanisms of *Moringa oleifera* Leaf extract. *Eur. Rev. Med. Pharmacol. Sci.* **2011**, *15*, 803–808. [PubMed]

61. Toma, A.; Makonnen, E.; Debella, A.; Tesfaye, B. Antihyperglycemic Effect on Chronic Administration of Butanol Fraction of Ethanol Extract of *Moringa stenopetala* Leaves in Alloxan Induced Diabetic Mice. *Asian Pac. J. Trop. Biomed.* **2012**, *2*, S1606–S1610. [CrossRef]

62. Hassarajani, S.; Souza, T.D.; Mengi, S.A. Efficacy study of the bioactive fraction (F-3) of Acorus calamus in hyperlipidemia. *Indian J. Pharmacol.* **2007**, *39*, 196–200.

63. Mensah, J.K.; Ikhajiagbe, B.; Edema, N.E.; Emokhor, J. Phytochemical, nutritional and antibacterial properties of dried leaf powder of *Moringa oleifera* (Lam.) from Edo Central Province, Nigeria. *J. Nat. Prod. Plant Resour.* **2012**, *2*, 107–112.

64. Bamishaiye, E.I.; Olayemi, F.F.; Awagu, E.F.; Bamshaiye, O.M. Proximate and phytochemical composition of *Moringa oleifera* leaves at three stages of maturation. *Adv. J. Food. Sci. Technol.* **2011**, *3*, 233–237.

65. Posmontie, B. The medicinal qualities of *Moringa oleifera. Holist. Nurs. Pract.* **2011**, *25*, 80–87. [CrossRef] [PubMed]

66. Mishra, G.; Singh, P.; Verma, R.; Kumar, R.S.; Srivastava, S.; Khosla, R.L. Traditional uses, phytochemistry and pharmacological properties of *Moringa oleifera* plant: An overview. *Der Pharmacia Lettre* **2011**, *3*, 141–164.

67. Tejas, G.H.; Umang, J.H.; Payal, B.N.; Tusharbinu, D.R.; Pravin, T.R. A panoramic view on pharmacognostic, pharmacological, nutritional, therapeutic and prophylactic values of *Moringa olifera* Lam. *Int. Res. J. Pharm.* **2012**, *3*, 1–7.

68. Lopez-Teros, V.; Ford, J.L.; Green, M.H.; Tang, G.; Grusak, M.A.; Quihui-Cota, L.; Muzhingi, T.; Paz-Cassini, M.; Astiazaran-Garcia, H. Use of a "Super-child" Approach to Assess the Vitamin A Equivalence of *Moringa oleifera* Leaves, Develop a Compartmental Model for Vitamin A Kinetics, and Estimate Vitamin A Total Body Stores in Young Mexican Children. *J. Nutr.* **2017**. [CrossRef] [PubMed]

69. Ayoola, G.A.; Coker, H.A.B.; Adesegun, S.A.; Adepoju-Bello, A.A.; Obaweya, K.; Ezennia, E.C. Phytochemical screening and antioxidant activities of some selected medicinal plants used for malaria therapy in southwestern Nigeria. *Trop. J. Pharm. Res.* **2008**, *7*, 1019–1024.

70. Davinelli, S.; Bertoglio, J.C.; Zarrelli, A.; Pina, R.; Scapagnini, G. A randomized clinical trial evaluating the efficacy of an anthocyanin-maqui berry extract (Delphinol®) on oxidative stress biomarkers. *J. Am. Coll. Nutr.* **2015**, *34* (Suppl. 1), 28–33. [CrossRef] [PubMed]

71. Murillo, A.G.; Fernandez, M.L. The relevance of dietary polyphenols in cardiovascular protection. *Curr. Pharmacol. Rev.* **2017**, *23*, 2444–2452. [CrossRef] [PubMed]

72. Pokorny, J. Introduction. In *Antioxidant in Foods: Practical Applications*; Pokorny, J., Yanishlieva, N., Gordon, N.H., Eds.; Woodhead Publishing Limited: Cambridge, UK, 2001; pp. 1–3.

73. Zheng, W.; Wang, S.Y. Antioxidant activity and phenolic compounds in selected herbs. *J. Agric. Food Chem.* **2001**, *49*, 5165–5170. [CrossRef] [PubMed]

74. Siddhuraju, P.; Becker, K. Antioxidant properties of various solvent extracts of total phenolic constituents from three different agroclimatic origins of drumstick tree (*Moringa oleifera* Lam.) leaves. *J. Agric. Food Chem.* **2003**, *51*, 2144–2155. [CrossRef] [PubMed]

75. Iqbal, S.; Bhanger, M.I. Effect of season and production location on antioxidant activity of *Moringa oleifera* leaves grown in Pakistan. *J. Food Compos. Anal.* **2006**, *19*, 544–551. [CrossRef]

76. Kooltheat, N.; Sranujit, R.P.; Chumark, P.; Potup, P.; Laytragoon-Lewin, N.; Usuwanthim, K. An ethyl acetate fraction of *Moringa oleifera* Lam. Inhibits human macrophage cytokine production induced by cigarette smoke. *Nutrients* **2014**, *6*, 697–710. [CrossRef] [PubMed]

77. Waterman, C.; Cheng, D.M.; Rojas-Silva, P.; Poulev, A.; Dreifus, J.; Lila, M.A.; Raskin, I. Stable, water extractable isothiocyanates from *Moringa oleifera* leaves attenuate inflammation in vitro. *Phytochemistry* **2014**, *103*, 114–122. [CrossRef] [PubMed]

78. Sudha, P.; Asdaq, S.M.; Dhamingi, S.S.; Chandrakala, G.K. Immunomodulatory activity of methanolic leaf extract of *Moringa oleifera* in animals. *Indian J. Physiol. Pharmacol.* **2010**, *54*, 133–140. [PubMed]

79. Gupta, A.; Gautam, M.K.; Singh, R.K.; Kumar, M.V.; Rao, C.H.V.; Goel, R.K.; Anupurba, S. Immunomodulatory effect of *Moringa oleifera* Lam. extract on cyclophosphamide induced toxicity in mice. *Indian J. Exp. Biol.* **2010**, *48*, 1157–1160. [PubMed]

80. Das, N.; Sikder, K.; Ghosh, S.; Fromenty, B.; Dey, S. *Moringa oleifera* Lam. leaf extract prevents early liver injury and restores antioxidant status in mice fed with high-fat diet. *Indian J. Exp. Biol.* **2012**, *50*, 404–412. [PubMed]

81. Joung, H.; Kim, B.; Park, H.; Lee, K.; Kim, H.H.; Sim, H.C.; Do, H.J.; Hyun, C.K.; Do, M.S. Fermented *Moringa oleifera* Decreases Hepatic Adiposity and Ameliorates Glucose Intolerance in High-Fat Diet-Induced Obese Mice. *J. Med. Food* **2017**, *20*, 439–447. [CrossRef] [PubMed]

82. Sharifudin, S.A.; Fakurazi, S.; Hidayat, M.T.; Hairuszah, I.; Moklas, M.A.; Arulselvan, P. Therapeutic potential of *Moringa oleifera* extracts against acetaminophen-induced hepatotoxicity in rats. *Pharm. Biol.* **2013**, *51*, 279–288. [CrossRef] [PubMed]

83. Ouedraogo, M.; Lamien-Sanou, A.; Ramde, N.; Ouédraogo, A.S.; Ouédraogo, M.; Zongo, S.P.; Goumbri, O.; Duez, P.; Guissou, P.I. Protective effect of *Moringa oleifera* Leaves against gentamicin-induced nephrotoxicity in rabbits. *Exp. Toxicol. Pathol.* **2013**, *65*, 335–339. [CrossRef] [PubMed]

84. Adeyemi, O.S.; Elebiyo, T.C. *Moringa oleifera* supplemented diets prevented nickel-induced nephrotoxicity in Wistar rats. *J. Nutr. Metab.* **2014**, *2014*, 958621. [CrossRef] [PubMed]

85. Oyagbemi, A.A.; Omobowale, T.O.; Azeez, I.O.; Abiola, J.O.; Adedokun, R.A.; Nottidge, H.O. Toxicological evaluations of methanolic extract of *Moringa oleifera* Leaves in liver and kidney of male Wistar rats. *J. Basic Clin. Physiol. Pharmacol.* **2013**, *24*, 307–312. [CrossRef] [PubMed]

86. Asiedu-Gyekye, I.J.; Frimpong-Manso, S.; Awortwe, C.; Antwi, D.A.; Nyarko, A.K. Micro- and macroelemental composition and safety evaluation of the nutraceutical *Moringa oleifera* Leaves. *J. Toxicol.* **2014**, *2014*, 786979. [CrossRef] [PubMed]

87. Almatrafi, M.M.; Vergara-Jimenez, M.; Murillo, A.G.; Norris, G.H.; Blesso, C.N.; Fernandez, M.L. *Moringa* leaves prevent hepatic lipid accumulation and inflammation in guinea pigs by reducing the expression of genes involved in lipid Metabolism. *Int. J. Mol. Sci.* **2017**, *18*, 1330. [CrossRef] [PubMed]

88. Almatrafi, M.M.; Vergara-Jimenez, M.; Smyth, J.A.; Medina-Vera, I.; Fernandez, M.L. *Moringa olifeira* leaves do not alter adipose tissue colesterol accumulation or inflammation in guinea pigs fed a hypercholesterolemic diet. *EC Nutr.* **2017**, *18*, 1330.

89. Waterman, C.; Rojas-Silva, P.; Tumer, T.; Kuhn, P.; Richard, A.J.; Wicks, S.; Stephens, J.M.; Wang, Z.; Mynatt, R.; Cefalu, W.; et al. Isothiocyanate-rich *Moringa oleifera* extract reduces weight gain, insulin resistance and hepatic gluconeogenesis in mice. *Mol. Nutr. Food Res.* **2015**, *59*, 1013–1024. [CrossRef] [PubMed]

90. Fabio, G.D.; Romanucci, V.; De Marco, A.; Zarrelli, A. Triterpenoids from Gymnema sylvestre and their pharmacological activities. *Molecules* **2014**, *19*, 10956–10981. [CrossRef] [PubMed]

91. Oh, Y.S.; Jun, H.S. Role of bioactive food components in diabetes prevention: Effects on Beta-cell function and preservation. *Nutr. Metab. Insights* **2014**, *7*, 51–59. [CrossRef] [PubMed]

92. Oboh, G.; Agunloye, O.M.; Adefegha, S.A.; Akinyemi, A.J.; Ademiluyi, A.O. Caffeic and chlorogenic acids inhibit key enzymes linked to type 2 diabetes (in vitro): A comparative study. *J. Basic Clin. Physiol. Pharmacol.* **2015**, *26*, 165–170. [CrossRef] [PubMed]

93. Manohar, V.S.; Jayasree, T.; Kishore, K.K.; Rupa, L.M.; Dixit, R.; Chandrasekhar, N. Evaluation of hypoglycemic and antihyperglycemic effect of freshly prepared aqueous extract of *Moringa oleifera* leaves in normal and diabetic rabbits. *J. Chem. Pharmacol. Res.* **2012**, *4*, 249–253.

94. Omodanisi, E.I.; Aboua, Y.G.; Oguntibeju, O.O. Assessment of the Anti-Hyperglycaemic, Anti-Inflammatory and Antioxidant Activities of the Methanol Extract of *Moringa oleifera* in Diabetes-Induced Nephrotoxic Male Wistar Rats. *Molecules* **2017**, *22*, 439. [CrossRef] [PubMed]

95. Dubey, D.K.; Dora, J.; Kumar, A.; Gulsan, R.K. A Multipurpose Tree-*Moringa oleifera*. *Int. J. Pharm. Chem. Sci.* **2013**, *2*, 415–423.

96. Randriamboavonjy, J.I.; Rio, M.; Pacaud, P.; Loirand, G.; Tesse, A. *Moringa oleifera* seeds attenuate vascular oxidative and nitrosative stresses in spontaneously hypertensive rats. *Oxid. Med. Cell. Longev.* **2017**, *2017*, 4129459. [CrossRef] [PubMed]

97. Karim, N.A.; Ibrahim, M.D.; Kntayya, S.B.; Rukayadi, Y.; Hamid, H.A.; Razis, A.F. *Moringa oleifera* Lam: Targeting Chemoprevention. *Asian Pac. J. Cancer Prev.* **2016**, *17*, 3675–3686, Review. [PubMed]

98. Sidker, K.; Sinha, M.; Das, N.; Das, D.K.; Datta, S.; Dey, S. *Moringa oleifera* Leaf extract prevents in vitro oxidative DNA damage. *Asian J. Pharm. Clin. Res.* **2013**, *6*, 159–163.

99. Khalafalla, M.M.; Abdellatef, E.; Dafalla, H.M.; Nassrallah, A.; Aboul-Enein, K.M.; Lightfoot, D.A.; El-Deeb, F.E.; El-Shemyet, H.A. Active principle from *Moringa oleifera* Lam leaves effective against two leukemias and a hepatocarcinoma. *Afr. J. Biotechnol.* **2010**, *9*, 8467–8471.

100. Abdull Razis, A.F.; Ibrahim, M.D.; Kantayya, S.B. Health benefits of *Moringa oleifera*. *Asian Pac. J. Cancer Prev.* **2014**, *15*, 8571–8576. [CrossRef] [PubMed]

101. Berkovich, L.; Earon, G.; Ron, I.; Rimmon, A.; Vexler, A.; Lev-Ari, S. *Moringa oleifera* aqueous leaf extract down-regulates nuclear factor-kB and increases cytotoxic effect of chemotherapy in pancreatic cancer cells. *BMC Complement. Altern. Med.* **2013**, *13*. [CrossRef] [PubMed]

102. Adebayo, I.A.; Arsad, H.; Samian, M.R. Antiproliferative effect on breast cancer (MCF7) of *Moringa oleifera* seed extracts. *Afr. J. Tradit. Complement. Altern. Med.* **2017**, *14*, 282–287. [CrossRef] [PubMed]

103. Abd-Rabou, A.A.; Abdalla, A.M.; Ali, N.A.; Zoheir, K.M. *Moringa oleifera* root induces cancer apoptosis more effectively than leave nanocomposites and its free counterpart. *Asian Pac. J. Cancer Prev.* **2017**, *18*, 2141–2149. [PubMed]

104. Sadek, K.M.; Abouzed, T.K.; Abouelkhair, R.; Nasr, S. The chemo-prophylactic efficacy of an ethanol *Moringa oleifera* leaf extract against hepatocellular carcinoma in rats. *Pharm. Biol.* **2017**, *55*, 1458–1466. [CrossRef] [PubMed]

105. Budda, S.; Butryee, C.; Tuntipopipat, S.; Rungsipipat, A.; Wangnaithum, S.; Lee, J.S.; Kupradinun, P. Suppressive effects of *Moringa oleifera* Lam pod against mouse colon carcinogenesis induced by azoxymethane and dextran sodium sulfate. *Asian Pac. J Cancer Prev.* **2011**, *12*, 3221–3228. [PubMed]

106. Obulesu, M.; Rao, D.M. Effect of plant extracts on Alzheimer's disease: An insight into therapeutic avenues. *J. Neurosci. Rural Pract.* **2011**, *2*, 56–61. [CrossRef] [PubMed]

107. Ganguly, R.; Hazra, R.; Ray, K.; Guha, D. Effect of *Moringa oleifera* in experimental model of Alzheimer's disease: Role of antioxidants. *Ann. Neurosci.* **2005**, *12*, 36–39. [CrossRef]

108. Ganguly, R.; Guha, D. Alteration of brain monoamines and EEG wave pattern in rat model of Alzheimer's disease and protection by *Moringa oleifera*. *Indian J. Med. Res.* **2008**, *128*, 744–755. [PubMed]

109. Giacoipo, S.; Rajan, T.S.; De Nicola, G.R.; Iori, R.; Rollin, P.; Bramanti, P.; Mazzon, E. The Isothiocyanate Isolated from *Moringa oleifera* Shows Potent Anti-Inflammatory Activity in the Treatment of Murine Subacute Parkinson's Disease. *Rejuvenation Res.* **2017**, *20*, 50–63. [CrossRef] [PubMed]

antioxidants

MDPI

Article

In Vitro and In Vivo Antioxidant and Anti-Hyperglycemic Activities of Moroccan Oat Cultivars

Ilias Marmouzi [1,*], El Mostafa Karym [2], Nezha Saidi [3], Bouchra Meddah [1], Mourad Kharbach [4,5], Azlarab Masrar [6], Mounya Bouabdellah [6], Layachi Chabraoui [6], Khalid El Allali [7], Yahia Cherrah [1] and My El Abbes Faouzi [1]

[1] Laboratoire de Pharmacologie et Toxicologie, équipe de Pharmacocinétique, Faculté de Médecine et Pharmacie, University Mohammed V in Rabat, BP 6203, Rabat Instituts, Rabat 10100, Morocco; b.meddah@um5s.net.ma (B.M.); cherrahy@yahoo.fr (Y.C.); myafaouzi@yahoo.fr (M.E.A.F.)

[2] Laboratoire de Biochimie et Neurosciences, FST, Université Hassan I, BP 577, Settat 26000, Morocco; karímex100@hotmail.com

[3] Regional Office of Rabat, National Institute for Agricultural Research, P.O. Box 6570, Rabat Institutes, Rabat 10101, Morocco; nezsaidi@yahoo.fr

[4] Pharmaceutical and Toxicological Analysis Research Team, Laboratory of Pharmacology and Toxicology, Faculty of Medicine and Pharmacy, University Mohammed V, Rabat 10100, Morocco; mourad.kharbach@hotmail.fr

[5] Department of Analytical Chemistry, Applied Chemometrics and Molecular Modelling, CePhaR, Vrije Universiteit Brussel (VUB), Laarbeeklaan 103, B-1090 Brussels, Belgium

[6] Central Laboratory of Biochemistry, Ibn Sina Hospital, Rabat 10100, Morocco; amasrar@yahoo.fr (A.M.); monybouabdellah@yahoo.com (M.B.); lchabraoui@yahoo.fr (L.C.)

[7] Comparative Anatomy Unit-URAC-49, Hassan II Agronomy and Veterinary Institute, Rabat 10101, Morocco; khalid_elallali@yahoo.fr

* Correspondence: ilias.marmouzi@um5s.net.ma; Tel.: +212-6-6282-7643; Fax: +212-037-77-3701

Received: 13 September 2017; Accepted: 24 October 2017; Published: 6 December 2017

Abstract: Improvement of oat lines via introgression is an important process for food biochemical functionality. This work aims to evaluate the protective effect of phenolic compounds from hybrid Oat line (F11-5) and its parent (Amlal) on hyperglycemia-induced oxidative stress and to establish the possible mechanisms of antidiabetic activity by digestive enzyme inhibition. Eight phenolic acids were quantified in our samples including ferulic, p-hydroxybenzoic, caffeic, salicylic, syringic, sinapic, p-coumaric and chlorogenic acids. The Oat extract (2000 mg/kg) ameliorated the glucose tolerance, decreased Fasting Blood Glucose (FBG) and oxidative stress markers, including Superoxide dismutase (SOD), Catalase (CAT), Glutathione peroxidase (GPx), Glutathione (GSH) and Malondialdehyde (MDA) in rat liver and kidney. Furthermore, Metformin and Oat intake prevented anxiety, hypercholesterolemia and atherosclerosis in diabetic rats. In vivo anti-hyperglycemic effect of Oat extracts has been confirmed by their inhibitory activities on α-amylase (723.91 µg/mL and 1027.14 µg/mL) and α-glucosidase (1548.12 µg/mL & 1803.52 µg/mL) enzymes by mean of a mixed inhibition.

Keywords: streptozotocin-nicotinamide; anti-hyperglycemic; hybrid Oat; digestive enzyme

1. Introduction

Oxidative stress (OS) has been implicated as a contributor to both the onset and progression of diabetes [1]. In physiologic concentrations, endogenous reactive oxygen species (ROS) help to maintain homeostasis. However, when ROS accumulate in excess for prolonged periods of time, they cause chronic oxidative stress and adverse effects [2]. The mitochondrial overproduction of ROS in hyperglycemia has been postulated to cause redox imbalance, oxidative insults, mitochondrial dysfunction, and cell death. The overwhelmed free radicals can damage DNA integrity, membrane lipids and protein function by oxidation and lead to functional abnormalities, apoptosis or necrosis [3]. Generally, many of the common risk factors, such as obesity, increased age, and unhealthy eating habits, enhance pro-oxidative milieu, and may contribute to the development of insulin resistance [4,5], β-cell dysfunction, impaired glucose tolerance, and mitochondrial dysfunction [1], which can ultimately lead to the diabetic disease state. Data from experimental and clinical studies suggest an inverse association between insulin sensitivity and radical oxygenated species (ROS) levels [6]. OS also contributes in the development of diabetic complications, including diabetic retinopathy, nephropathy, peripheral neuropathy, and cardiovascular disease [7].

As can be expected, nutritional therapies that alter or disrupt OS mechanisms may serve to reduce the risk of development of diabetes [1]. Especially, cereal foods such as oats are recommended for healthy diets as being recognized sources of antioxidants [8], contributing to the management of the oxidative stress and consequently the amelioration of the diabetic status [9]. Above all, the phenolic content in Oat reveals that it may serve as an excellent dietary source of natural antioxidants [10]. Antioxidant therapy can protect pancreatic cells from apoptosis and preserve their functions [11]. Therefore, the higher antioxidant effects a compound might have, the higher the positive effects in diabetes prevention. Many previous works has reported anti-hyperglycemic effect of whole Oat products and β-glucans [12,13]. However few papers described the protective effect of Oat phenolics against oxidative damage in diabetic and cellular models and simultaneously evaluated the cognitive impairments that result from diabetic and oxidative environment. Accordingly, diabetes has been shown to be strongly implicated in Alzheimer, memory loss, depression and anxiety [14,15]. Thus, there is a need to evaluate the protective effect of nutritional intervention against cognitive diabetic complications. Moreover, it is established that Oat mechanisms of antidiabetic activity are diverse and complementary; however Oat interactions with digestive enzymes are not very well described, especially α-amylase and α-glucosidase [16,17], unless recently for some disaccharides [18].

Nowadays, the food industry is more and more directed to minor crops that have been underestimated and mainly used in animal feeding in the past decades [10,19]. In this regard, it has been shown that improvement of oat lines via introgression from tetraploids has a promising effect on Oat nutritional characteristics [20]. In our previous investigation [10], we have described the nutritional characteristics of Moroccan Oat varieties; in this focus the current work aims to compare the therapeutic and preventive effect of a Moroccan hybrid Oat (F11-5) and its parent (Amlal), on oxidative damage and antioxidants enzymes under diabetic status, also to evaluate their antioxidant effects in *Tetrahymena* model, and to characterize the inhibitory properties on digestive enzymes (α-amylase and α-glucosidase).

2. Material and Methods

2.1. Oat Material Hybridization and Extraction

Oat material was obtained from the National Institute for Agricultural Research (INRA) in Rabat, Morocco. Accessions of tetraploid *A. murphyi dom* collected in different regions of Morocco, were involved in interspecific crosses with the Moroccan cultivar of *A. sativa* (Amlal). Amlal (Am) was used as female parent in the first crossing cycle. The yielded hybrids were backcrossed to their hexaploid parents respectively (*A. sativa* × *A. murphyi dom* × *A. sativa*). Ploidy analysis of the derivative hybrids was analyzed and only hexaploid hybrids ($2n = 6 \times = 42$) were selected and had been subjected to pedigree selection until reaching genetic stability. The Oat varieties (Amlal and F11-5) have been grown under the same conditions in Marchouch experimental station (68 km from Rabat), at 410 m of altitude, 498.3 mm of average annual rainfall, and a black crumbling soil. Oat varieties have been harvested in May 2013. The grain of each variety was cleaned and stored for evaluation. Oats phenolic extraction and quantification has been performed as previously described [10].

2.2. Chromatographic Analysis

α-Tocopherol analysis was performed using HPLC-FD (High Performance Liquid Chromatography-Fluorescence detector) equipped with a Zorbax SB-C18 column (Agilent Technologies, Palo Alto, CA, USA), using a fluorescence detector (excitation wavelength 290 nm, detection wavelength 330 nm) (Perkin Elmer, Monza, Italy).

The LC–DAD/ESI-MS (liquid chromatography-diode array detection/electrospray ionization mass spectrometry) system consisted of a binary pump (G1312A, Agilent Technologies, Inc., Wilmington, DE, USA) and an autosampler (G1330B, Agilent Technologies, Inc., Wilmington, DE, USA) coupled to a mass spectrometer equipped with an electrospray ionizer source (MS; ESI-; Micromass Quattro Micro; Waters, Milford, MA, USA). Reversed phase HPLC separation was carried out using a zorbax C18 column Zorbax (100 mm × 2.1 mm × 1.7 μm, Agilent Technologies, Santa Clara, CA, USA). The mass spectrometer was operated in negative ion mode with the following parameters: capillary voltage, 3.0 kV; cone voltage, 20 V; and extractor, 2 V. Source temperature was 100 °C, desolvation temperature was 350 °C, cone gas flow was 30 L/h, and desolvation gas flow was 350 L/h. The mobile phase components were 0.1% formic acid (A) and acetonitrile with 0.1% formic acid (B). The mobile phase gradient was: 0 min, 90% A; 0–18 min, 30% A; 18–20 min, 30% A; 20–23 min, 30% A; 23–25 min, 90% A; 25–30 min, 90% A. The injection volume was 10 μL and the column temperature was 35 °C. The flow rate of the mobile phase was 0.5 mL/min. The phenolic acids were identified on the basis of their retention times, MS spectra and molecular-ion identification.

2.3. Antioxidant Effect in Tetrahymena pyriformis Cell Culture

T. pyriformis is a suitable experimental organism for wide spectrum of functional, pharmacological studies allowing the use of indices common in animal studies, such as growth arrest and metabolic inhibition, kinetics and synthesis of specific molecules or enzymes. In this assay *T. pyriformis* was used as a cellular model to follow the protective effect of natural products against Hydrogen peroxide (H_2O_2)-induced OS. The ciliated protozoa *T. pyriformis* was grown axenically, without shaking, in the Proteose-peptone yeast Glucose defined medium (PPYG), as described by Mori [21]. The cells were incubated at 28 °C in capped 500 mL Fernbach flasks containing 100 mL of PPYG medium. For the bioassays, the *T. pyriformis* cultures were always in exponential growth phase, and they were adjusted to a density of 10^4 cells/mL in fresh PPYG medium just before treatment with the tested chemicals. Protective antioxidant effect has been evaluated on *T. pyriformis* under H_2O_2 treatments in the presence of Oats extracts (controls were performed without H_2O_2). For this purpose, the 50% inhibitory concentration values (IC_{50}) of tested substances (H_2O_2, Am and F11) were determined previously on *T. pyriformis* by cell counting. Treatments solutions were prepared in deionized water, just before testing; pH was checked, and readjusted to 6.5 if necessary. The IC_{10} of Oat extracts has been used for protective effects of Oat extracts simultaneously with H_2O_2 cytotoxicity under 75 μM doses. To evaluate cell viability, aliquots of 1 mL were taken from untreated and treated *T. pyriformis*. The samples were diluted in distilled water, and fixed with neutral buffered formalin containing 10% (*v/v*) formalin in Isoton buffer for 1 h. The cell number was determined with a hemocytometer under an optical microscope. To determine cell viability, a counting without fixation was realized in order to identify immobile and mobile cells (dead and live cells, respectively). The MTT assay [22] using the MTT reagent (3-(4,5-dimethylthiazol-2-yl)-2,5-diphenyltetrazolium bromide) was used to evaluate the effects of treatments on cell proliferation and/or mitochondrial activity. Four hours later the MTT salts were reduced to formazan blue in the metabolic active cells by mitochondrial enzyme succinate dehydrogenase to form NADH (Nicotinamide Adenine Dinucleotide Hydrogen) and NADPH (Nicotinamide Adenine Dinucleotide Phosphate). Absorbance was read at 570 nm. Oat antioxidant protective effect has been evaluated by measuring the protein content and two antioxidant enzymes: total superoxide dismutase (T-SOD) and Catalase (CAT) in the *T. pyriformis* homogenate [23–25].

2.4. Digestive Enzymes Inhibition and Kinetics

The α-amylase inhibition assay was conducted according to Kee [26] work, with slight modifications. Briefly, 250 μL of sample was mixed with 250 μL of α-amylase (240 U/mL, in 0.02 M phosphate buffer solution (PBS), pH 6.9, with 0.006 M NaCl). After incubating at 37 °C for 10 min, 250 μL of 1% (w/v) soluble starch (in 0.02 M PBS, pH 6.9) was added and the mixture was further incubated at 37 °C for 30 min, followed by adding 250 μL of dinitrosalicylic acid color reagent (DNS 96 mM, 30% Na-K tartrate, 0.4 M NaOH), and stopped by heating in a boiling water bath for 10 min. After cooling to room temperature, the mixture has been diluted with 2 mL of PBS and the absorbance was measured at 540 nm.

The α-glucosidase enzyme (0.1 U/mL) and substrate *p*-Nitrophenyl-α-D-glucopyranoside (*p*-NPG, 1 mM) were dissolved in PBS (0.1 M, pH 6.7), and all samples were dissolved in distilled water. The inhibitor (150 μL) was pre-incubated with the enzyme (100 μL) at 37 °C for 10 min, and then the substrate (200 μL) was added to the reaction mixture. The enzymatic reaction was performed at 37 °C for 30 min. The reaction was then terminated by the addition of Na_2CO_3 (1 M, 1 mL). All samples were analyzed in triplicate with different concentrations to determine the IC_{50} values, and the absorbance was recorded at 405 nm. The inhibition percentage (%) was calculated by the following equation for both assays:

$$\text{Inhibition (\%)} = \frac{(\text{AC} - \text{AC}_b) - (\text{AS} - \text{AS}_b)}{\text{AC} - \text{AC}_b} \times 100$$

Absorbances are abbreviated as follows: AC (control), AC_b (control blank), AS (sample) and AS (sample blank).

The mode of inhibition of α-amylase and α-glucosidase was investigated with increasing concentrations of the substrate (Starch and *p*-NPG) in the presence of different concentrations of extracts. Then, the type of inhibition was determined by Lineweaver–Burk plot analysis of the data, calculated from the results according to Michaelis–Menten kinetics.

2.5. Anti-Hyperglycemic and Antioxidant Effect in Diabetic Model

2.5.1. Animals

Wistar rats weighing (250–300 g) and Swiss albino mice (20–25 g), are bred in the central animal facility of the Faculty of Medicine and Pharmacy of Rabat, Morocco. Animals were kept in cages under standard laboratory conditions with tap water and standard diet *ad libitum*, in a 12 h light/12 h dark cycle at a temperature of 21 to 23 °C.

Ethics approval was also obtained from Mohammed V University in Rabat, under the responsibility of the Central Animal Facility and the Laboratory of Pharmacology and Toxicology at the Faculty of Medicine and Pharmacy of Rabat (01DEC2015). The experiments were conducted in accordance with the accepted principles outlined in the "Guide for the Care and Use of Laboratory Animals" prepared by the National Academy of Sciences and published by the National Institutes of Health and all efforts were made to minimize animal suffering and the number of animals used.

2.5.2. Acute Toxicity

Acute oral toxicity study was performed as per 425 guidelines (OECD) from the Organization of Economic Co-operation and Development. Four groups of male Swiss albino mice (*n* = 6), each one selected by random sampling technique, were used for acute toxicity study. The animals were kept fasting overnight providing them only with water. Subsequent to the administration of extracts (2000 mg/kg), the animals were observed closely for the first 3 h in order to detect any toxic manifestations such as increased locomotors activity, salivation, clonic convulsion, coma and death.

Subsequent observations were made at regular intervals for 24 h. The animals were observed for a further two weeks.

2.5.3. Experimental Diabetes

Overnight fasted rats (OFR) were treated with nicotinamide (110 mg/kg, i.p.). Streptozotocin (65 mg/kg, i.p.) was injected 15 min after nicotinamide administration, in all groups except for normal control [27]. Animals were fed with glucose solution (5%) for 12 h to avoid hypoglycemia. Hyperglycemia was confirmed three days later, and the steady state of hyperglycemia was reached after 10 days. Serum glucose was determined by the glucose oxidase peroxidase method using a glucometer (One Touch Ultra, LifeScan, Milpitas, CA, USA). Animals having serum glucose between 200–300 mg/dL were selected for the study.

2.5.4. Experimental Design

After establishment of the diabetic rats model, animals were divided into five treatment groups: Normal control (NC; 1 mL DW/200 g), Diabetic control (DC; 1 mL DW/200 g), Oat treatment (Am and F11; 2000 mg/kg) and Metformin (Met; 300 mg/kg). The dose was selected based on the results of our preliminary oral glucose tolerance tests (OGTTs) assays, in which we found that the effect of the 2000 mg/kg treatment was significantly higher than 500 mg/kg ($p < 0.05$) (data not shown). The animals received the respective treatment for 42 days (6 weeks). Blood was collected at the end of the experiment for hematological and biochemical analysis. Liver and kidney homogenate were analyzed for oxidative stress markers. Pancreas has been used for the histological characterization. OGTTs were performed on day 1 and 40 from the beginning of the experiment. During the study period of 42 days, the rats were weighed daily using electronic balance, and glycaemia levels were recorded weekly. Food and water intake, and urinary volume were determined for the first and last day before OGTT using metabolic cages.

2.5.5. Oral Glucose Tolerance Test

OFR were administered with glucose (2000 mg/kg) orally by means of gastric intubation. Animals in Am and F11 groups were administered orally with Oat extracts, at a dose of 2000 mg/kg, 30 min before the oral administration of glucose. Whereas animals in group Met were given metformin (300 mg/kg). The animals of control i.e. group NC and DC were given orally equal volume of water only. Blood samples were collected from the tail vein at 0, 30, 60, 90, 120 and 150 min. Total glycemic responses to OGTT were calculated from respective areas under curves (AUC) of glycaemia during the 150 min observation period. The Δ variation of glycaemia was defined as the difference between glycaemia at t_0 and a following time point.

2.5.6. Behavioral Assays

The elevated plus maze (EPM) is an ethological model of anxiety in rodents. After 34 days of treatments, OFR have received their treatments at the fixed doses. Thirty minutes after administration, each rat was placed in the central square facing an open arm and allowed to freely explore the maze for 5 min. The following measurements were recorded; arm entries, total time spent in open arms and total number of entries in open arms [28]. Two days later (day 36), the OFR have received their treatments at the fixed doses, and thirty minutes later, the animals' spontaneous activity was evaluated in an open field test (OFT). In individual tests, the rats were placed at the same point and allowed to freely explore the apparatus for 10 min. The following measurements were then recorded; total squares entries, central squares entries and time spent in central squares [29].

2.5.7. Hematological and Biochemical Analysis

On day 42, blood samples were collected from the Jugular vein by using capillary tubes containing Ethylenediaminetetraacetic acid (EDTA) (anti-coagulant). The following hematological parameters were evaluated in the collected blood samples: Total hemoglobin (HGB), red (RBC) and white (WBC) blood corpuscles count, neutrophils, lymphocytes, eosinophils, monocytes, basophils and platelet count using fully automated analyzer (Architect c8000, Clinical Chemistry System, Chicago, IL, USA). Biochemical parameters *viz.* aspartate aminotransferase (AST), alanine aminotransferase (ALT), Total proteins, urea, uric acid, creatinine, cholesterol, triacylglycerols (TG), high density lipoprotein (HDL), low density lipoprotein (LDL), lactate dehydrogenase (LDH), Sodium, Potassium and Chlorine were determined using the same analyzer. The atherosclerosis index (AI) was calculated as LDL/HDL ratio.

2.5.8. Key Enzymes and Markers of Oxidative Stress

After sacrifice, liver and kidney were dissected out, and a 10% of organs homogenate was prepared in ice-cold PBS 50 mM using Teflon glass homogenizer. The homogenate was centrifuged at 3500 rpm for 10 min (at 4 °C) using cooling centrifuge (Mikro 220R, Hettick Lab Technology, Tuttlingen, Germany). The pellet was discarded and supernatant obtained was used for the estimation of antioxidant enzymes, oxidative stress markers and protein content in the homogenate.

The protein content was estimated following the method of Lowry [23]. The antioxidant enzymes were expressed as μmol/min/mg protein. Catalase (CAT) activity was determined according to the method of Aebi [30]. T-SOD activity was determined by the method of Beauchamp and Fridovich [22]. The MnSOD activity was measured after addition of 2 mM KCN in the solution. The CuZn-SOD activity was calculated by subtraction as follows:

$$CuZn - SOD = T\text{-}SOD - Mn\text{-}SOD$$

Activities of Glutathione peroxidase (GPx) were estimated according to the method of Mannervik [25]. Glutathione (GSH) was estimated by a previously described method [31], and expressed as μmol GSH/mg protein using GSH standard. Thiobarbituric acid reactive substances were estimated as malondialdehyde (MDA) equivalent (nmol MDA/mg protein), from the calibration curve [32].

2.5.9. Histological Study of Pancreas

The pancreas from each rat was fixed in 10% buffered formalin and processed via classical histology method using the paraffin wax embedding techniques (dehydration, clearing and embedding). The paraffin embedded-sections were cut at 4 μm thickness using microtome (Shandon Hypercut, Runcorn, UK), and then stained by hematoxylin and eosin staining method. Stained sections of pancreas were qualitatively (morphological) analyzed on microscope (Leica Microsystems DM2500, Wetzlar, Germany), and photomicrographs of histological alterations have been taken at ×400 magnification.

2.6. Statistical Analysis

Data were expressed as the mean values ± standard deviation (SD) for each measurement. The data were also analyzed by one-way analysis of variance (one-way ANOVA). Post Hoc procedure was used for significance of difference ($p < 0.05$). Analysis was performed with Graph pad prism 6.0.

3. Results

3.1. Chemical Analysis

Chemical analysis of Oat varieties (Table 1) revealed the phenolic composition of Oat extracts. The *p*-Hydroxybenzoic acid constitute the major phenolic compound in both extracts with 1840.34 ± 30.45 mg/Kg for Amlal and 1270.02 ± 38.34 mg/Kg for F11, followed by the Syringic acid (1830.66 ± 90.21 mg/Kg for Amlal and 310.41 ± 33.09 mg/Kg for F11), and the Caffeic acid (250.67 ± 32.11 mg/Kg for Amlal and 421.54 ± 12.32 mg/Kg for F11). Significant contents of Sinapic, Ferulic, Gallic, *p*-Coumaric and Chlorogenic acids were found. α-Tocopherol contents in Amlal and F11 were 1.65 ± 0.22 and 1.82 ± 0.12 mg/kg in Amlal and F11 respectively.

Table 1. Phenolic and tocopherol composition of hybrid Oat and parent in mg/Kg.

Compounds	Amlal	F11-5
Gallic acid	41.08 ± 2.32 [b]	13.34 ± 2.29 [a]
Chlorogenic acid	2.56 ± 0.45 [a]	1.78 ± 0.12 [a]
p-Hydroxybenzoic acid	1840.34 ± 30.45 [b]	1270.02 ± 38.34 [a]
Caffeic acid	250.67 ± 32.11 [a]	421.54 ± 12.32 [b]
Syringic acid	1830.66 ± 90.21 [b]	310.41 ± 33.09 [a]
p-Coumaric acid	26.67 ± 3.22 [b]	16.43 ± 1.90 [a]
Ferulic acid	70.45 ± 1.87 [b]	1.98 ± 0.49 [a]
Sinapic acid	17.10 ± 2.09 [b]	11.62 ± 2.78 [a]
Salicilyc acid	4.67 ± 0.07 [b]	1.78 ± 0.16 [a]
α-Tocopherol	1.65 ± 0.22 [a]	1.82 ± 0.12 [b]

Data are reported to mean (n = 2) ± SD. Values in the same raw not sharing a common letter ([a,b]) differs significantly at $p < 0.05$.

3.2. Antioxidant Effect in T. pyriformis

Different concentrations of Oat extracts and H_2O_2 were screened to determine their cytotoxicity (IC_{50}) on *T. pyriformis*. The determined IC_{50} of H_2O_2 (0.75 mM) was used to establish the oxidative stress environment and the IC_{10} of Oat varieties (1 µg/mL) were used for antioxidant treatment in comparison with vitamin C at the same dose (Figure 1).

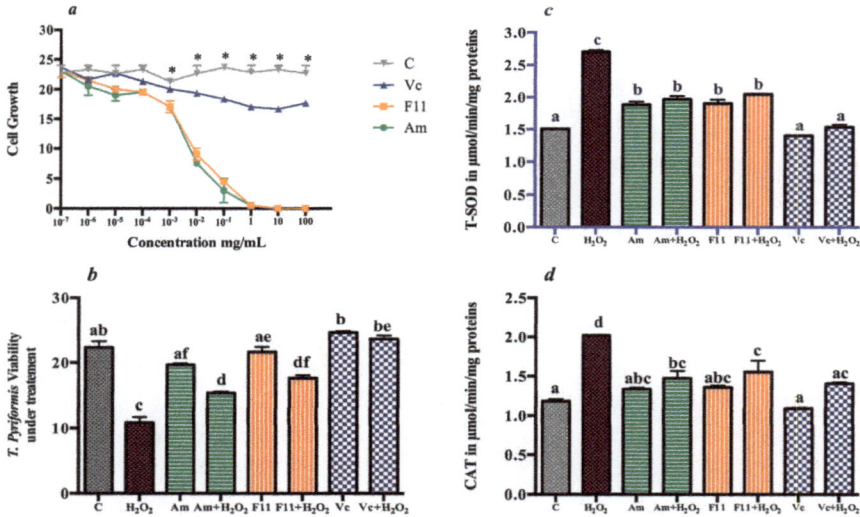

Figure 1. Oat protective effect on antioxidant enzymes under oxidative stress in *T. pyriformis*. (**a**) Oat and Vitamin C effect on *T. pyriformis* growth. (**b**) *T. pyriformis* viability under H_2O_2 treatment in the presence of extracts. (**c**) T-SOD levels *T. pyriformis* cultures. (**d**) CAT levels *T. pyriformis* cultures, Data are reported to mean (n = 3) ± SD. Values not sharing a common letter (a–d) differs significantly at $p < 0.05$. (*) differ significantly from the control.

H$_2$O$_2$-induced stress reduced *T. pyriformis* viability by 48.49% compared to the control group. Cell viability in treated groups was improved by 20.15% in Am, 30.60% in F11 and 57.47% in Vc compared to H$_2$O$_2$ group. Antioxidant enzymes, including T-SOD and CAT levels increased significantly by 38.92% and 36.04% in H$_2$O$_2$ compared to the non-treated control ($p < 0.05$). However Oat treatment reduced the level of the expression of those enzymes. T-SOD release was reduced by −20.16% in Am, −17.80% in F11, and −37.77% in the positive control (Vc). Similarly, the treatments decreased CAT levels by −20.14% in Am, −15.80% in F11 and −23.77% in Vc.

3.3. In Vitro Inhibition of Digestive Enzymes

In an array to explore the in vitro antidiabetic activity, Moroccan Oat extracts were screened for the α-amylase and α-glucosidase inhibitory properties. Effects were compared with the commercially available α-glucosidase inhibitor, acarbose (Ac).

Inhibitory activities of the extracts were evaluated at different concentrations and results were given in Figure 2. Acarbose and extracts showed a dose dependent inhibitory effect on enzymes. The IC$_{50}$ values for anti-amylase and anti-glucosidase, activity of Am and F11 were lower than that of acarbose, a very well known drug with well-established activity (Figure 2). IC$_{50}$ for amylase inhibition were 723.91 µg/mL and 1027.14 µg/mL for Am and F11 respectively. While, acarbose inhibitory concentration was 396.42 µg/mL. α-glucosidase inhibitory effect of Oat extracts was much more lower than its equivalent of acarbose. In term of IC$_{50}$, significant differences have been registered between Oat varieties and acarbose: Am (1548.12 µg/mL), F11 (1803.52 µg/mL) and Ac (199.53 µg/mL).

A Lineweaver-Burk double reciprocal plot of enzymes activity in the presence of Oat extracts as inhibitor was plotted. Km and Vmax values increased with increasing Oat concentrations. The kinetic study (Figure 2) suggests that Oat extracts metabolites intersected in the second quadrant indicating a mixed-type of inhibition in both enzymes (α-amylase and α-glucosidase).

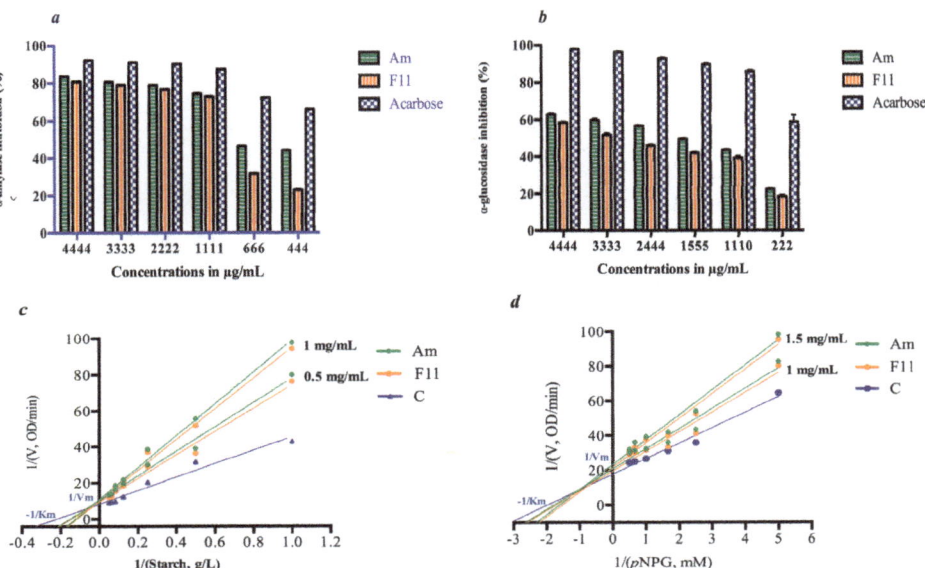

Figure 2. α-amylase and α-glucosidase inhibitory activities. (**a**) α-amylase inhibitory activities of Oat extracts. (**b**) α-glucosidase inhibitory activities of Oat extracts. (**c**) Lineweaver burk plot of Oat α-amylase inhibition. (**d**) Lineweaver burk plot of Oat α-glucosidase inhibition, Data are reported to mean ($n = 3$) ± SD.

3.4. Anti-Hyperglycemic and Antioxidant Effect in Diabetic Model

3.4.1. Acute Oral Toxicity and Metabolic Parameters

Animals treated with Oat extracts did not show any change in their behavioral pattern during the acute toxicity study. There was no significant difference in the body weight and food consumption when compared to the vehicle treated group. Also, no apparent pathological changes were seen. Thus, it was concluded that Oat extracts were safe at 2000 mg/kg.

Table 2 shows the effect of daily administration of Oat extracts on body weight, urinary volume, and food and water intakes in diabetic rats. After six weeks of treatment, DC rats showed a considerably reduction of body weight by 50.91% compared to the initial value. The administration of Oat extracts (Am and F11) resulted in significant weight gain as compared to DC (40.47%, 30.07% and 39.01% for Met, Am and F11 respectively). In the other hand, food intake increased by 29.47% in DC, 11.99% in Met, 19.29% in Am and 26.58% in F11. Also, the elevated water intake and urinary volume (139.43–56.11%) in DC was significantly reduced to (−18.95%, −18.68%) in Met, (−45.96%, −39.33%) in Am and to (−62.57%, −39.20%) in F11. Compared to DC, urinary volume, food and water intake in treated groups has decreased, while body weight has increased significantly ($p < 0.05$).

Table 2. Changes in fasting blood glucose and metabolic parameters.

Measure		NC	DC	Met	Am	F11
FBG (mg/dL)	D1	89.15 ± 16.48 [a]	249.85 ± 28.05 [a]	246.16 ± 52.60 [a]	236.71 ± 70.16 [a]	215.80 ± 48.83 [a]
	D14	86.83 ± 9.64 [a]	295.50 ± 28.98 [b]	107.66 ± 7.14 [b]	216.42 ± 25.26 [a]	193.14 ± 34.78 [a]
	D28	91.83 ± 10.04 [a]	381.80 ± 38.10 [c]	100.50 ± 3.50 [b]	189.85 ± 33.30 [a]	176.28 ± 34.51 [a]
	D42	92.16 ±7.19 [a]	412.20 ± 49.87 [c]	92.83 ± 9.78 [b]	152.50 ± 22.38 [b]	137.57 ± 33.57 [b]
Food intake (g)	D1	16.85 ± 1.57 [a]	19.14 ± 3.02 [a]	18.85 ± 2.60 [a]	19.28 ± 2.13 [a]	18.28 ± 2.87 [a]
	D41	19.42 ± 3.10 [a]	27.14 ± 2.19 [b]	21.42 ± 1.98 [a]	23.00 ± 1.41 [a]	23.14 ± 2.91 [a]
Water intake (mL)	D1	24.71 ± 4.46 [a]	71 ± 11.71 [a]	57.28 ± 7.69 [a]	67.14 ± 3.71 [a]	63.00 ± 7.16 [a]
	D41	25.14 ± 5.61 [a]	170 ± 14.54 [b]	68.14 ± 10.17 [a]	98.00 ± 7.50 [b]	102.42 ± 15.05 [b]
Urinary volume (mL)	D1	10.57 ± 1.98 [a]	45.57 ± 6.87 [a]	43.57 ± 2.69 [a]	37.42 ± 5.42 [a]	39.71 ± 5.46 [a]
	D41	11.14 ± 1.77 [a]	71.14 ± 2.60 [b]	51.71 ± 5.61 [a]	52.14 ± 2.41 [b]	55.28 ± 5.58 [b]
Body Weight (g)	D1	208.57 ± 25.33 [a]	241.85 ± 10.52 [a]	221.71 ± 56.29 [a]	221.42 ± 26.45 [a]	214.85 ± 57.01 [a]
	D41	236.16 ± 25.89 [a]	118.71 ± 21.62 [b]	198.57 ± 13.50 [a]	175.28 ± 14.39 [a]	189.28 ± 58.01 [a]

Data are reported to mean ($n = 8$) ± SD. Values in the same column and category not sharing a common letter ([a–c]) differs significantly at $p < 0.05$.

3.4.2. Glucose Tolerance and Anti-Hyperglycemic Effect

In order to evaluate the effectiveness of each treatment, OGTT was performed at the beginning and the end of the study. OGTT revealed that blood glucose reached its peak level at 30 min after the glucose load in all groups before and after treatments (Figure 3). The initial administration of Oat extracts (Am and F11) at 2000 mg/kg did not significantly reduce the AUC of blood glucose levels compared to the vehicle in DC. However, as expected Metformin show decrease (−52.48%) in glucose levels and stabilize glycaemia to normal values in diabetic rats, 90 min after administration.

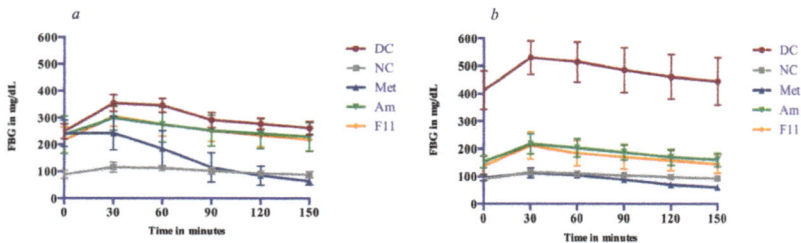

Figure 3. *Cont.*

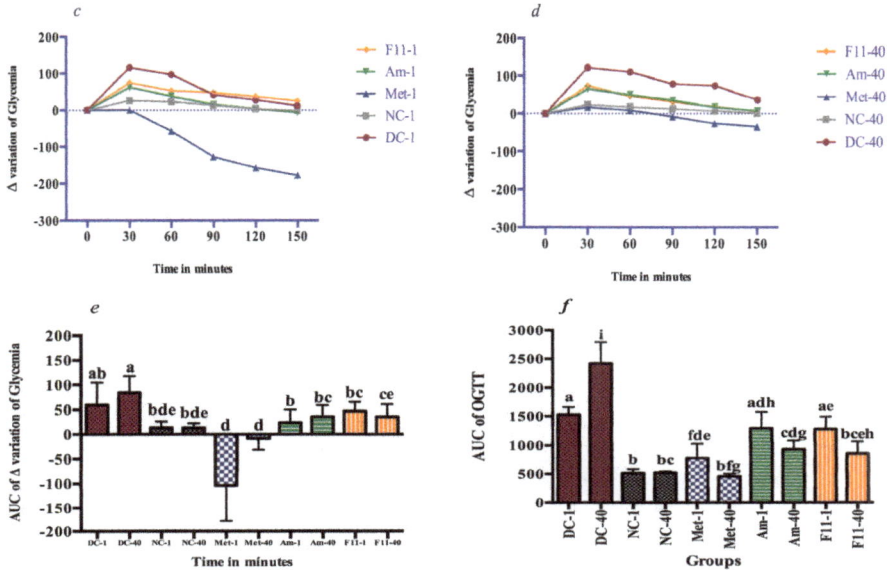

Figure 3. Oral glucose tolerance before and after treatment period. (**a**) OGTT of the first day 1; (**b**) OGTT of the day 40; (**c**) Δ variation of glycaemia during OGTT-Day 1; (**d**) Δ variation of glycaemia during OGTT-Day 40; (**e**) AUC of Δ variation of glycaemia during OGTT challenge; (**f**) AUC of OGTT, Data are reported to mean (n = 8) ± SD. Values not sharing a common letter (a–i) differs significantly at $p < 0.05$.

Across the six weeks of treatment period, FBG has been shown to be decreased gradually in Am, F11, and Met groups (Table 2). Glycemic levels in Oat-treated groups decreased significantly (−35.57% and −36.25%) compared to the dramatically increase (64.97%) in DC. Metformin has been shown to possess largely stronger effect (−60.28%), stabilizing FBG to its normal levels (Table 1).

At the end of treatment, the beneficial effects of Oat extracts on glucose metabolism of diabetic rats become evident. In the OGTT, after 30 min of the glucose overload, treated groups (Am, F11 and Met) were able to reduce glycaemia (Δ variation of glycaemia) more efficiently (65.33, 73.14 and 17.14 mg/dL) than DC (118 mg/dL). To the end of the challenge test, Oat extracts reduced blood glucose level significantly compared to DC. Hence, the glucose tolerance was significantly improved after Oat supplementation and metformin treatment.

3.4.3. Hematological and Biochemical Indices

In order to evaluate the protective effect of the above treatments on physiological functions, the hematological and biochemical parameters were analyzed (Table 3). The total hemoglobin, red blood corpuscles count, neutrophils, lymphocytes, monocytes and eosinophils were not significantly different between the evaluated groups (Table 3). However, the white blood corpuscles count increased significantly in DC ($p < 0.05$). Moreover, the ALT and AST in Am, F11 and Met treated rats decreased significantly ($p < 0.05$) as compared to diabetic animals (Table 3). Nevertheless, total serum protein did not show any significant change in DC compared to NC. In the other hand LDH and urea in Oat and Met has shown decreased levels as compared to DC ($p < 0.01$). Serum creatinine did not show significant variations among groups. In the lipid profile analysis, the diabetic animals showed significant increase ($p < 0.01$) in the total cholesterol and TG; simultaneously a decrease in HDL compared to normal and treated rats. LDL levels did not show significant variations between DC and NC. Atherosclerosis index demonstrated significant differences in all groups ($p < 0.05$).

Table 3. Hematological and biochemical analysis of diabetic animals.

Parameters	Unit	NC	DC	Met	Am	F11
			Hematology			
HGB	g/dL	11.25 ± 0.49 [a]	14.32 ± 1.27 [a]	12.93 ± 0.98 [a]	11.1 ± 3.39 [a]	12.02 ±2.04 [a]
RBC	10^{-6}/uL	6.87 ± 0.02 [a]	8.36 ± 0.59 [b]	7.84 ± 0.21 [ab]	12.76 ± 0.50 [c]	6.76 ± 1.05 [a]
WBC	10^{-3}/uL	4.23 ± 0.89 [a]	21.31 ± 3.69 [c]	10.43 ± 5.43 [ab]	18.05 ± 1.18 [b]	12.83 ± 3.90 [ab]
Neutrophils	10^{-3}/uL	15.50 ± 4.65 [a]	21.63 ± 6.29 [ab]	16.43 ± 3.47 [b]	15.83 ± 3.44 [b]	17.01 ± 0.10 [b]
Lymphocytes	10^{-3}/uL	65.25 ± 15.34 [abc]	54.05 ± 1.62 [c]	45.65 ± 9.89 [b]	73.93 ± 4.20 [b]	55.33 ± 4.90 [c]
Monocytes	10^{-3}/uL	4.95 ± 0.91 [a]	9.86 ± 1.77 [d]	5.24 ± 1.69 [c]	6.71 ± 1.65 [b]	5.73 ± 0.49 [bcd]
Eosinophils	10^{-3}/uL	2.80 ± 0.97 [ab]	2.77 ± 0.73 [a]	1.98 ± 0.59 [c]	3.91 ± 1.60 [abc]	2.10 ± 0.81 [bc]
Basophils	10^{-3}/uL	n.d	n.d	n.d	n.d	n.d
Platelet count	10^{-3}/uL	508 ± 89.78 [ab]	516.75 ± 93.11 [ab]	534.33 ± 51.18 [a]	548.86 ± 41.71 [ab]	584.50 ± 61.51 [b]
			Liver function			
ALT	IU/L	66.66 ± 12.01 [ab]	130.05 ± 4.56 [c]	84.66 ± 17.62 [b]	68.28 ± 37.66 [ab]	72.40 ± 18.14 [a]
AST	IU/L	104.5 ± 14.86 [ac]	225.66 ± 53.65 [d]	118.33 ± 33.26 [c]	139.14 ± 29.53 [a]	141.33 ± 19.75 [bc]
Total proteins	g/L	62.00 ± 5.32 [a]	63.00 ± 1.87 [abcd]	68.28 ± 3.98 [def]	67.11 ± 4.82 [ce]	64.76 ± 5.28 [bf]
			Renal function			
LDH	U/L	327.33 ± 26.00 [ac]	693.00 ± 78.23 [b]	351.88 ± 91.34 [ac]	569.00 ± 89.50 [a]	426.26 ± 49.25 [c]
Urea	g/L	0.22 ± 0.04 [ac]	0.62 ± 0.12 [b]	0.43 ± 0.09 [c]	0.49 ± 0.19 [ab]	0.28 ± 0.05 [a]
Creatinine	mg/L	4.61 ± 0.28 [abcd]	7.70 ± 1.49 [a]	4.18 ± 0.22 [b]	4.95 ± 0.35 [c]	4.68 ± 0.32 [d]
			Lipid profile			
Cholesterol	g/L	0.44 ± 0.16 [ac]	1.09 ± 0.20 [b]	0.46 ± 0.17 [ad]	0.66 ± 0.19 [c]	0.63 ± 0.12 [cd]
TG	g/L	0.58 ± 0.09 [acde]	4.51 ± 0.26 [b]	0.41 ± 0.16 [c]	0.72 ± 0.21 [d]	1.01 ± 0.22 [e]
HDL	g/L	0.22 ± 0.04 [a]	0.14 ± 0.04 [b]	0.21 ± 0.05 [a]	0.17 ± 0.05 [a]	0.17 ± 0.01 [a]
LDL	g/L	0.20 ± 0.03 [a]	0.26 ± 0.05 [ab]	0.21 ± 0.09 [ab]	0.21 ± 0.08 [b]	0.21 ± 0.12 [a]
AI	ratio	0.91 ± 0.05 [a]	2.36 ± 0.03 [d]	0.99 ± 0.05 [b]	1.23 ± 0.04 [c]	1.23 ± 0.10 [bc]
			Minerals			
Sodium	mmol/L	141.28 ± 1.60 [a]	134.66 ± 9.71 [ab]	141.14 ± 2.11 [ab]	140.14 ± 1.57 [b]	139.80 ± 2.48 [ab]
Potassium	mmol/L	5.42 ± 0.54 [ac]	4.41 ± 0.60 [bc]	5.4 ± 1.15 [ab]	5.39 ± 0.35 [abc]	6.31 ± 1.33 [c]
Chlore	mmol/L	106.71 ± 2.62 [a]	97.00 ± 7.93 [ab]	103.85 ± 1.95 [ac]	101.42 ± 1.98 [b]	98.60 ± 3.78 [bc]

Data are reported to mean ($n = 8$) ± SD. Values in the same row and category not sharing a common letter ([a–d]) differs significantly at $p < 0.05$.

3.4.4. Oxidative Stress Markers and Antioxidants Enzymes

By the end of the treatment period, oxidative stress markers (MDA and GSH) and antioxidant enzymes (T-SOD, Mn-SOD, CuZn-SOD, CAT and GPx) were analyzed and results are presented in Table 4. Lipid peroxidation products as expressed in MDA equivalents in diabetic non-treated animals were much higher than those found in NC and treated rats. MDA levels at the liver was significantly higher compared to NC ($p < 0.001$), while treatment registered a lesser increase in Met (30.95%), Am (154.76%) and F11 (407.14%). Likewise, at the kidney level, similar increase in MDA has been registered (405.49%). Treatments of diabetic groups decreased the MDA production in Met (−274.72%), Am (−378.01%) and F11 (−256.58%) compared to DC. The T-SOD, Mn-SOD and CuZn-SOD levels in DC have increased significantly compared to NC and treated groups ($p < 0.05$). T-SOD increased by 124.51% in DC compared to NC at the liver, while in the kidney the increase was less marked by 52.16%. Am and Met treatments decreased T-SOD and similarly Mn-SOD and CuZn-SOD in both organs ($p < 0.05$). Also, compared to NC, diabetic animals have shown a significant increase ($p < 0.05$) of CAT by 218.68% in the liver and 50.05% in the kidney (Table 4). The treated groups exhibited a significant decrease ($p < 0.05$) in CAT levels except for the F11 group. Oat extract and metformin induced a reduction of CAT peaks in treated diabetic animals (−194.63%, −170.41% and −63.66% for Met, Am and F11 respectively) in the liver and similarly in the kidney. Moreover, DC rats were also characterized by an increase in GSH by 85.59% in the liver. Administration of the treatments induced a significant ($p < 0.05$) decrease of GSH in the liver for Met (−37.28%), Am (−55.08%) and F11 (−42.36%), compared to DC. The kidney levels of GSH followed the same tendency. GPx Levels in the other hand increased significantly in liver of the diabetic controls (123.60%). Treated groups (Am and Met) show decreased values compared to diabetic control at the liver level, except for F11 group who did not show significant variations in comparison with DC. GPx levels in the kidney did not show significant differences.

Table 4. Levels of antioxidant enzymes and oxidative stress markers in liver and kidney.

	MDA	T-SOD	Mn-SOD	CuZn-SOD	CAT	GSH	GPx
				Liver			
NC	0.42 ± 0.05 [a]	7.22 ± 1.57 [a]	1.66 ± 0.44 [a]	3.58 ± 0.24 [ab]	5.78 ± 1.44 [a]	2.36 ± 0.25 [a]	3.22 ± 0.27 [ac]
DC	6.37 ± 0.60 [d]	16.21 ± 2.86 [c]	3.64 ± 0.63 [d]	11.80 ± 2.06 [c]	18.42 ± 1.81 [c]	4.38 ± 0.99 [c]	7.20 ± 0.42 [d]
Met	0.55 ± 0.06 [b]	7.25 ± 2.46 [a]	1.96 ± 0.46 [c]	4.33 ± 1.30 [b]	7.17 ± 0.77 [a]	3.50 ± 0.97 [b]	2.05 ± 0.71 [a]
Am	1.07 ± 0.17 [c]	8.30 ± 0.48 [abc]	2.72 ± 0.54 [b]	3.60 ± 1.34 [a]	8.57 ± 1.71 [b]	3.08 ± 0.35 [abc]	2.75 ± 0.23 [b]
F11	2.13 ± 0.98 [abc]	14.52 ± 3.00 [b]	2.93 ± 0.75 [abc]	9.25 ± 0.50 [c]	14.74 ± 3.88 [abc]	3.38 ± 1.60 [abc]	5.91 ± 0.94 [cd]
				Kidney			
NC	0.182 ± 0.02 [ab]	8.30 ± 0.65 [ab]	2.64 ± 0.38 [a]	4.32 ± 0.98 [a]	8.57 ± 0.96 [a]	1.46 ± 0.18 [a]	3.94 ±0.68 [ae]
DC	0.92 ± 0.05 [e]	12.63 ± 1.02 [c]	4.01 ± 0.12 [b]	9.99 ± 1.07 [b]	12.86 ± 0.81 [c]	2.13 ± 0.31 [d]	7.53 ± 1.41 [e]
Met	0.42 ± 0.03 [cd]	10.04 ± 0.63 [a]	2.79 ± 0.36 [a]	5.39 ± 0.87 [a]	10.32 ± 0.72 [b]	1.48 ± 0.20 [ab]	5.34 ± 1.26 [bde]
Am	0.232 ± 0.09 [ac]	7.93 ± 0.95 [ab]	2.83 ± 0.59 [a]	5.61 ± 0.54 [a]	7.99 ± 1.69 [ab]	1.14 ± 0.11 [bc]	6.52 ± 1.18 [cde]
F11	0.453 ± 0.09 [bd]	7.16 ± 0.85 [b]	3.44 ± 0.72 [a]	4.75 ± 0.71 [a]	7.82 ± 1.24 [a]	1.22 ± 0.18 [c]	5.45 ± 0.57 [d]

Data are reported to mean (n = 3) ± SD. Values in the same row and category not sharing a common letter ([a–e]) differs significantly at $p < 0.05$. CAT, Mn-SOD, CuZn-SOD and GPx are expressed in in µmol/min/mg protein, GSH in µmol GSH/mg protein, MDA in nmol MDA/mg protein.

3.4.5. Behavioral Analysis

As depicted in Table 5 the behavioral analysis of DC based on the EPM demonstrated a significant decrease in arm entries (−90.99%), open arms entries (−89.18%), and in time spent in open arms (−87.29%) compared to NC. Similarly, the OFT parameters of DC show significant decrease in total squares entries (−71.14%), central squares entries (−89.10%) and time spent in central squares (−62.21%), compared to NC. Metformin and Oat treatments resulted in significant increase of locomotor and exploratory activities in both assays ($p < 0.05$).

Table 5. Behavioral analysis using OFT and EPM.

	NC	DC	Met	Am	F11
		Open Field test			
Total squares entries	42.32 ± 5.32 [c]	12.21 ± 2.11 [a]	29.21 ± 3.21 [b]	25.43 ± 2.42 [b]	27.43 ± 5.32 [b]
Central squares entries	19.45 ± 3.23 [d]	2.12 ± 0.98 [a]	9.43 ± 1.21 [b]	5.32 ± 1.05 [ab]	4.98 ± 1.43 [ab]
Time spent in central squares (%)	45.95 ± 4.23 [c]	17.36 ± 3.32 [a]	32.28 ± 2.79 [b]	20.92 ± 2.56 [a]	18.15 ± 3.26 [a]
		Elevated Plus Maze test			
Arm entries	35.60 ± 4.56 [d]	3.21 ± 0.45 [a]	29.32 ± 2.22 [d]	7.75 ± 1.45 [b]	12.25 ± 1.67 [c]
Open arms entries	23.12 ± 3.21 [d]	2.50 ± 0.32 [a]	21.32 ± 3.17 [c]	4.75 ± 0.89 [a]	7.75 ± 1.22 [b]
Time spent in open arms (%)	26.29 ± 5.43 [d]	3.34 ± 1.12 [a]	16.78 ± 5.69 [c]	5.88 ± 3.45 [a]	9.43 ± 4.01 [b]

Data are reported to mean (n = 8) ± SD. Values in the same column and category not sharing a common letter ([a–d]) differs significantly at $p < 0.05$.

3.4.6. Pancreas Histopathology

The pathologic features of pancreas tissues in each group are shown in Figure 4. Normal control showed normal pancreatic parenchyma cells and islet cells (round or oval-shaped cell mass) whereas DC showed focal necrosis and dilated acini. Metformin and Oat-treated groups showed minimal pathological changes.

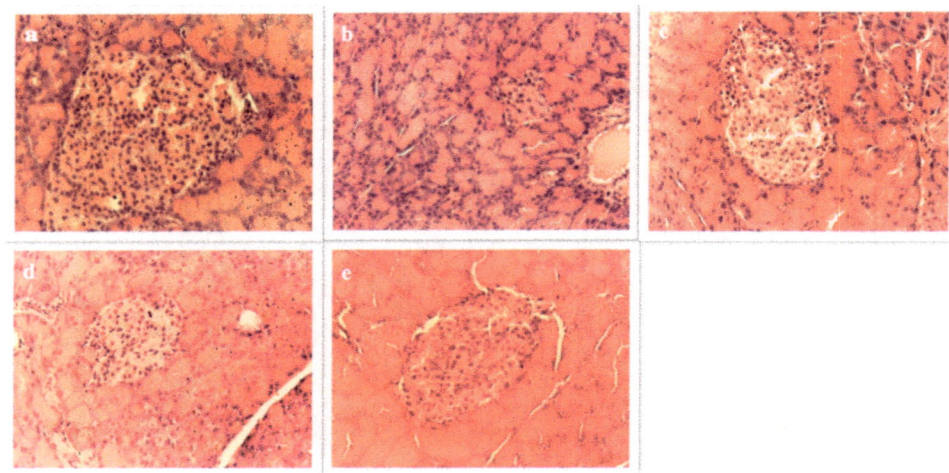

Figure 4. Effect of diabetes and oat treatment on histological alterations in the pancreas. Representative photomicrographs of histological alterations stained with hematoxylin and eosin at ×400 magnification. Normal control (**a**); Diabetic control (**b**); Metformin (**c**); Amlal (**d**); F11-5 (**e**).

4. Discussion

4.1. In Vitro Antidiabetic Activity

This study reports the inhibitory kinetics of Oat aqueous extracts against key enzymes linked to hyperglycemia. The in vitro assay demonstrated a moderate inhibitory effect compared to the positive control. Regarding the safe profile of Oat extract, it could be of interest to consider Oat formulation for inhibition of digestive enzymes in the small intestine. Oat mixture can display an antidiabetic effect if used as a capsulated complement. The presence of numerous phenolic compounds among the extract mixture e.g., *p*-Hydroxybenzoic, Syringic and Caffeic acids, resulted in a mixed type of inhibition. Previously, the association of several bioactive compounds revealed to be much more effective and safe in therapeutic strategies [33].

In fact, the heterogeneous character of type 2 diabetes results from the dynamic interaction between defects in insulin secretion and insulin action. Such a deficiency results in increased concentrations of blood glucose, which in turn damages many of the body's systems [34]. Therefore, there is a need to control postprandial blood glucose for diabetes management. One of the strategies in glucose control is the inhibition of digestive enzymes [17], such as α-glucosidase and α-amylase, which results in a significant reduction of the post-prandial blood glucose conferring an important target for diabetes management [35]. Although the wide availability of glucosidase inhibitors, many works sought natural sources in the hope to present nutritional alternatives with minimal side effects and low therapy costs [36]. Actually, a pseudo-tetrasaccharide of microbial origin namely acarbose, inhibits the brush-border enzymes glucoamylase, dextrinase, maltase and sucrase as well as the pancreatic α-amylase [17]. Additionally, other digestive enzymes inhibitors [37] are available (miglitol, emiglitate voglibose). Despite the effectiveness of acarbose as antidiabetic drug with glucosidase inhibitory properties, food alternatives are indeed needed because of acarbose side effects such as diarrhea and flatulence that occurs when colonic bacteria ferment the undigested carbohydrates, resulting in gas formation [38]. Also uncommon hepatotoxicity cases have been registered following long-term intake [39]. Currently, there is renewed interest in nutritional therapies and functional foods with preventive effects on diabetes and obesity [40]. Few papers described the Oat glucosidase and amylase inhibitory activities [16,17]. Nevertheless, none of the previous studies has determined the inhibitory concentrations and the kinetic of inhibition. Also the reported works focused on whole oatmeal and did not especially study the aqueous phenolic rich extract. Oat anti-enzymatic activity is probably linked in part to its phenolic content. As we have adopted a complementary nutritional therapy with multi-target bioactivities (antioxidant and antidiabetic), the safe Oat phenolic extracts have been chosen for in vivo and in vitro activities. Inhibitory effect of hybrid and parent Oat varieties was lower than acarbose, especially for glucosidase inhibition. As a complex mixture, Oat extracts exhibited a mixed inhibition, which is an intermediate mode between the competitive and uncompetitive inhibition [41]. The extracts metabolites were able to bind either to the free enzyme or the enzyme-substrate complex. Phenolic compounds were reported to exhibit a glucosidase inhibitory effect. Especially cereal grains, such as Barley [42], Wheat [43] and Quinoa [44], were reported to inhibit intestinal glucosidase.

4.2. Oxidative Damage Prevention and Anti-Hyperglycemic Effect

Recently, we have reported the nutritional characteristics and biochemical composition of Moroccan Oat varieties [10]. This assay is an attempt to investigate the in vivo antidiabetic and antioxidant properties of the Moroccan Oat variety (Amlal) and its derivative new line (F11-5).

Oat plays a role in modulating the metabolic effects observed after fiber-rich meals. As a soluble fiber with viscous characteristics, Oat β-glucans modifies properties of chyme in the upper part of the gastrointestinal tract affecting gastric emptying, gut motility, and nutrient absorption, which are reflected in lower postprandial glycemic and insulin responses [45]. Thus, oat β-glucan intake is beneficial for healthy subjects and patients with type-2 diabetes [46]. Manifestly, many papers described the Oat antidiabetic action linked to its fiber content and viscosity [47]. Moreover, Oat antioxidant effect is well described in literature using in vitro methods [10]. However few papers discussed the preventive effect of Oat extracts on hyperglycemia-induced OS. In our multi-target Oat-therapy approach, we tried to focus on both diabetes and oxidative stress. In this regard we have decided to use the aqueous phenolic extracts instead of other Oat extracts. The in vivo antioxidant and antidiabetic activities were evaluated in male wistar rats, following acute toxicity study in female Swiss mice. The use of female in toxicity studies is mainly due to their higher sensitivity. According to the OECD 425, the preferred rodent species is the rat, although other rodent species may be used. On the other hand, for diabetes models, several animal species, including the mouse and rat are sensitive to the pancreatic β-cell cytotoxic effects of Streptozotocin (STZ). However, some mice strains are less sensitive to this toxin.

In summary, the results from in vivo study of Oat extract intake, revealed an amelioration of glucose tolerance, a decrease in FBG and an antioxidant effect based on oxidative stress markers expression compared to diabetic stressed control. Furthermore, Oat intake prevented anxiety like behavior, hypercholesterolemia and atherosclerosis in diabetic rats.

Animal models of diabetes are an important tool for in vivo screening of antidiabetic compounds. The fungal compound streptozotocin is a hydrophilic nitrosourea analogue with antibiotic and chemotherapeutic activities. Owing to its glucose like similar structure, it enters β-cells via GLUT2 (Glucose transporter 2) transporters in a similar way to glucose [48], resulting in DNA alkylation, nitric oxide release and ROS generation leading to insulin reduced synthesis and diabetogenic state. In the STZ-NA (Streptozotocin-Nicotinamide) model of diabetes the use of NA a poly-ADP-ribose (Poly-Adenosine diphosphate-ribose) synthetase inhibitor, is mainly for its protective effect on β-cells function via the prevention of reduction in the level of nicotinamide adenine dinucleotide; thereby it results in a reverses of the insulin secretion inhibition lowering the degenerescence in the experimental model following the STZ administration, which show similar characteristics to T2DM (Type 2 Diabetes Mellitus) [49].

Actually, we have demonstrated that Oat extracts at 2000 mg/kg reduced significantly the FBG in STZ-NA-induced diabetic rats after 6 weeks of treatment. The OGTT performed at the beginning and the end of the treatment period show significant improvement in impaired glucose tolerance. Body weight loss in diabetic animals has been attributed to the waste in muscle mass [50]. Oat treatment and anti-hyperglycemic effect resulted in a decrease of body weight loss frequency. In the other hand, the increase in urinary volume and water and food intake in diabetic animals is comparable to polyphagia and polydipsia observed in diabetic patients [51]. Oat treatment prevented those manifestations and decreased diabetic symptoms impact. The decreased serum glucose levels after treatment period can probably be linked to Oat ability to inhibit digestive enzymes at the intestinal level. Based upon histopathological results it can be also hypothesized that Oat may probably exhibit protective effect on pancreatic β-cells against STZ toxic effects. Manifestly, β-cells number and size in treated animals support this hypothesis. Furthermore, diabetes complications such as hypercholesteremia and hypertriacylglycerolemia are primary factors involved in the development of atherosclerosis and coronary heart disease [52]. Oat treatment significantly reduced serum TG and total cholesterol in STZ diabetic rats. Consequently, it is suggested that Oat intake and Met treatment could modulate blood lipid abnormalities and reduce atherosclerosis risk. As a vital organ the liver metabolizes nutrients and detoxifies harmful substances. Accordingly, ALT and AST are reliable markers of liver function [53]. In the STZ-NA model of diabetes the increase in ALT and AST activities in plasma indicates the STZ hepatotoxic effect and the liver necrosis [54]. Am and F11 treatments reduced these enzymes levels in plasma compared to the DC and consequently prevented the liver damage; which indicates their hepatoprotective effect. The positive control group has been treated by metformin, which is a guanidine-containing compound currently recommended as first line therapy for all newly diagnosed T2DM patients [51]. Metformin decrease glucose production in the liver via the suppression of gluconeogenesis. Moreover, enhance the insulin suppression of endogenous glucose production and, to a lesser extent, reduce the intestinal glucose absorption and probably improve the glucose uptake and utilization by peripheral tissues [55]. Indeed, there is a growing evidence that hyperglycemia causes OS in a variety of tissues through ROS production [56]. At the same time, free radicals and ROS are implicated in the physiological gluco-regulatory systems. For instance, insulin release that is stimulated by glucose production in β-cells has been linked to H_2O_2 regulatory action [57]. Recent mathematical modeling described the OS dynamics to be of oscillatory nature [58] and tissue depending [59], which may explain the inconsistency of experimental data related to antioxidant enzymes measurements. Alternatively, focusing on stable OS markers such as MDA may reflect more reliable information on OS damage under long-term dynamics. Correspondingly, antioxidant enzymes and thiols form the first line of defense against ROS in the cells. In fact, SOD activity enhances the spontaneous dismutation of superoxide radicals to H_2O_2, before it is removed by CAT activity. In contrast to CAT, the GPx act by reducing lipid hydroperoxides to the corresponding alcohols and reduce free H_2O_2 to H_2O even at low concentrations. Moreover, the major endogenous antioxidant GSH directly neutralizes the free radicals and ROS, and maintains the exogenous antioxidants such as vitamins C and E in their reduced forms [60]. The increase in antioxidant enzymes activity in various tissues could reflect an adaptation to diabetes-induced OS. Our results showed that the level of MDA and GSH and the activities of T-SOD, Mn-SOD, CuZn-SOD, and CAT in the liver and kidney of diabetic rats significantly increased when compared with the normal control animals. However GPx increased only in the liver. The change in the antioxidant enzymes activities at the kidney and liver, show similarities to those seen in rats subjected to food deprivation-induced weight loss [61]. The reduced lipid peroxidation products and antioxidant enzymes after treatment likely indicates that Oat extracts might be a good source of metabolites with protective effect against diabetic-induced oxidative stress complications. Similarly, our results demonstrated clearly the protective effect of Oat extract against H_2O_2 oxidative stress in the cellular model of *T. pyriformis*. Both models indicate the in vivo antioxidant effect of Oat extracts, which can be attributed to its main phenolic compounds. Many phenolics with reported antioxidant effect has been isolated and identified in Oat such as ferulic acid and avenanthramides derivatives [62].

The radical scavenging effect of those Oat metabolites is highly linked to their bioavailability [63] and biotransformation [64].

Diabetes and hyperglycemia-induced oxidative stress can result in neurological complications such as dementia and depression [14]. Also the diabetogenic agent streptozotocin when injected at the intracerebro-ventricular level in rats can produce an Alzheimer disease (AD) model that triggers an insulin resistant brain state [65]. The plausible explanation suggests that low metabolism of brain glucose associated with oxidative stress and neuro-inflammation can result in mitochondrial dysfunction and apoptosis at the hippocampus, leading to accumulation of β-amyloid plaques and neurofibrillary tangles. This oxidative, neuro-inflammatory and degenerative environment induces progressively the memory impairment [66]. Moreover, in previous reports [67], STZ-treated rats displayed increased anxiety-like behavior in different paradigms, such as the open-field test, and the elevated plus maze. In fact, metformin administration significantly attenuated the anxiety-like behavior [68]. In our assays, the Oat and metformin treatment significantly improved cognitive status and prevented anxiety-like behavior in diabetic animals.

5. Conclusions

The present study suggests that Oat extracts are able to prevent oxidative stress in the diabetic liver and kidney. Also it has been demonstrated that Oat antidiabetic activity is mediated by digestive enzymes inhibition including (α-amylase and α-glucosidase). Oat extracts have been shown to protect antioxidant systems under diabetic stress and to prevent lipid peroxidation. Furthermore, Oat treatment preserved body weight and ameliorated liver functions (ALT and AST), lipid profile (cholesterol, TG) and reduced atherosclerosis index in diabetic animals. Finally, Oat and metformin intake significantly improved the anxiety-like behavior in diabetic animals.

Acknowledgments: Ilias Marmouzi wishes to thank Mohammed Achaaban for advices, guidance and help with dissection, also Laila Jahidi Ghazali Laboratory for histological experiments and comments. Many thanks to Ali Berraaouan for pancreas protocols, to Fatima Affifi and Violet Kasabri for enzymes protocols, and to Abdellatif Bouayyadi for providing reagents. I would also like to thank my colleagues Miloud El Karbane, Najmouddine, Karima Sayah and Hanae Naceiri Mrabeti for help in experimental work, and to Hayat Aouari, Ahmed Lafrouhi and Hassan Mounakhil, for technical support. This work was financially supported by the CNRST (Centre National de Recherche Scientifique et Technique), Morocco (RS/2011/27).

Author Contributions: Ilias Marmouzi , My El Abbes Faouzi, Bouchra Meddah and Yahia Cherrah conceived and designed the experiments; Ilias Marmouzi performed and participated in all of the experiments; Ilias Marmouzi analyzed the data and wrote the paper; El Mostafa Karym helped performing cell assays and enzyme measurement; Mourad Kharbach helped performing chemical analysis; Khalid El Allali supervised histological experiments, provided the platform for the assay and revised the manuscript; Mounya Bouabdellah, Layachi Chabraoui and Azlarab Masrar performed hematological analysis; Nezha Saidi performed plant hybridization and provided the plant material.

Conflicts of Interest: The authors declare no conflict of interest.

Abbreviations

AD	Alzheimer disease
AI	Atherosclerosis index
AST	Aspartate aminotransferase
ALT	Alanine aminotransferase
FBG	Fasting blood glucose
AUC	Areas under curve
CAT	Catalase
DNS	Dinitrosalicylic acid
EPM	Elevated plus maze
GPx	Glutathione Peroxidase
GSH	Glutathione
HDL	High density lipoprotein
HGB	Hemoglobin
LDH	Lactate dehydrogenase
LDL	Low density lipoprotein

MDA	Malondialdehyde
p-NPG	*p*-Nitrophenyl-α-D-glucopyranoside
OFR	Overnight fasted rats
OFT	Open field test
OGTT	Oral glucose tolerance test
OS	Oxidative stress
PBS	Phosphate buffer solution
PPYG	Proteose-peptone yeast Glucose defined medium
RBC	Red blood corpuscles count
ROS	Radical oxygenated species
SOD	Superoxide dismutase
STZ-NA	Streptozotocine nicotinamide
TG	Triacylglycerols
HDL	High density lipoprotein
T2DM	Type 2 diabetes mellitus
WBC	White blood corpuscles count

References

1. Rains, J.L.; Jain, S.K. Oxidative stress, insulin signaling, and diabetes. *Free Radic. Biol. Med.* **2011**, *50*, 567–575. [CrossRef] [PubMed]
2. Robertson, R.P. Chronic oxidative stress as a central mechanism for glucose toxicity in pancreatic islet beta cells in diabetes. *J. Biol. Chem.* **2004**, *279*, 42351–42354. [CrossRef] [PubMed]
3. Wold, L.E.; Asli, F.C.; Jun, R. Oxidative stress and stress signaling: Menace of diabetic cardiomyopathy. *Acta Pharmacol. Sin.* **2005**, *26*, 908–917. [CrossRef] [PubMed]
4. Evans, J.L.; Goldfine, I.D.; Maddux, B.A.; Grodsky, G.M. Are Oxidative Stress—Activated Signaling Pathways Mediators of Insulin Resistance and β-Cell Dysfunction? *Diabetes* **2003**, *52*, 1–8. [CrossRef] [PubMed]
5. Henriksen, J.E.; Diamond-Stanic, M.K.; Marchionne, E.M. Oxidative stress and the etiology of insulin resistance and type 2 diabetes. *Free Radic. Biol. Med.* **2011**, *51*, 993–999. [CrossRef] [PubMed]
6. Houstis, N.; Rosen, E.D.; Lander, E.S. Reactive oxygen species have a causal role in multiple forms of insulin resistance. *Nature* **2006**, *440*, 944–948. [CrossRef] [PubMed]
7. Ceriello, A. New Insights on Oxidative Stress and Diabetic Complications May Lead to a "Causal" Antioxidant Therapy. *Diabetes Care* **2003**, *26*, 1589–1596. [CrossRef] [PubMed]
8. Truswell, A.S. Cereal grains and coronary heart disease. *Eur. J. Clin. Nutr.* **2002**, *56*, 1–14. [CrossRef] [PubMed]
9. Marventano, S.; Vetrani, C.; Vitale, M.; Godos, J.; Riccardi, G.; Grosso, G. Whole Grain Intake and Glycaemic Control in Healthy Subjects: A Systematic Review and Meta-Analysis of Randomized Controlled Trials. *Nutrients.* **2017**, *9*, 769. [CrossRef] [PubMed]
10. Marmouzi, I.; Saidi, N.; Meddah, B.; Bouksaim, M.; Gharby, S.; El Karbane, M.; Serragui, S.; Cherrah, Y.; Faouzi, M.E.A. Nutritional characteristics, biochemical composition and antioxidant activities of Moroccan Oat varieties. *Food Meas.* **2016**, *10*, 156–165. [CrossRef]
11. Rahimi, R.; Nikfar, S.; Larijani, B.; Abdollahi, M. A review on the role of antioxidants in the management of diabetes and its complications. *Biomed. Pharmacother.* **2005**, *59*, 365–373. [CrossRef] [PubMed]
12. Shen, R.L.; Cai, F.L.; Dong, J.L.; Hu, X.Z. Hypoglycemic Effects and Biochemical Mechanisms of Oat Products on Streptozotocin-Induced Diabetic Mice. *J. Agric. Food Chem.* **2011**, *59*, 8895–8900. [CrossRef] [PubMed]
13. Zhao, Q.; Hu, X.; Guo, Q.; Cui, S.W.; Xian, Y.; You, S.; Chen, X.; Xu, C.; Gao, X. Physicochemical properties and regulatory effects on db/db diabetic mice of β-glucans extracted from oat, wheat and barley. *Food Hydrocoll.* **2014**, *37*, 60–68. [CrossRef]
14. Detka, J.; Kurek, A.; Basta-Kaim, A.; Kubera, M.; Lasoń, W.; Budziszewska, B. Neuroendocrine link between stress, depression and diabetes. *Pharmacol. Rep.* **2013**, *65*, 1591–1600. [CrossRef]
15. Butterfield, D.A.; Domenico, F.D.; Barone, E. Elevated risk of type 2 diabetes for development of Alzheimer disease: A key role for oxidative stress in brain. *Biochim. Biophys. Acta* **2014**, *1842*, 1693–1706. [CrossRef] [PubMed]
16. Rocher, A.; Colilla, F.; Ortiz, M.L.; Mendez, E. Identification of the three major coeliac immunoreactive proteins and one α-amylase inhibitor from oat endosperm. *FEBS Lett.* **1992**, *310*, 37–40. [CrossRef]
17. Bischoff, B.A.G.H. Pharmacology of α-glucosidase inhibition. *Eur. J. Clin. Investig.* **1994**, *24* (Suppl. 3), 3–10.

18. Dong, J.; Cai, F.; Shen, R.; Liu, Y. Hypoglycaemic effects and inhibitory effect on intestinal disaccharidases of oat beta-glucan in streptozotocin-induced diabetic mice. *Food Chem.* **2011**, *129*, 1066–1071. [CrossRef] [PubMed]

19. Marmouzi, I.; El Madani, N.; Charrouf, Z.; Cherrah, Y.; Faouzi, M.E.A. Proximate analysis, fatty acids and mineral composition of processed Moroccan *Chenopodium quinoa Willd* and antioxidant properties according to the polarity. *Phytothérapie* **2015**, *13*, 110–117. [CrossRef]

20. Saidi, N.; Saidi, S.; Hilali, A.; Benchekroun, M.; Al Faiz, C.; Bouksaim, M.; Shaimi, N.; Souihka, A.; Idrissi, S.A.; Gaboune, F.; et al. Improvement of oat hexaploid lines's groat nutritive value via hybridisation with tetraploid oat *A. magna. Am. J. Res. Commun.* **2013**, *1*, 126–135.

21. Mori, K.; Kashiwagi, A.; Yomo, T. Single-Cell Isolation and Cloning of *Tetrahymena thermophila* Cells with a Fluorescence-Activated Cell Sorter. *J. Eukaryot. Microbiol.* **2011**, *58*, 37–42. [CrossRef] [PubMed]

22. Lizard, G.; Gueldry, S.; Deckert, V.; Gambert, P.; Lagrost, L. Evaluation of the cytotoxic effects of some oxysterols and of cholesterol on endothelial cell growth: Methodological aspects. *Pathol. Biol.* **1997**, *45*, 281–290. [PubMed]

23. Lowry, O.H.; Rosebrough, H.N.; Farr, A.L.; Randall, R.J. Protein measurement with the folin phenol reagent. *J. Biol. Chem.* **1951**, *193*, 265–275. [PubMed]

24. Beauchamp, C.; Fridovich, I. Superoxide dismutase: Improved assays and an assay applicable to acrylamide gels. *Anal. Biochem.* **1971**, *44*, 276–287. [CrossRef]

25. Mannervik, B. Glutathione peroxidase. *Methods Enzymol.* **1985**, *113*, 490–495. [PubMed]

26. Furman, B.L. Streptozotocin-induced diabetic models in mice and rats. *Curr. Protoc. Pharmacol.* **2015**, *70*, 1–20.

27. Kee, K.T.; Koh, M.; Oong, L.X.; Ng, K. Screening culinary herbs for antioxidant and α-glucosidase inhibitory activities. *Int. J. Food Sci. Technol.* **2013**, *48*, 1884–1891. [CrossRef]

28. Walf, A.A.; Frye, C.A. The use of the elevated plus maze as an assay of anxiety-related behavior in rodents. *Nat. Protoc.* **2007**, *2*, 322–328. [CrossRef] [PubMed]

29. Prut, L.; Belzung, C. The open field as a paradigm to measure the effects of drugs on anxiety-like behaviors: A review. *Eur. J. Pharmacol.* **2003**, *463*, 3–33. [CrossRef]

30. Aebi, H. Catalase in vitro. *Methods Enzymol.* **1984**, *105*, 121–126. [PubMed]

31. Moron, M.S.; Depierre, J.W.; Mannervik, B. Levels of glutathione, glutathione reductase and glutathione S-transferase activities in rat lung and liver. *Biochim. Biophys. Acta* **1979**, *582*, 67–78. [CrossRef]

32. Ohkawa, H.; Ohishi, N.; Yagi, K. Assay of lipid peroxides in animal tissue by thio barbituric acid reaction. *Anal. Biochem.* **1979**, *95*, 351–358. [CrossRef]

33. Efferth, T.; Egon, K. Complex interactions between phytochemicals. The multi-target therapeutic concept of phytotherapy. *Curr. Drug Targets* **2011**, *12*, 122–132. [CrossRef] [PubMed]

34. Lin, Y.; Sun, Z. Current views on type 2 diabetes. *J. Endocrinol.* **2010**, *204*, 1–11. [CrossRef] [PubMed]

35. Lebovitz, H.E. Alpha-Glucosidase Inhibitors. *Endocrinol. Metab. Clin. N. Am.* **1997**, *26*, 539–551. [CrossRef]

36. Van de Laar, F.A.; Lucassen, P.L.B.J.; Akkermans, R.P.; Van de Lisdonk, E.H.; Rutten, G.E.H.M.; Van Weel, C. Alpha-glucosidase inhibitors for type 2 diabetes mellitus. *Cochrane Database Syst. Rev.* **2005**, *18*, CD003639.

37. Krause, H.P.; Ahr, H.J. Pharmacokinetics and Metabolism of Glucosidase Inhibitors. In *Oral Antidiabetics*; Handbook of Experimental Pharmacology; Springer: Berlin/Heidelberg, Germany, 1996; Volume 119, pp. 541–555.

38. Balfour, J.A.; McTavish, D. Acarbose: An Update of its Pharmacology and Therapeutic Use in Diabetes Mellitus. *Drugs* **1993**, *46*, 1025–1054. [CrossRef] [PubMed]

39. Wang, P.Y.; Kaneko, T.; Wang, Y.; Sato, A. Acarbose alone or in combination with ethanol potentiates the hepatotoxicity of carbon tetrachloride and acetaminophen in rats. *J. Hepatol.* **1999**, *29*, 161–165. [CrossRef] [PubMed]

40. Evert, A.B.; Boucher, J.L.; Cypress, M.; Dunbar, S.A.; Franz, M.J.; Mayer-Davis, E.J.; Neumiller, J.J.; Nwankwo, R.; Verdi, C.L.; Urbanski, P.; et al. Nutrition Therapy Recommendations for the Management of Adults With Diabetes. *Diabetes Care* **2014**, *37*, S120–S143. [CrossRef] [PubMed]

41. Thu Phan, M.A.; Wang, J.; Tang, J.; Lee, Y.Z.; Ng, K. Evaluation of α-glucosidase inhibition potential of some flavonoids from Epimedium brevicornum. *LWT Food Sci. Technol.* **2013**, *53*, 492–498. [CrossRef]

42. Ramakrishna, R.; Sarkar, D.; Schwarz, P.; Shetty, K. Phenolic linked anti-hyperglycemic bioactives of barley (*Hordeum vulgare* L.) cultivars as nutraceuticals targeting type 2 diabetes. *Ind. Crops Prod.* **2017**, *107*, 509–517. [CrossRef]

43. Malunga, L.N.; Eck, P. Inhibition of intestinal α-glucosidase and glucose absorption by feruloylated arabinoxylan mono-and oligosaccharides from corn bran and wheat aleurone. *J. Nutr. Metab.* **2016**, *2016*, 1932532. [CrossRef] [PubMed]

44. Hemalatha, P.; Bomzan, D.P.; Rao, B.S.; Sreerama, Y.N. Distribution of phenolic antioxidants in whole and milled fractions of quinoa and their inhibitory effects on α-amylase and α-glucosidase activities. *Food Chem.* **2016**, *199*, 330–338. [CrossRef] [PubMed]

45. Daou, C.; Zhang, H. Oat Beta-Glucan: Its Role in Health Promotion and Prevention of Diseases. *Compr. Rev. Food Sci. Food Saf.* **2012**, *11*, 355–365. [CrossRef]

46. Shen, X.L.; Zhao, T.; Zhou, Y.; Shi, X.; Zou, Y.; Zhao, G. Effect of oat β-glucan intake on glycaemic control and insulin sensitivity of diabetic patients: A meta-analysis of randomized controlled trials. *Nutrients* **2016**, *8*, 39. [CrossRef] [PubMed]

47. Wanga, Q.; Ellisa, P.R. Oat β-glucan: Physico-chemical characteristics in relation to its blood-glucose and cholesterol-lowering properties. *Br. J. Nutr.* **2014**, *112*, S4–S13. [CrossRef] [PubMed]

48. Radenković, M.; Stojanović, M. Milica Prostran Experimental diabetes induced by alloxan and streptozotocin: The current state of the art. *J. Pharmacol. Toxicol. Methods* **2016**, *78*, 13–31. [CrossRef] [PubMed]

49. Masiello, P.; Broca, C.; Gross, R.; Roye, M.; Manteghetti, M.; Hillaire-Buys, D.; Novelli, M.; Ribes, G. Experimental NIDDM: Development of a New Model in Adult Rats Administered Streptozotocin and Nicotinamide. *Diabetes* **1998**, *47*, 224–229. [CrossRef] [PubMed]

50. Swanston-Flat, S.K.; Day, C.; Bailey, C.J.; Flatt, P.R. Traditional plant treatment for diabetes: Studies in normal and streptozotocin diabetic mice. *Diabetologia* **1990**, *33*, 462–464. [CrossRef]

51. American Diabetes Association. 2. Classification and Diagnosis of Diabetes. *Diabetes Care* **2015**, *38*, S8–S16.

52. Taylor, F.; Huffman, M.D.; Macedo, A.F.; Moore, T.H.M.; Burke, M.; Davey Smith, G.; Ward, K.; Ebrahim, S. Statins for the primary prevention of cardiovascular disease. *Cochrane Database Syst. Rev.* **2013**, CD004816. [CrossRef]

53. Ohaeri, O.C. Effect of garlic oil on the levels of various enzymes in the serum and tissue of streptozotocin diabetic rats. *Biosci. Rep.* **2001**, *21*, 19–24. [CrossRef] [PubMed]

54. Palsamy, P.; Subramanian, S. Resveratrol, a natural phytoalexin, normalizes hyperglycemia in streptozotocin-nicotinamide induced experimental diabetic rats. *Biomed. Pharmacother.* **2008**, *62*, 598–605. [CrossRef] [PubMed]

55. Natali, A.; Ferrannini, E. Effects of metformin and thiazolidinediones on suppression of hepatic glucose production and stimulation of glucose uptake in type 2 diabetes: A systematic review. *Diabetologia* **2006**, *49*, 434–441. [CrossRef] [PubMed]

56. King, G.L.; Loeken, M.R. Hyperglycemia-induced oxidative stress in diabetic complications. *Histochem. Cell. Biol.* **2004**, *122*, 333–338. [CrossRef] [PubMed]

57. Mahadev, K.; Zilbering, A.; Zhu, L.; Goldstein, B.J. Insulin-stimulated Hydrogen Peroxide Reversibly Inhibits Protein-tyrosine Phosphatase 1B in Vivo and Enhances the Early Insulin Action Cascade. *J. Biol. Chem.* **2001**, *276*, 21938–21942. [CrossRef] [PubMed]

58. Porokhovnik, L.N.; Passekov, V.P.; Gorbachevskaya, N.L.; Sorokin, A.B.; Veiko, N.N.; Lyapunova, N.A. Active ribosomal genes, translational homeostasis and oxidative stress in the pathogenesis of schizophrenia and autism. *Psychiatr. Genet.* **2015**, *25*, 79–87. [CrossRef] [PubMed]

59. Genet, S.; Kale, R.K.; Baquer, N.Z. Alterations in antioxidant enzymes and oxidative damage in experimental diabetic rat tissues: Effect of vanadate and fenugreek (*Trigonella foenum graecum*). *Mol. Cell. Biochem.* **2002**, *236*, 7–12. [CrossRef] [PubMed]

60. Rajendran, P.; Nandakumar, N.; Rengarajan, T.; Palaniswami, R.; Gnanadhas, E.N.; Lakshminarasaiah, U.; Gopas, J.; Nishigaki, I. Antioxidants and human diseases. *Clin. Chim. Acta* **2014**, *436*, 332–347. [CrossRef] [PubMed]

61. Godin, D.V.; Wohaieb, S.A.; Garnett, M.E.; Goumeniouk, A.D. Antioxidant enzyme alterations in experimental and clinical diabetes. *Mol. Cell. Biochem.* **1988**, *84*, 223–231. [CrossRef] [PubMed]

62. Peterson, D.M. Oat Antioxidants. *J. Cereal Sci.* **2001**, *33*, 115–129. [CrossRef]

63. Chen, C.Y.; Milbury, P.E.; Kwak, H.K.; Collins, F.W.; Samuel, P.; Blumberg, J.B. Avenanthramides and Phenolic Acids from Oats Are Bioavailable and Act Synergistically with Vitamin C to Enhance Hamster and Human LDL Resistance to Oxidation. *J. Nutr.* **2004**, *134*, 1459–1466. [PubMed]

64. Wang, P.; Chen, H.; Zhu, Y.; McBride, J.; Fu, J.; Sang, S. Oat Avenanthramide-C (2c) Is Biotransformed by Mice and the Human Microbiota into Bioactive Metabolites. *J. Nutr.* **2015**, *145*, 239–245. [CrossRef] [PubMed]
65. Salkovic-Petrisic, M.; Knezovic, A.; Hoyer, S.; Riederer, P. What have we learned from the streptozotocin-induced animal model of sporadic Alzheimer's disease, about the therapeutic strategies in Alzheimer's research. *J. Neural Transm.* **2013**, *120*, 233–252. [CrossRef] [PubMed]
66. Samy, D.M.; Ismail, C.A.; Nassr, R.A.; Zeitoun, T.M.; Nomair, A.M. Down stream modulation of extrinsic apoptotic pathway in streptozotocin-induced Alzheimer's dementia in rats: Erythropoietin versus curcumin. *Eur. J. Pharmacol.* **2016**, *770*, 52–60. [CrossRef] [PubMed]
67. Rebolledo-Solleiro, D.; Crespo-Ramírez, M.; Roldán-Roldán, G.; Hiriart, M.; Pérez de la Mora, M. Role of thirst and visual barriers in the differential behavior displayed by streptozotocin-treated rats in the elevated plus-maze and the open field test. *Physiol. Behav.* **2013**, *120*, 130–135. [CrossRef] [PubMed]
68. Garabadu, D.; Krishnamurthy, S. Diazepam Potentiates the Antidiabetic, Antistress and Anxiolytic Activities of Metformin in Type-2 Diabetes Mellitus with Cooccurring Stress in Experimental Animals. *BioMed Res. Int.* **2014**, *2014*, 693074. [CrossRef] [PubMed]

antioxidants

MDPI

Review

Do Coffee Polyphenols Have a Preventive Action on Metabolic Syndrome Associated Endothelial Dysfunctions? An Assessment of the Current Evidence

Kazuo Yamagata

Laboratory of Molecular Health Science of Food, Department of Food Bioscience and Biotechnology, Nihon University College of Bioresource Sciences (NUBS), 1866, Kameino, Fujisawa, Kanagawa 252-8510, Japan; kyamagat@brs.nihon-u.ac.jp; Tel.: +81-466-84-3986

Received: 9 January 2018; Accepted: 1 February 2018; Published: 4 February 2018

Abstract: Epidemiologic studies from several countries have found that mortality rates associated with the metabolic syndrome are inversely associated with coffee consumption. Metabolic syndrome can lead to arteriosclerosis by endothelial dysfunction, and increases the risk for myocardial and cerebral infarction. Accordingly, it is important to understand the possible protective effects of coffee against components of the metabolic syndrome, including vascular endothelial function impairment, obesity and diabetes. Coffee contains many components, including caffeine, chlorogenic acid, diterpenes and trigonelline. Studies have found that coffee polyphenols, such as chlorogenic acids, have many health-promoting properties, such as antioxidant, anti-inflammatory, anti-cancer, anti-diabetes, and antihypertensive properties. Chlorogenic acids may exert protective effects against metabolic syndrome risk through their antioxidant properties, in particular toward vascular endothelial cells, in which nitric oxide production may be enhanced, by promoting endothelial nitric oxide synthase expression. These effects indicate that coffee components may support the maintenance of normal endothelial function and play an important role in the prevention of metabolic syndrome. However, results related to coffee consumption and the metabolic syndrome are heterogeneous among studies, and the mechanisms of its functions and corresponding molecular targets remain largely elusive. This review describes the results of studies exploring the putative effects of coffee components, especially in protecting vascular endothelial function and preventing metabolic syndrome.

Keywords: antioxidative effects; chlorogenic acid; endothelial dysfunction; metabolic syndrome

1. Introduction

Metabolic syndrome is a combination of medical conditions, including dyslipidemia, elevated blood pressure, insulin resistance, and excess body weight. An increase in the rate of adult obesity has led to increases in obesity-associated metabolic disorders, such as insulin resistance, glucose intolerance, dyslipidemia, and hypertension, which are major risk factors for cardiovascular disease (CVD) [1]. While the mechanism underlying metabolic syndrome is poorly understood, insulin resistance appears to be an important factor [2]. The occurrence of hyperglycemia, dyslipidemia, and hypertension in metabolic syndrome is associated with endothelial dysfunction and promotion of atherogenesis [3]. These changes show that patients with metabolic syndrome have an increased risk of myocardial infarction and stroke [4,5], which, together, contribute to the burden of non-communicable diseases [6].

Various components of the common diet have been suggested to potentially affect endothelial function, including fruits, vegetables, olive oil and nuts [7–10]. Several epidemiological studies have shown that coffee consumption is associated with a lower risk of cardiovascular disease, metabolic disorders and certain cancers [11]. Among the observed results, evidence suggests that individuals consuming moderate amounts of coffee might be less likely to be affected by metabolic syndrome [12,13]. Evidence has been increasing recently supporting the protective effects of coffee and its components, such as caffeine, chlorogenic acids and diterpenes, against oxidative stress and related metabolic syndrome risk [14,15]. Results from a recent meta-analysis indicated that the coffee intake was inversely related with the risk of type 2 diabetes (T2DM) [16]. These effects are likely due to the presence of chlorogenic acids and caffeine [17]. Moreover, other reports described the effect of coffee intake on lipid, protein and DNA damage, and the modulation of antioxidant capacity and antioxidant enzymes in human studies [18,19]. These results suggested that coffee consumption enhances glutathione levels and offers protection against DNA damage [19,20]. In contrast, caffeine consumption of ≥6 units/day during pregnancy is related with impaired fetal length growth [21]. Also, this report indicated that a higher caffeine intake might preferentially adversely affect fetal skeletal growth.

In metabolic syndrome, inflammation is promoted, which strongly increases the risk of systemic CVD [22,23]. Endothelial dysfunction reduces vasodilation, enhances proinflammatory status and thrombogenesis, and is the first step in the development of CVD [24]. The development of arteriosclerosis that occurs in metabolic syndrome is attributed to endothelial dysfunction. Namely, these events of endothelial dysfunction strongly induce CVD [25]. On the other hand, the intake of coffee contributes to the prevention of endothelial dysfunction and decrease in CVD. A recent study demonstrated that the inverse relationship between coffee consumption and metabolic syndrome may reflect coffee's content of antioxidants that offer cardiovascular protection [26]. This review considers the effects of coffee polyphenols on vascular endothelial function in preventing metabolic syndrome. The association of coffee intake with metabolic syndrome and with vascular disorders induced in metabolic syndrome will be discussed.

2. Antioxidant Effects of Coffee Components

Coffee is a major source of antioxidant polyphenols in the Japanese diet [27]. As shown in Table 1, coffee has the highest total polyphenol content in beverages, followed by green tea.

Table 1. Total polyphenol intake from beverages in the Japanese diet [a].

Main Food Sources	Consumption of Beverages		Average Total Polyphenol Content	Daily Total Polyphenol Consumption	
Beverages	(mL/day)	(%)	(mg/100 mL)	(mL/day)	(%)
Green tea	353 ± 337	23	115	292 ± 398	34
Coffee	213 ± 213	19	200	426 ± 424	50
Barley tea	174 ± 325	16	9	15 ± 28	2
Oolong tea	76 ± 214	7	39	30 ± 84	4
Fresh milk	60 ± 127	5			
Black tea	59 ± 146	5	96	57 ± 140	7
Other tea	53 ± 182	5	8	4 ± 15	1
Sports drinks	52 ± 180	3			
Carbonated drinks	37 ± 127	3			
Mineral water	35 ± 136	3			
Fruit juice	32 ± 71	3	34	11 ± 24	1
Tomato/vegetable juice	14 ± 56	1	69	9 ± 38	1
Cocoa/chocolate malt drinks	10 ± 49	1	62	6 ± 30	1
Soy milk	6 ± 30	0	36	2 ± 11	0
Others	40 ± 147	4			
Total	1113 ± 512	100		853 ± 512	100

[a] Adapted from reference Fukushima et al., 2009 [27].

Coffee contains over 1500 chemicals, most of which are formed during the roasting process. The total polyphenol consumption from coffee is 426 mg/day, which makes up 50% of the amount of polyphenols consumed daily in the Japanese population [27]. Typically found polyphenols include caffeine, chlorogenic acid, diterpenes and trigonelline [28]. The chemical composition of green coffee beans is shown in Table 2 [29]. The main water-soluble constituents in coffee are phenolic polymers, polysaccharides, chlorogenic acids, minerals, caffeine, organic acids, sugars, and lipids. Furthermore, the lipid-soluble constituents of coffee include mainly triacylglycerols, tocopherols, and esters of diterpene alcohols with fatty acids.

Table 2. Comparison of chemical components of green beans in *Coffea arabica* and *Coffea canephora*.

Component	C. arabica	C. canephora
Minerals *	3.5–4.5	3.9–4.5
Lipids *	13–17	7.2–11
Caffeine *	0.7–2.2 (average 1.4)	1.5–2.8 (average 2.2)
Chlorogenic acid *	4.80–6.14	5.34–6.41
Trigonelline *	1–1.2	0.6–1.7
Oligosaccharides *	6–8	5–7
Total polysaccharides *	50–55	37–47

* % dry matter. Adapted from reference Stamach et al., 2006 [29].

Coffee is a major source of antioxidant polyphenols in the Japanese diet [28]. Coffee has antioxidative components and contributes to oxidative stress prevention [30]. The antioxidative compounds inhibit nicotinamide adenine dinucleotide phosphate (NADPH) oxidase in the mitochondria and decrease production of reactive oxygen species (ROS) [31]. For this reason, coffee is understood to be a diet beverage that can inhibit oxidative stress. The coffee component induces antioxidant activity and endothelial nitric oxide (NO) production. For example, intake of caffeoylquinic (CQA, a representative of chlorogenic acid) for 8 weeks decreased NADPH-dependent ROS production and enhanced NO production in the aorta of spontaneously hypertensive rats [32]. In addition, CQA was directly related to the blocking of gene expression of the NADPH-oxidase component, p22phox [32]. These results indicate that the CQA might decrease oxidative stress and induce NO production and help to prevent endothelial dysfunction, such as vascular hypertrophy and hypertension, in spontaneously hypertensive rats.

The main antioxidants in coffee are the chlorogenic acids, caffeine, and melanoidins [17]. Chlorogenic acids consist of a family of esters formed between quinic acid and caffeic acid. The subclasses of chlorogenic acids are CQA, feruloylquinic (FQA) and dicaffeoylquinic (diCQA) acid. The main chlorogenic acid subclass in coffee is CQA [17,33] (Figure 1). Following acute consumption of coffee, a significant 5.5% rise in plasma antioxidant activity in human has been demonstrated [34], including inhibition of low density lipoprotein (LDL) oxidation [35,36]. Therefore, the effect of CQA may contribute to an arteriosclerotic preventive mechanism. Also, caffeine has antioxidant properties, such as the ability to scavenge hydroxyl radicals [37]. Caffeine has been shown to scavenge superoxide radicals by measurements of O_2^- after reaction with caffeine using electron paramagnetic resonance (EPR) [38,39]. These results support the findings of another study in which caffeine had antioxidative effects and inhibited peroxidation of LDL [40]. Finally, the browned material in coffee also has antioxidative effects. The taste and the color of the coffee are produced mainly by a roasting process with the Maillard reaction. These substances contribute to the antioxidation effect and oxidative stress of the coffee [41,42]. In the roasted coffee bean, melanoidin is produced from nonenzymatic browning and accounts for the coffee bean's antioxidant activity [43]. Chlorogenic acid decreases in roasting, but melanoidins increase and may make up for the decrease in antioxidation from a loss of chlorogenic acid [44]. The antioxidative difference of the coffee changes by a roast. Differences in reactivity for roasts on antioxidants, such as chlorogenic acid, may influence the antioxidation characteristics of the coffee.

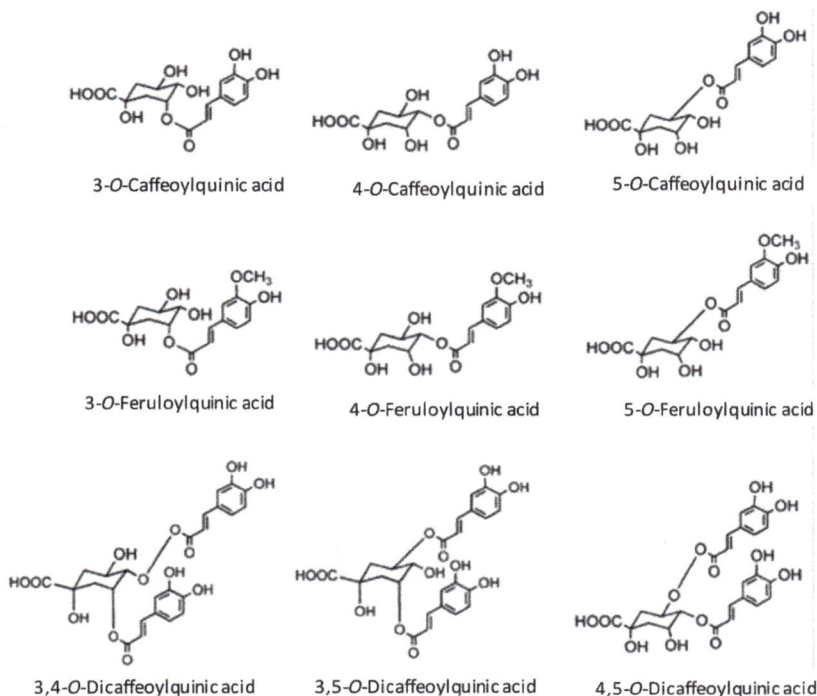

Figure 1. Structure of main chlorogenic acids in coffee (Adapted from reference Stamach et al., 2006) [29].

3. Epidemiological Studies of Coffee Consumption and the Metabolic Syndrome

3.1. Coffee Intake and Metabolic Syndrome

Regular consumption of coffee has been associated with lower odds of having metabolic syndrome [12,13]. Clinical evidence of the effects of coffee consumption on various components of the metabolic syndrome has been provided by a cross-over, randomized controlled study, investigating men and women with normal cholesterol levels (*n* = 25) and those with hypercholesterolemia (*n* = 27) aged 18–45 years with body mass index (BMI) ranging from 18–25 kg/m². For 8 weeks, the study subjects consumed three servings/day of a blend providing coffee polyphenols, hydroxycinnamic acids (510.6 mg) and caffeine (121.2 mg), or a control drink. In the coffee consumption groups, blood pressure, body fat percentage, and levels of leptin, plasminogen activator inhibitor-1 (PAI-1) and resistin were reduced. In addition, glucose concentration, insulin resistance and triglyceride levels were reduced. Notably, these reductions were much greater in the group with hypercholesterolemia compared with the controls. These results suggest that regular coffee consumption can improve the pathologic condition of patients with metabolic syndrome-associated hypercholesterolemia.

Most of the existing evidence relies on the results of two meta-analyses showing an association between coffee consumption and metabolic syndrome in observational studies [12,13]. The meta-analyses included 13 studies with a total of 159,805 participants and showed an inverse association between regular coffee consumption and metabolic syndrome, despite evidence of heterogeneity between results of the studies included. The causes for these differences may have been variations across the studies in terms of lifestyle and the percentages of patients with metabolic syndrome as well as in coffee-drinking habits, such as adding milk, full-fat cream or sugar [45]. Several other factors may have accounted for the heterogeneity among results. For instance, most of the studies did not consider methods of preparation, type, and roasting process of coffee, which have been shown to influence the phytochemical component of the beverage [46]. Moreover, collinearity may exist with the intake of certain foods, sugar, and the presence or absence of milk [47]. Other sources of heterogeneity may be derived from lifestyle differences among individuals, for instance, the level of smoking, which has been shown to be an effect modifier of the association between coffee intake and health outcomes [48]. Third, none of the studies took into account genetic factors, which have been reported to affect the relationship between coffee and cardiovascular outcomes due to polymorphisms related to caffeine metabolization [49,50].

Collectively, it was shown that the association between coffee consumption and occurrence of metabolic syndrome varied greatly across studies (Table 3). In a large general population cohort study, high coffee consumption was associated with low risks of obesity, metabolic syndrome and T2DM [51]. The study results indicated that high coffee consumption was associated with decrease in obesity, metabolic syndrome and T2DM. Moreover, high coffee consumption was associated with low BMI, weight, height, systolic/diastolic blood pressure, triglycerides and cholesterol. In another study, the effects of coffee consumption in metabolic syndrome were investigated in healthy subjects: 174 men and 194 women were followed from the age of 27 years onwards [52]. This study began in 1977, along with an observational longitudinal study that examined 600 girls and boys. The strongest evidence supporting a positive health effect of coffee consumption has been for diabetes. However, this study demonstrated that long-term coffee consumption was not associated with metabolic syndrome. While coffee consumption appeared to be significantly reversely correlated with blood pressure, the relationship was no longer significant after adjustment for lifestyle covariates. In a Mendelian randomization study, we examined the relationship between coffee intake and obesity, metabolic syndrome, and T2DM in 93,179 people with T2DM in two large cohorts. A high intake of coffee was associated with a reduced risk of obesity, metabolic syndrome, and type II diabetes mellitus. Furthermore, higher coffee consumption was associated with reduced BMI, weight, waist circumference, blood pressure, triglycerides and total cholesterol and increased high-density lipoprotein cholesterol [51]. In addition, a study of metabolic syndrome in Poland investigated the association of tea and coffee consumption with the prevalence of metabolic syndrome in 8821 subjects aged 20 years and older. The relationship of coffee intake with metabolic syndrome suggested a role for coffee intake in cardiovascular prevention. The effect of the coffee may have been caused by antioxidant action [53]. Still another study showed that the risk of metabolic syndrome was associated with coffee intake in 15,691 Korean women, indicating that coffee intake might be related to a decreased occurrence of metabolic syndrome in this population [54].

Table 3. Characteristics of studies investigating the relationship between coffee consumption and metabolic syndrome and its components.

Design	Population Characteristics	Cases	Diagnosis Criteria	Adjustments	Results	Country	Reference
Cross-sectional	1889 (760 M, 1129 F, mean age 50.2 ± 16.3)	226 (91 M, 135 F)	IDF-MetS	Gender, age, BMI, educational level, socio-economic status, alcohol drinking, smoking status, alcohol drinking, physical activity level, MedDietScore, caffeine, source of caffeine.	Coffee, but not caffeine, was inversely associated with MetS and triglycerides.	Italy	[54]
	8821 (4291 M, 4530 F, mean age 56.8 ± 7)	2461 (1126 M, 1335 F)	IDF-MetS	Gender, age, educational level, occupational level, physical activity, smoking status, alcohol drinking, total energy intake, tea consumption.	Coffee was negatively associated with MetS, WC, hypertension and triglycerides.	Poland	[55]
	17,953 (6879 M, 11,074 F, mean age 39.7, range 19–65)	na	NCEP ATPIII	Age, gender, smoking status, physical activity, alcohol, total energy, education, income.	Comparing ≥3 times/day consumers with those who consumed coffee <1 time/week, the OR for MetS was 1.37, 95% CI 1.10–1.72. In addition, coffee drinkers had an elevated risk of obesity, abdominal obesity and low HDL.	Republic of Korea	[56]
	19,839 (all male, age range 30–79)	3957 (all male)	NCEP APTIII	Age, education level, physical activity, occupation, smoking habits, alcohol habits, dietary factors, and family history of diabetes, hypertension, and cerebrovascular and CVD in second-degree relatives.	Regular drinking of coffee was not associated with MetS.	China	[57]
	554 (409 M, 145 F, mean age 52.2 ± 9.3)	114 (NCEP ATPIII), 77 (JASSO)	NCEP ATPIII/JASSO	Age, gender, total energy intake, physical activity, and smoking and drinking habits.	NCEP ATPIII criteria: Coffee was associated with a lower prevalence of MetS and drinkers of ≥3 cups/day had a lower OR for triglycerides. JASSO criteria: MetS prevalence was not associated with coffee consumption. However 1.5 to 3 cups/day drinkers registered a lower OR for high FPG.	Japan	[58]
	361 (all male, mean age 74.7 ± 6.1)	132 (all male)	Modified NCEP ATPIII	Age, BMI, UA, HOMA-IR, hsCRP, physical activity, psycho-social factors (occupational status, marital status, educational status), alcohol habits, coffee drinking habits.	Coffee drinking was not associated with MetS (OR 0.92, 95% CI 0.27–3.14).	Taiwan	[59]
	3283 (2335 M and 948 F, mean age 46.4, range 20–65)	406 (374 M and 32 F)	JASSO	Age, alcohol drinking, smoking, physical activity.	Coffee consumption of 4 cups or more was protective for MetS (OR 0.61, 95% CI 0.39–0.95), high blood pressure and high triglycerides, when compared with non-coffee drinkers in men. In women, coffee consumption was not associated with the prevalence of MetS or its components.	Japan	[60]

Table 3. *Cont.*

Design	Population Characteristics	Cases	Diagnosis Criteria	Adjustments	Results	Country	Reference
Cross-sectional/prospective	83,436	26,046	Not standard criteria	Age, gender, smoking status, physical inactivity and use of antihypertensive and lipid-lowering medication.	A high coffee intake was associated with low risk of MetS (OR 0.89, 95% CI 0.83–0.95), obesity, type 2 diabetes, high BMI, WC, total cholesterol and low HDL.	Denmark	[51]
	9514 (4497 M and 5317 F, mean age 53.6 ± 5.7)	3782	AHA	Age, gender, race, education, center, total calories, smoking status, pack-years, physical activity, and intakes of meat, dairy, fruits and vegetables, whole grains, and refined grains.	No relationship was observed between coffee and MetS.	USA	[61]
	17,014 (age range: 20–56)	1942	modified NCEP ATPIII	Age, baseline examination, alcohol intake, coffee consumption, number of cigarettes smoked, years of education, leisure-time physical activity.	Coffee intake was not associated with MetS, both in men and women.	Norway	[62]
Prospective	368 (174 M and 194 F, mean age 36)	37	NCEP ATPIII	Gender, physical activity, energy intake, smoking behavior, alcohol consumption.	Coffee consumption was not associated with MetS or its components.	Netherlands	[45]
	1902 (785 M and 1117 F, mean age 62.7 ± 11)	188 (137 M and 51 F)	JASSO	Age, gender, total energy intake, alcohol intake, current smoking, and habitual physical activity.	In those with lower coffee consumption there was a higher MetS prevalence, with an inverse relationship between the number of components and coffee consumption. All components of MetS except for HDL-cholesterol were directly associated with coffee.	Japan	[63]
Case-control	250 (103 M and 147 F, age range: 18–81)	74 (27 M, 47 F)	NCEP ATPIII	Age, gender, education level, socio-economic status, marital status, hyperglycaemia, chocolate, coffee, milk, sleep.	Coffee was inversely associated with metabolic syndrome.	Brazil	[47]

Abbreviations: AHA: American Heart Association; BMI: body mass index; CI: confidence interval; CVD: cardiovascular disease; HDL: high-density lipoprotein; HMW-Ad: high-molecular-weight serum adiponectin; HOMA-IR: homeostasis model-insulin resistance index; hsCRP: high-sensitivity C-reactive protein; IDF: International Diabetes Federation; JASSO: Japan Society for the Study of Obesity; MetS: metabolic syndrome; na: not available; NCEP ATPIII: National Cholesterol Education Program Adult Treatment Panel III; OR: odds ratio; WC: waist circumference. Adapted from reference Marventano et al., 2016 [12].

Furthermore, the report showed a relationship between coffee intake and metabolic syndrome in overweight and normal individuals. The studies included a questionnaire-style interview, blood pressure measurements and examination of fasting blood samples. In the obese and overweight groups, lower coffee intake compared with higher intake was associated with a higher risk of abdominal obesity, hypertension, abnormal glucose concentration, triglycerides and metabolic syndrome [55]. Another study examined the relationship between dietary lifestyle factors with metabolic syndrome. Daily drinking of 2–3 cups of coffee was inversely related to metabolic syndrome, and sleeping 7–8 h per night was related to decreased odds of metabolic syndrome [15]. In a cross-sectional study, coffee intake was inversely associated with metabolic syndrome and triglyceride levels in 1886 Italian subjects [55]. Also, in 8821 Italian subjects, coffee consumption was negatively associated with metabolic syndrome, waist circumference, hypertension and triglyceride levels [55]. The report also demonstrated that coffee intake was negatively associated with metabolic syndrome, waist circumference, hypertension and triglycerides [56]. Furthermore, in 17,953 Korean adults, when comparing those who consumed instant coffee >3 times/day with those who consumed instant coffee <1 time/week, the odds ratio for metabolic syndrome was 1.37. In addition, coffee drinkers had an increased risk of obesity, abdominal obesity and low levels of high-density lipoprotein [61]. In these studies, most of the subjects consumed milk or consumed an instant coffee mix containing sugar and powdered creamer. These results indicate that instant coffee drinkers have increased risks of these metabolic conditions, and that the consumption of the instant coffee mixture may have a noxious effect on metabolic syndrome. Therefore, the increased risk of metabolic syndrome may be attributed in part to the excessive intake of sugar and powdered creamer. Furthermore, the association between coffee and metabolic syndrome was evaluated by several large-scale prospective studies, despite results being mostly contrasting (9514 in the United States, [61]; 17,014 in Norway, [62]; and 368 in the Netherlands, [46]).

3.2. Coffee Intake and Obesity

The rate of obesity has increased on a global scale in both adults and children, is related to several comorbidities, such as hypertension and T2DM, and is strongly related to the onset of CVD [64]. In the United States, prospective studies examined the interaction of habitual coffee consumption with the genetic predisposition to obesity in relation to BMI in 5116 men and 9841 women [65]. Higher coffee consumption appeared to reduce the genetic association between obesity and BMI—individuals with greater genetic predisposition to obesity were observed to have a lower BMI, related to higher coffee consumption. Furthermore, a randomized clinical trial was performed with obese women aged 20–45 years in which an intervention group received 400 mg of coffee [66]. The body weight, body mass and fat mass indices, and waist-to-hip circumference ratio of the intervention group decreased, compared to the control group. Furthermore, the intervention group had decreased cholesterol and low-density lipoprotein (LDL) levels, compared to the control group. In contrast, the serum adiponectins increased in the intervention group. These results indicate that consumption of coffee may reduce obesity.

3.3. Coffee Intake and Type 2 Diabetes

A meta-analysis of prospective studies (10 articles involving 491,485 participants, including 29,165 with T2DM) was performed to evaluate the relationship between coffee and caffeine consumption and T2DM incidence [67]. The relative risk of T2DM decreased with coffee intake. In addition, a relationship between T2DM incidence and coffee intake was found among non-smokers with BMI <25 kg/m^2. In particular, the effect on T2DM risk was higher in women. Furthermore, a large-scale case-cohort study demonstrated evidence for an interaction of incretin-associated TCF7L2 genetic variants and an incretin-specific genetic risk score with coffee consumption in relation to T2DM risk [68]. The cohort study included 11,035 participants, among them, 8086 incident T2DM cases. However, none of these relationships were statistically significant.

In cohort study of 2332 Chinese subjects, coffee intake was indicated to be inversely related to T2DM [69]. Habitual coffee consumption was associated with a 38–46% reduced risk of T2DM, compared to than non-drinkers. In a prospective cohort study in 88,259 US women (younger and middle-aged, incident cases 1263) were examined for coffee and caffeine intake and risk of T2DM [70]. The relative risk of T2DM decreased depending on coffee intake. Coffee consumption may decrease the risk of T2DM in younger and middle-aged women. Furthermore, reports indicated that the consumption in adults of up to 3 cups a day of coffee decreased the risk of T2DM and of metabolic syndrome [71]. Recently, in a Mendelian randomization study, habitual coffee consumption was inversely associated with T2DM, along with depression and Alzheimer's disease onset [72]. These results indicate that coffee consumption may contribute to decreases in T2DM.

3.4. Coffee Intake and Non-Alcoholic Steatohepatitis

The relationship between non-alcoholic steatohepatitis with coffee intake was explored by several epidemiologic studies. One prospective cohort study indicated an inverse association between coffee consumption and liver cirrhosis [73,74]. This study involved a cohort of 63,275 Chinese subjects (middle-aged and older) in the Singapore Chinese Health Study [75]. Compared to non-daily coffee drinkers, those who drank two or more cups per day had a 66% reduction in mortality risk. Coffee intake was related to a decrease in the risk of mortality, except coffee intake was not associated with hepatitis B-related cirrhosis mortality. Furthermore, a cross-sectional study (n = 347) showed an inverse association between coffee consumption and liver fibrosis [76]. In the study, high coffee consumption was related to a lower occurrence of clinically significant fibrosis. This result suggests that coffee consumption may exert beneficial effects on fibrosis progression [77].

However, neither the occurrence of fatty liver, nor the prevalence of fatty liver, as assessed by ultrasonography (SteatoTest) and the hepatorenal index, were related to coffee consumption. In a cross-sectional study, the effects of coffee consumption were investigated in patients (n = 1018) with non-alcoholic fatty liver disease (n = 155), hepatitis C virus (n = 378), and hepatitis B virus (n = 485). Drinking two or more cups of coffee per day was related to improvements in pathologic conditions [77]. In an epidemiological study on the association of coffee intake with chronic liver disease, 286 patients in a liver outpatient department in a hospital in Scotland completed a questionnaire regarding coffee consumption and lifestyle factors. The results indicated that coffee intake may be related to a reduced prevalence of cirrhosis in patients with chronic liver disease [78].

3.5. Coffee Intake and Atherosclerosis

A cohort study in Tokushima Prefecture, Japan, investigated the relationship between coffee intake and arterial stiffness [79]. A report indicated that the intake of coffee was inversely related to arterial stiffness in 540 Japanese men [79]. Coffee intake was inversely related to arterial stiffness, independent of atherosclerotic risk factors. This result was related partially to a decrease in circulating triglycerides. In addition, other studies suggest that the addition of milk may affect coffee's preventive action on arteriosclerosis [46,51]. On the other hand, in a cohort study, no association was observed between coffee or caffeine intake and coronary and carotid atherosclerosis. The Coronary Artery Risk Development in Young Adults (CARDIA) study examined the relationship between coffee intake and atherosclerosis in 5115 young adults [80]. No relationship was observed between atherosclerosis and the intake of average coffee, decaffeinated coffee, or caffeine intake. Furthermore, in 6508 ethnically diverse participants, coffee intake (>1 cup per day) was not associated with coronary artery calcification or cardiovascular events [81]. In a cross-sectional study, 1929 participants without known coronary heart disease, coffee intake and calcified atherosclerotic plaques in the coronary arteries were examined [82]. The results did not support a relationship between coffee intake and coronary-artery calcification in men and women. On the other hand, caffeine consumption was marginally inversely related to coronary artery calcification. As for the intake of more than 1 cup of coffee per day, caffeine may be related to cardiovascular events. In addition to epidemiological studies, further interventional studies may be needed to confirm the causal association.

3.6. Coffee Intake and Hypertension

In a meta-analysis of seven cohort studies, including 205,349 individuals and 44,120 cases of hypertension, an increase in 1 cup/day of coffee consumption was associated with a 1% decreased risk of hypertension [83]. Results from individual cohort studies suggest that the risk of hypertension depends on the coffee intake level. In a prospective cohort study of 24,710 Finnish subjects with no history of antihypertensive drug treatment, coronary heart disease, or stroke at baseline, the association between coffee intake and the incidence of antihypertensive drug treatment was investigated. The multivariate-adjusted hazard ratios for the amount of coffee consumed daily were marginally significant for baseline systolic blood pressure [84]. Low-to-moderate coffee intake appeared to increase the risk of antihypertensive drug treatment. In a cross-sectional population-based study including 8821 adults (51.4% female) in Poland, coffee consumption was negatively related to hypertension [85]. More details on the relationship of coffee intake with hypertension will have to be determined in future studies.

4. Coffee Composition and Features

Commercially important coffee comes from the species *Coffea arabica* and *Coffea robusta*. Coffee from the arabica species has good flavor and constitutes approximately 80% of the coffee consumed in the world [14]. The composition of coffee changes with the coffee bean species and roast process conditions [41]. Specifically, *Coffea arabica* differs from *Coffea robusta*, and the roast condition differs with time or temperature. Chemical components of green beans in *Coffea arabica* and *Coffea canephora* are shown in Table 2 [16]. Preparations include boiled unfiltered coffee, filtered coffee, and decaffeinated coffee. The compositions differ across the species, degree of roasts and preparation of the coffee. A large number of different compositions are present in coffee, but caffeine, diterpene, kahweol and chlorogenic acid and other phenols are the basic components (Figure 2) [86]. Components with which metabolic syndrome prevention is expected are caffeine, diterpenes, kahweol and polyphenols. In particular, chlorogenic acid has many beneficial effects [69].

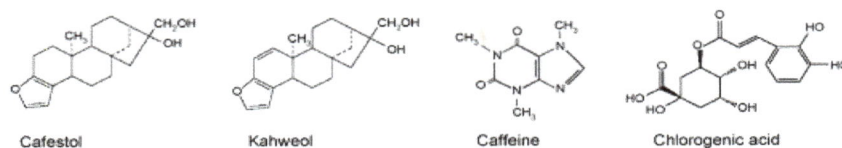

| Cafestol | Kahweol | Caffeine | Chlorogenic acid |

Figure 2. Chemical structures of proposed bioactive compounds in coffee (Adapted from reference Bonita et al., 2007) [86].

Chlorogenic acids are a family of molecules formed between quinic and cinnamic acids and metabolized to several molecules in the body (Figure 3) [87]. The most common chlorogenic acid is 5-*O*-caffeoylquinic acid (CQA), but it is called simply chlorogenic acid. It was reported that the chlorogenic acid content of one 200-mL cup of coffee ranges from 70 to 350 mg [88]. Trigonelline is a niacin-related compound and is another component of coffee. Trigonelline was observed to alter induction of estrogen-dependent growth through the estrogenic action in human breast cancer cells [89]. This activity indicates that trigonelline has an estrogenic effect.

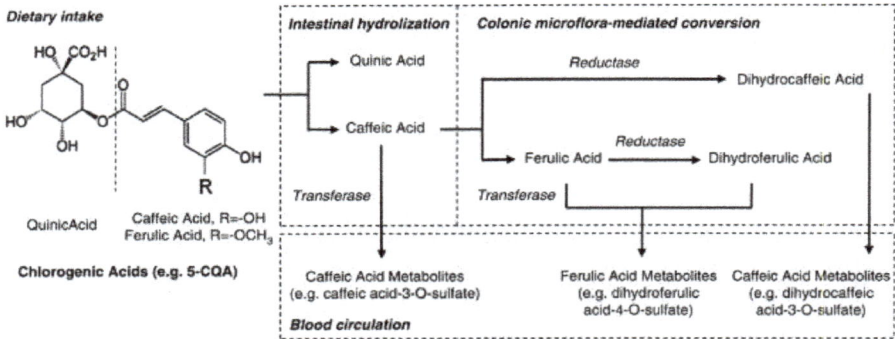

Figure 3. Main metabolic pathway of chlorogenic acids. Dietary chlorogenic acids is hydrolyzed into quinic acid, caffeic and ferulic acid, and further metabolized in small intestine and colon before entering into blood stream (Adapted from reherence Zhao et al., 2012) [87].

5. Chlorogenic Acid and Metabolic Syndrome Associated-Endothelial Dysfunction

The increase of markers of inflammation enhances global cardiovascular risk. The inflammatory response is enhanced early in adipose expansion and chronic obesity during metabolic syndrome onset [27]. Increasing evidence suggests that chronic subclinical inflammation is part of the metabolic syndrome. For example, increased serum concentrations of tumor necrosis factor-α (TNF-α) and IL-6 might attenuate insulin action by inhibiting insulin signaling [88]. A number of features of metabolic syndrome are related to low-grade inflammatory pathological conditions (Table 4) [89]. Increased plasma C-reactive protein has been observed in insulin-resistant and obese subjects and is a surrogate marker for both coronary heart disease and diabetes [90]. The adipose tissue secretes several adipocytokines and induces inflammation and oxidative stress in vascular tissue. In particular, adiponectin and resistin regulate monocyte adherence to vascular endothelial cells [91]. Subsequently, enhancing monocytic migration to the subendothelial space is one of the key events in the development of atherosclerosis. Specifically, the metabolic syndrome induces vascular endothelial cell disorder and induces arteriosclerosis, and arteriosclerosis, in turn increasing the risk for conditions such as myocardial infarction or cerebral infarction. Chlorogenic acid appears to protect normal endothelial activity [92]. On the other hand, chlorogenic acid inhibited interleukin 1 beta (IL-1β)-induced gene expression of vascular cell adhesion molecule-1, intercellular cell adhesion molecule-1 and endothelial cell selectin in human umbilical vein endothelial cells [59]. Also, chlorogenic acid blocked IL-1β-induced nuclear translocation of nuclear factor-kappaB subunits p50 and p65 and suppressed the adhesion of human lymphoma cell line, U937 cells. In addition, a recent study demonstrated the protective effects of chlorogenic acid on human umbilical vein endothelial cells [93]. Namely, chlorogenic acid induced a cell growth higher than those stimulated with inflammatory TNF-α only. Furthermore, chlorogenic acid reduced reactive oxygen species and xanthine oxidase-1 levels, and enhanced superoxide dismutase and heme oxygenase-1 levels in endothelial cells. A study described the effects of chlorogenic acid on endothelial function with oxidant-enhanced damage in isolated aortic rings from mice. Chlorogenic acid reduced HOCl-induced oxidative damage in endothelial cells. The mechanism of the beneficial effect of chlorogenic acid was associated with the production of NO and induction of heme oxygenase-1 [94]. Consumption of coffee with a high content of chlorogenic acids repaired endothelial dysfunction by decreasing oxidative stress [95]. A previous study indicated that oxidative stress has been demonstrated to play a important role in the development of endothelial dysfunction [96]. Chlorogenic acid appears to protect against endothelial dysfunction due to its antioxidant activity. Also, chlorogenic acid inhibited TNFα-induced intercellular adhesion molecule-1, vascular cell adhesion molecule-1, and monocyte chemotactic protein-1 expression in human endothelial cells. In addition, it has been reported that chlorogenic acid blocks α-glucosidase activities, and may thus prevent T2DM [97]. It is suggested that chlorogenic acid prevents vascular endothelial disorder through this inhibitory activity. These results suggest that chlorogenic acid prevents induced atherosclerosis that would otherwise stimulate inflammation. Lysophosphatidylcholine (LPC) is a major phospholipid component of oxidized LDL and is associated with atherogenic induction [98]. A recent report indicated the effects of chlorogenic acid on intracellular calcium control in LPC-treated endothelial cells [99]. Namely, the gene expression of the transient receptor potential canonical (TRPC) channel 1 was enhanced significantly by LPC treatment and inhibited by chlorogenic acid. Thus, chlorogenic acid may protect endothelial cells against LPC injury and inhibit atherosclerosis.

Table 4. The inflammatory component of the metabolic syndrome [a].

Vascular dysfunction	Endothelial dysfunction Microalbuminuria
Proinflammatory state	Elevated high sensitivity C-reactive protein and serum amyloid A Elevated inflammatory cytokines (TNF-α, IL-6) Decreased adiponectin levels
Prothrombotic state	
Insulin resistance	
Visceral adiposity	

Abbreviations: IL-6, interleukin 6; TNF-α, tumor necrosis factor-α. [a] Adapted from reherence Paoletti et al., 2006 [90].

Furthermore, the effects of the representative chlorogenic acid CQA on vascular function and blood pressure were evaluated in normotensive Wistar–Kyoto rats and spontaneously hypertensive rats [32]. CQA increased NO production and decreased ROS production. In addition, CQA reduced hypertension, and prevented the impairment of endothelial function in spontaneously hypertensive rats. NADPH oxidase-derived superoxide had a important role in the control of vascular tone in health and disease [100]. In endothelial cells, heme oxygenase-1 is induced in response to oxidative stress, which may play a role in vascular prevention. Furthermore, a report demonstrated that the pretreatment of cultured human aortic endothelial cells with 10 μM chlorogenic acid prevented endothelial cell viability following exposure to hypochlorous acid [94]. Chlorogenic acid enhanced endothelial nitric oxide synthase dimerization and induced heme oxygenase-1 protein expression in human aortic endothelial cells. These results are consistent with the endothelial protective effects of coffee consumption. These reports suggest that the coffee component, chlorogenic acid, can protect cultured endothelial cells against inflammation-enhanced endothelial dysfunction and play an important role in the prevention of atherosclerotic complications.

6. Conclusions

Metabolic syndrome is a strong risk factor for atherosclerosis-associated CVD and T2DM. Obesity due to excess energy intake strongly enhances the metabolic syndrome—concomitant obesity is the major driver of the syndrome. However, coffee polyphenols can reverse the metabolic risk factors of metabolic syndrome. Coffee polyphenols inhibit atherosclerosis-related CVD and T2DM, respectively. Coffee has many health-promoting properties, and chlorogenic acid appears to protect against metabolic syndrome through its antioxidant activity. The antioxidative effects of coffee components may be a basic feature of prevention.

Acknowledgments: The authors received no specific funding for this work.

Conflicts of Interest: The author declares no conflict of interest.

Abbreviations

BMI	body mass index
CVD	cardiovascular disease
CQA	5-*O*-caffeoylquinic acid
EPR	electron paramagnetic resonance
IL-1β	interleukin 1 beta
LDL	low-density lipoprotein
LPC	lysophosphatidylcholine
NADPH	nicotinamide adenine dinucleotide phosphate
NAFLD	non-alcoholic fatty liver disease
NO	nitric oxide
ROS	reactive oxygen species
T2DM	type 2 diabetes
TNF-α	tumor necrosis factor-α

References

1. Fuster, J.J.; Ouchi, N.; Gokce, N.; Walsh, K. Obesity-induced changes in adipose tissue microenvironment and their impact on cardiovascular disease. *Circ. Res.* **2016**, *118*, 1786–1807. [CrossRef] [PubMed]
2. Eckel, R.H.; Grundy, S.M.; Zimmet, P.Z. The metabolic syndrome. *Lancet* **2005**, *365*, 1415–1428. [CrossRef]
3. Vykoukal, D.; Davies, M.G. Biology of metabolic syndrome in a vascular patient. *Vascular* **2012**, *20*, 156–165. [CrossRef] [PubMed]
4. Maruyama, K.; Uchiyama, S.; Iwata, M. Metabolic syndrome and its components as risk factors for first-ever acute ischemic noncardioembolic stroke. *J. Stroke Cerebrovasc. Dis.* **2009**, *18*, 173–177. [CrossRef] [PubMed]
5. Mottillo, S.; Filion, K.B.; Genest, J.; Joseph, L.; Pilote, L.; Poirier, P.; Rinfret, S.; Schiffrin, E.L.; Eisenberg, M.J. The metabolic syndrome and cardiovascular risk a systematic review and meta-analysis. *J. Am. Coll. Cardiol.* **2010**, *56*, 1113–1132. [CrossRef] [PubMed]
6. Low, W.Y.; Lee, Y.K.; Samy, A.L. Non-communicable diseases in the Asia-Pacific region: Prevalence, risk factors and community-based prevention. *Int. J. Occup. Med. Environ. Health* **2015**, *28*, 20–26. [PubMed]
7. Landberg, R.; Naidoo, N.; van Dam, RM. Diet and endothelial function: From individual components to dietary patterns. *Curr. Opin. Lipidol.* **2012**, *23*, 147–155. [CrossRef] [PubMed]
8. Davinelli, S.; Scapagnini, G. Polyphenols: A Promising nutritional approach to prevent or reduce the progression of prehypertension. *High Blood Press Cardiovasc. Prev.* **2016**, *23*, 197–202. [CrossRef] [PubMed]
9. Katz, D.L.; Doughty, K.; Ali, A. Cocoa and chocolate in human health and disease. *Antioxid. Redox Signal.* **2011**, *15*, 2779–27811. [CrossRef] [PubMed]
10. Garcia-Vilas, J.A.; Quesada, A.R.; Medina, M.A. Hydroxytyrosol targets extracellular matrix remodeling by endothelial cells and inhibits both ex vivo and in vivo angiogenesis. *Food Chem.* **2017**, *221*, 1741–1746. [CrossRef] [PubMed]
11. Grosso, G.; Godos, J.; Galvano, F.; Giovannucci, E.L. Coffee, Caffeine, and Health Outcomes: An Umbrella Review. *Annu. Rev. Nutr.* **2017**, *37*, 131–156. [CrossRef] [PubMed]
12. Marventano, S.; Salomone, F.; Godos, J.; Pluchinotta, F.; Del Rio, D.; Mistretta, A.; Grosso, G. Coffee and tea consumption in relation with non-alcoholic fatty liver and metabolic syndrome: A systematic review and meta-analysis of observational studies. *Clin. Nutr.* **2016**, *35*, 1269–1281. [CrossRef] [PubMed]
13. Shang, F.; Li, X.; Jiang, X. Coffee consumption and risk of the metabolic syndrome: A meta-analysis. *Diabetes Metab.* **2016**, *42*, 80–87. [CrossRef] [PubMed]
14. Di Lorenzo, A.; Curti, V.; Tenore, G.C.; Nabavi, S.M.; Daglia, M. Effects of tea and coffee consumption on cardiovascular diseases and relative risk factors: An update. *Curr. Pharm. Des.* **2017**, *23*, 2474–2487. [CrossRef] [PubMed]
15. Martini, D.; Del Bo', C.; Tassotti, M.; Riso, P.; Del Rio, D.; Brighenti, F.; Porrini, M. Coffee consumption and oxidative stress: A review of human intervention studies. *Molecules* **2016**, *21*, 979. [CrossRef] [PubMed]
16. Ding, M.; Bhupathiraju, S.N.; Chen, M.; van Dam, R.M.; Hu, F.B. Caffeinated and decaffeinated coffee consumption and risk of type 2 diabetes:a systematic review and a dose-response meta-analysis. *Diabetes Care* **2014**, *37*, 569–586. [CrossRef] [PubMed]

17. Santos, R.M.; Lima, D.R. Coffee consumption, obesity and type 2 diabetes: A mini-review. *Eur. J. Nutr.* **2016**, *55*, 1345–1358. [CrossRef] [PubMed]

18. Bloomer, R.J.; Trepanowski, J.F.; Farney, T.M. Influence of acute coffee consumption on postprandial oxidative stress. *Nutr. Metab. Insights* **2013**, *6*, 35–42. [CrossRef] [PubMed]

19. Bakuradze, T.; Boehm, N.; Janzowski, C.; Lang, R.; Hofmann, T.; Stockis, J.P.; Albert, F.W.; Stiebitz, H.; Bytof, G.; Lantz, I.; et al. Antioxidant-rich coffee reduces DNA damage, elevates glutathione status and contributes to weight control: Results from an intervention study. *Mol. Nutr. Food Res.* **2011**, *55*, 793–797. [CrossRef] [PubMed]

20. Kotyczka, C.; Boettler, U.; Lang, R.; Stiebitz, H.; Bytof, G.; Lantz, I.; Hofmann, T.; Marko, D.; Somoza, V. Dark roast coffee is more effective than light roast coffee in reducing body weight, and in restoring red blood cell vitamin E and glutathione concentrations in healthy volunteers. *Mol. Nutr. Food Res.* **2011**, *55*, 1582–1586. [CrossRef] [PubMed]

21. Bakker, R.; Steegers, E.A.; Obradov, A.; Raat, H.; Hofman, A.; Jaddoe, V.W. Maternal caffeine intake from coffee and tea, fetal growth, and the risks of adverse birth outcomes: the Generation R Study. *Am. J. Clin. Nutr.* **2010**, *91*, 1691–1698. [CrossRef] [PubMed]

22. Hotamisligil, G.S. Inflammation and metabolic disorders. *Nature* **2006**, *444*, 860–867. [CrossRef] [PubMed]

23. Saltiel, A.R.; Olefsky, J.M. Inflammatory mechanisms linking obesity and metabolic disease. *J. Clin. Investig.* **2017**, *127*, 1–4. [CrossRef] [PubMed]

24. Vanhoutte, P.M.; Shimokawa, H.; Tang, E.H.; Feletou, M. Endothelial dysfunction and vascular disease. *Acta Physiol.* **2009**, *196*, 193–222. [CrossRef] [PubMed]

25. Grassi, D.; Desideri, G.; Ferri, C. Cardiovascular risk and endothelial dysfunction: The preferential route for atherosclerosis. *Curr. Pharm. Biotechnol.* **2011**, *12*, 1343–1353. [CrossRef] [PubMed]

26. Micek, A.; Grosso, G.; Polak, M.; Kozakiewicz, K.; Tykarski, A.; Puch Walczak, A.; Drygas, W.; Kwasniewska, M.; Pajak, A. Association between tea and coffee consumption and prevalence of metabolic syndrome in Poland—Results from the WOBASZ II study (2013–2014). *Int. J. Food Sci. Nutr.* **2017**, *9*, 1–11. [CrossRef] [PubMed]

27. Fukushima, Y.; Ohie, T.; Yonekawa, Y.; Yonemoto, K.; Aizawa, H.; Mori, Y.; Watanabe, M.; Takeuchi, M.; Hasegawa, M.; Taguchi, C.; et al. Coffee and green tea as a large source of antioxidant polyphenols in the Japanese population. *J. Agric. Food Chem.* **2009**, *57*, 1253–1259. [CrossRef] [PubMed]

28. Yesil, A.; Yilmaz, Y. Review article: Coffee consumption, the metabolic syndrome and non-alcoholic fatty liver disease. *Aliment. Pharmacol. Ther.* **2013**, *38*, 1038–1044. [CrossRef] [PubMed]

29. Stalmach, A.; Mullen, W.; Nagai, C.; Crozier, A. On-line HPLC analysis of the antioxidant activity of phenolic compounds in brewed paper-filtered coffee. *Braz. J. Plant Physiol.* **2006**, *18*, 253–262. [CrossRef]

30. Serafini, M.; Testa, M.F. Redox ingredients for oxidative stress prevention: The unexplored potentiality of coffee. *Clin. Dermatol.* **2009**, *27*, 225–229. [CrossRef] [PubMed]

31. Andriantsitohaina, R.; Auger, C.; Chataigneau, T.; Etienne-Selloum, N.; Li, H.; Martinez, M.C.; Schini-Kerth, V.B.; Laher, I. Molecular mechanisms of the cardiovascular protective effects of polyphenols. *Br. J. Nutr.* **2012**, *108*, 1532–1549. [CrossRef] [PubMed]

32. Suzuki, A.; Yamamoto, N.; Jokura, H.; Yamamoto, M.; Fujii, A.; Tokimitsu, I.; Saito, I. Chlorogenic acid attenuates hypertension and improves endothelial function in spontaneously hypertensive rats. *J. Hypertens.* **2006**, *24*, 1065–1073. [CrossRef] [PubMed]

33. Stalmach, A.; Steiling, H.; Williamson, G.; Crozier, A. Bioavailability of chlorogenic acids following acute ingestion of coffee by humans with an ileostomy. *Arch. Biochem. Biophys.* **2010**, *501*, 98–105. [CrossRef] [PubMed]

34. Natella, F.; Nardini, M.; Giannetti, I.; Dattilo, C.; Scaccini, C. Coffee drinking influences plasma antioxidant capacity in humans. *J. Agric. Food Chem.* **2002**, *50*, 6211–6216. [CrossRef] [PubMed]

35. Andreasen, M.F.; Landbo, A.K.; Christensen, L.P.; Hansen, A.; Meyer, A.S. Antioxidant effects of phenolic rye (*Secale cereale* L.) extracts, monomeric hydroxycinnamates, and ferulic acid dehydrodimers on human low-density lipoproteins. *J. Agric. Food Chem.* **2001**, *49*, 4090–4096. [CrossRef] [PubMed]

36. Godos, J.; Pluchinotta, F.R.; Marventano, S.; Buscemi, S.; Volti, G.L.; Galvano, F.; Grosso, G. Coffee components and cardiovascular risk: Beneficial and detrimental effects. *Int. J. Food Sci. Nutr.* **2014**, *65*, 925–936. [CrossRef] [PubMed]

37. Leon-Carmona, J.R.; Galano, A. Is caffeine a good scavenger of oxygenated free radicals? *J. Phys. Chem. B* **2011**, *115*, 4538–4546. [CrossRef] [PubMed]

38. Kumar, S.S.; Devasagayam, T.P.; Jayashree, B.; Kesavan, P.C. Mechanism of protection against radiation-induced DNA damage in plasmid pBR322 by caffeine. *Int. J. Radiat. Biol.* **2001**, *77*, 617–623. [CrossRef] [PubMed]

39. Brezova, V.; Slebodova, A.; Stasko, A. Coffee as a source of antioxidants: An EPR study. *Food Chem.* **2009**, *114*, 859–868. [CrossRef]

40. Lee, C. Caffein may antioxidant ability and oxygen radical absorbing capacity and inhibition of LDL peroxidation. *Clin. Chim. Acta* **2000**, *295*, 141–154. [CrossRef]

41. Yen, G.C.; Chung, D.Y. Antioxidant effects of extracts from Cassia tora L. prepared under different degrees of roasting on the oxidative damage to biomolecules. *J. Agric. Food Chem.* **1999**, *47*, 1326–1332. [CrossRef] [PubMed]

42. Edeas, M.; Attaf, D.; Mailfert, A.S.; Nasu, M.; Joubet, R. Maillard reaction, mitochondria and oxidative stress: Potential role of antioxidants. *Pathol. Biol.* **2010**, *58*, 220–225. [CrossRef] [PubMed]

43. Bekedam, E.K.; Loots, M.J.; Schols, H.A.; Van Boekel, M.A.; Smit, G. Roasting effects on formation mechanisms of coffee brew melanoidins. *J. Agric. Food Chem.* **2008**, *56*, 7138–7145. [CrossRef] [PubMed]

44. Opitz, S.E.; Smrke, S.; Goodman, B.A.; Keller, M.; Schenker, S.; Yeretzian, C. Antioxidant generation during coffee roasting: A comparison and Interpretation from three complementary assays. *Foods* **2014**, *3*, 586–604. [CrossRef] [PubMed]

45. Driessen, M.T.; Koppes, L.L.; Veldhuis, L.; Samoocha, D.; Twisk, J.W. Coffee consumption is not related to the metabolic syndrome at the age of 36 years: The Amsterdam Growth and Health Longitudinal Study. *Eur. J. Clin. Nutr.* **2009**, *63*, 536–542. [CrossRef] [PubMed]

46. Caprioli, G.; Cortese, M.; Sagratini, G.; Vittori, S. The influence of different types of preparation (espresso and brew) on coffee aroma and main bioactive constituents. *Int. J. Food Sci. Nutr.* **2015**, *66*, 505–513. [CrossRef] [PubMed]

47. Dos Santos, P.R.; Ferrari, G.S.; Ferrari, C.K. Diet, sleep and metabolic syndrome among a legal Amazon population. *Braz. Clin. Nutr. Res.* **2015**, *4*, 41–45. [CrossRef] [PubMed]

48. Grosso, G.; Micek, A.; Godos, J.; Sciacca, S.; Pajak, A.; Martinez-Gonzalez, M.A.; Giovannucci, E.L.; Galvano, F. Coffee consumption and risk of all-cause, cardiovascular, and cancer mortality in smokers and non-smokers: A dose-response meta-analysis. *Eur. J. Epidemiol.* **2016**, *31*, 1191–1205. [CrossRef] [PubMed]

49. Cornelis, M.C.; El-Sohemy, A.; Kabagambe, E.K.; Campos, H. Coffee, CYP1A2 genotype, and risk of myocardial infarction. *JAMA* **2006**, *295*, 1135–1141. [CrossRef] [PubMed]

50. Palatini, P.; Ceolotto, G.; Ragazzo, F.; Dorigatti, F.; Saladini, F.; Papparella, I.; Mos, L.; Zanata, G.; Santonastaso, M. CYP1A2 genotype modifies the association between coffee intake and the risk of hypertension. *J. Hypertens.* **2009**, *27*, 1594–1601. [CrossRef] [PubMed]

51. Nordestgaard, A.T.; Thomsen, M.; Nordestgaard, B.G. Coffee intake and risk of obesity, metabolic syndrome and type 2 diabetes: A Mendelian randomization study. *Int. J. Epidemiol.* **2015**, *44*, 551–565. [CrossRef] [PubMed]

52. Kim, K.; Kim, K.; Park, S.M. Association between the Prevalence of Metabolic Syndrome and the Level of Coffee Consumption among Korean Women. *PLoS ONE* **2016**, *11*, e0167007. [CrossRef] [PubMed]

53. Suliga, E.; Kozie, D.; Ciesla, E.; Rebak, D.; Gluszek, S. Coffee consumption and the occurrence and intensity of metabolic syndrome: A cross-sectional study. *Int. J. Food Sci. Nutr.* **2017**, *68*, 507–513. [CrossRef] [PubMed]

54. Grosso, G.; Marventano, S.; Galvano, F.; Pajak, A.; Mistretta, A. Factors associated with metabolic syndrome in a Mediterranean population: Role of caffeinated beverages. *J. Epidemiol.* **2014**, *24*, 327–333. [CrossRef] [PubMed]

55. Grosso, G.; Stepaniak, U.; Micek, A.; Topor-Madry, R.; Pikhart, H.; Szafraniec, K.; Pajak, A. Association of daily coffee and tea consumption and metabolic syndrome: Results from the Polish arm of the HAPIEE study. *Eur. J. Nutr.* **2015**, *54*, 1129–1137. [CrossRef] [PubMed]

56. Kim, H.J.; Cho, S.; Jacobs, D.R., Jr.; Park, K. Instant coffee consumption may be associated with higher risk of metabolic syndrome in Korean adults. *Diabetes Res. Clin. Pract.* **2014**, *106*, 145–153. [CrossRef] [PubMed]

57. Yen, A.M.; Chiu, Y.H.; Chen, L.S.; Wu, H.M.; Huang, C.C.; Boucher, B.J.; Chen, T.H. A population-based study of the association between betel-quid chewing and the metabolic syndrome in men. *Am. J. Clin. Nutr.* **2006**, *83*, 1153–1160. [PubMed]

58. Takami, H.; Nakamoto, M.; Uemura, H.; Katsuura, S.; Yamaguchi, M.; Hiyoshi, M.; Sawachika, F.; Juta, T.; Arisawa, K. Inverse correlation between coffee consumption and prevalence of metabolic syndrome: baseline survey of the Japan Multi-Institutional Collaborative Cohort (J-MICC) Study in Tokushima, Japan. *J. Epidemiol.* **2013**, *23*, 12–20. [CrossRef] [PubMed]

59. Chang, C.S.; Chang, Y.F.; Liu, P.Y.; Chen, C.Y.; Tsai, Y.S.; Wu, C.H. Smoking, habitual tea drinking and metabolic syndrome in elderly men living in rural community:the Tianliao old people (TOP) study 02. *PLoS ONE* **2012**, *7*, e38874.

60. Matsuura, H.; Mure, K.; Nishio, N.; Kitano, N.; Nagai, N.; Takeshita, T. Relationship between coffee consumption and prevalence of metabolic syndrome among Japanese civil servants. *J. Epidemiol.* **2012**, *22*, 160–166. [CrossRef] [PubMed]

61. Lutsey, P.L.; Steffen, L.M.; Stevens, J. Dietary intake and the development of the metabolic syndrome: The atherosclerosis risk in communities study. *Circulation* **2008**, *117*, 754–761. [CrossRef] [PubMed]

62. Wilsgaard, T.; Jacobsen, B.K. Lifestyle factors and incident metabolic syndrome. The Tromso Study 1979–2001. *Diabetes Res. Clin. Pract.* **2007**, *78*, 217–224. [CrossRef] [PubMed]

63. Hino, A.; Adachi, H.; Enomoto, M.; Furuki, K.; Shigetoh, Y.; Ohtsuka, M.; Kumagae, S.; Hirai, Y.; Jalaldin, A.; Satoh, A.; et al. Habitual coffee but not green tea consumption is inversely associated with metabolic syndrome: an epidemiological study in a general Japanese population. *Diabetes Res. Clin. Pract.* **2007**, *76*, 383–389. [CrossRef] [PubMed]

64. Lavie, C.J.; Milani, R.V.; Ventura, H.O. Obesity and cardiovascular disease: Risk factor, paradox, and impact of weight loss. *J. Am. Coll. Cardiol.* **2009**, *53*, 1925–1932. [CrossRef] [PubMed]

65. Wang, T.; Huang, T.; Kang, J.H.; Zheng, Y.; Jensen, M.K.; Wiggs, J.L.; Pasquale, L.R.; Fuchs, C.S.; Campos, H.; Rimm, E.B.; et al. Habitual coffee consumption and genetic predisposition to obesity: Gene-diet interaction analyses in three US prospective studies. *BMC Med.* **2017**, *15*, 97. [CrossRef] [PubMed]

66. Haidari, F.; Samadi, M.; Mohammadshahi, M.; Jalali, M.T.; Engali, K.A. Energy restriction combined with green coffee bean extract affects serum adipocytokines and the body composition in obese women. *Asia Pac. J. Clin. Nutr.* **2017**, *26*, 1048–1054. [PubMed]

67. Jiang, X.; Zhang, D.; Jiang, W. Coffee and caffeine intake and incidence of type 2 diabetes mellitus: A meta-analysis of prospective studies. *Eur. J. Nutr.* **2014**, *53*, 25–38. [CrossRef] [PubMed]

68. InterAct Consortium. Investigation of gene-diet interactions in the incretin system and risk of type 2 diabetes: The EPIC-InterAct study. *Diabetologia* **2016**, *59*, 2613–2621.

69. Lin, W.Y.; Pi-Sunyer, F.X.; Chen, C.C.; Davidson, L.E.; Liu, C.S.; Li, T.C.; Wu, M.F.; Li, C.I.; Chen, W.; Lin, C.C. Coffee consumption is inversely associated with type 2 diabetes in Chinese. *Eur. J. Clin. Investig.* **2011**, *41*, 659–666. [CrossRef] [PubMed]

70. Baspinar, B.; Eskici, G.; Ozcelik, A.O. How coffee affects metabolic syndrome and its components. *Food Funct.* **2017**, *8*, 2089–2101. [CrossRef] [PubMed]

71. Van Dam, R.M.; Willett, W.C.; Manson, J.E.; Hu, F.B. Coffee, caffeine, and risk of type 2 diabetes: A prospective cohort study in younger and middle-aged U.S. women. *Diabetes Care* **2006**, *29*, 398–403. [CrossRef] [PubMed]

72. Kwok, M.K.; Leung, G.M.; Schooling, C.M. Habitual coffee consumption and risk of type 2 diabetes, ischemic heart disease, depression and Alzheimer's disease: A Mendelian randomization study. *Sci. Rep.* **2016**, *6*. [CrossRef] [PubMed]

73. Ono, M.; Okamoto, N.; Saibara, T. The latest idea in NAFLD/NASH pathogenesis. *Clin. J. Gastroenterol.* **2010**, *3*, 263–270. [CrossRef] [PubMed]

74. Yki-Jarvinen, H. Non-alcoholic fatty liver disease as a cause and a consequence of metabolic syndrome. *Lancet Diabetes Endocrinol.* **2014**, *2*, 901–910. [CrossRef]

75. Goh, G.B.; Chow, W.C.; Wang, R.; Yuan, J.M.; Koh, W.P. Coffee, alcohol and other beverages in relation to cirrhosis mortality: The Singapore Chinese Health Study. *Hepatology* **2014**, *60*, 661–669. [CrossRef] [PubMed]

76. Zelber-Sagi, S.; Salomone, F.; Webb, M.; Lotan, R.; Yeshua, H.; Halpern, Z.; Santo, E.; Oren, R.; Shibolet, O. Coffee consumption and nonalcoholic fatty liver onset: A prospective study in the general population. *Transl. Res.* **2015**, *165*, 428–436. [CrossRef] [PubMed]

77. Hodge, A.; Lim, S.; Goh, E.; Wong, O.; Marsh, P.; Knight, V.; Sievert, W.; de Courten, B. Coffee Intake Is Associated with a Lower Liver Stiffness in Patients with Non-Alcoholic Fatty Liver Disease, Hepatitis C, and Hepatitis B. *Nutrients* **2017**, *9*, 56. [CrossRef] [PubMed]

78. Walton, H.B.; Masterton, G.S.; Hayes, P.C. An epidemiological study of the association of coffee with chronic liver disease. *Scott. Med. J.* **2013**, *58*, 217–222. [CrossRef] [PubMed]

79. Uemura, H.; Katsuura-Kamano, S.; Yamaguchi, M.; Nakamoto, M.; Hiyoshi, M.; Arisawa, K. Consumption of coffee, not green tea, is inversely associated with arterial stiffness in Japanese men. *Eur. J. Clin. Nutr.* **2013**, *67*, 1109–1114. [CrossRef] [PubMed]

80. Reis, J.P.; Loria, C.M.; Steffen, L.M.; Zhou, X.; van Horn, L.; Siscovick, D.S.; Jacobs, D.R., Jr.; Carr, J.J. Coffee, decaffeinated coffee, caffeine, and tea consumption in young adulthood and atherosclerosis later in life: The CARDIA study. *Arterioscler. Thromb. Vasc. Biol.* **2010**, *30*, 2059–2066. [CrossRef] [PubMed]

81. Miller, P.E.; Zhao, D.; Frazier-Wood, A.C.; Michos, E.D.; Averill, M.; Sandfort, V.; Burke, G.L.; Polak, J.F.; Lima, J.A.; Post, W.S.; et al. Associations of coffee, tea, and caffeine intake with coronary artery calcification and cardiovascular events. *Am. J. Med.* **2017**, *130*, 188–197. [CrossRef] [PubMed]

82. Patel, Y.R.; Gadiraju, T.V.; Ellison, R.C.; Hunt, S.C.; Carr, J.J.; Heiss, G.; Arnett, D.K.; Pankow, J.S.; Gaziano, J.M.; Djousse, L. Coffee consumption and calcified atherosclerotic plaques in the coronary arteries: The NHLBI Family Heart Study. *Clin. Nutr. ESPEN* **2017**, *17*, 18–21. [CrossRef] [PubMed]

83. Grosso, G.; Micek, A.; Godos, J.; Pajak, A.; Sciacca, S.; Bes-Rastrollo, M.; Galvano, F.; Martinez-Gonzalez, M.A. Long-Term Coffee Consumption Is Associated with Decreased Incidence of New-Onset Hypertension: A Dose—Response Meta-Analysis. *Nutrients* **2017**, *9*, 890. [CrossRef] [PubMed]

84. Hu, G.; Jousilahti, P.; Nissinen, A.; Bidel, S.; Antikainen, R.; Tuomilehto, J. Coffee consumption and the incidence of antihypertensive drug treatment in Finnish men and women. *Am. J. Clin. Nutr.* **2007**, *86*, 457–464. [PubMed]

85. Grosso, G.; Stepaniak, U.; Polak, M.; Micek, A.; Topor-Madry, R.; Stefler, D.; Szafraniec, K.; Pajak, A. Coffee consumption and risk of hypertension in the Polish arm of the HAPIEE cohort study. *Eur. J. Clin. Nutr.* **2016**, *70*, 109–115. [CrossRef] [PubMed]

86. Bonita, J.S.; Mandarano, M.; Shuta, D.; Vinson, J. Coffee and cardiovascular disease: In vitro, cellular, animal, and human studies. *Pharmacol. Res.* **2007**, *55*, 187–198. [CrossRef] [PubMed]

87. Zhao, Y.; Wang, J.; Ballevre, O.; Luo, H.; Zhang, W. Antihypertensive effects and mechanisms of chlorogenic acids. *Hypertens. Res.* **2012**, *35*, 370–374. [CrossRef] [PubMed]

88. Dandona, P.; Aljada, A.; Bandyopadhyay, A. Inflammation: The link between insulin resistance, obesity and diabetes. *Trends Immunol.* **2004**, *25*, 4–7. [CrossRef] [PubMed]

89. Paoletti, R.; Bolego, C.; Poli, A.; Cignarella, A. Metabolic syndrome, inflammation and atherosclerosis. *Vasc. Health Risk Manag.* **2006**, *2*, 145–152. [CrossRef] [PubMed]

90. Sjoholm, A.; Nystrom, T. Endothelial inflammation in insulin resistance. *Lancet* **2005**, *365*, 610–612. [CrossRef]

91. Kougias, P.; Chai, H.; Lin, P.H.; Yao, Q.; Lumsden, A.B.; Chen, C. Effects of adipocyte—Derived cytokines on endothelial functions: Implication of vascular disease. *J. Surg. Res.* **2005**, *126*, 121–129. [CrossRef] [PubMed]

92. Ochiai, R.; Sugiura, Y.; Otsuka, K.; Katsuragi, Y.; Hashiguchi, T. Coffee bean polyphenols ameliorate postprandial endothelial dysfunction in healthy male adults. *Int. J. Food Sci. Nutr.* **2015**, *66*, 350–354. [CrossRef] [PubMed]

93. Huang, W.Y.; Fu, L.; Li, C.Y.; Xu, L.P.; Zhang, L.X.; Zhang, W.M. Quercetin, Hyperin and chlorogenic acid improve endothelial function by antioxidant, antiinflammatory, and ACE inhibitory effects. *J. Food Sci.* **2017**, *82*, 1239–1246. [CrossRef] [PubMed]

94. Jiang, R.; Hodgson, J.M.; Mas, E.; Croft, K.D.; Ward, N.C. Chlorogenic acid improves ex vivo vessel function and protects endothelial cells against HOCl-induced oxidative damage, via increased production of nitric oxide and induction of Hmox-1. *J. Nutr. Biochem.* **2016**, *27*, 53–60. [CrossRef] [PubMed]

95. Kajikawa, M.; Maruhashi, T.; Hidaka, T.; Nakano, Y.; Kurisu, S.; Matsumoto, T.; Iwamoto, Y.; Kishimoto, S.; Matsui, S.; Aibara, Y.; et al. Coffee with a high content of chlorogenic acids and low content of hydroxyhydroquinone improves postprandial endothelial dysfunction in patients with borderline and stage 1 hypertension. *Eur. J. Nutr.* **2018**. [CrossRef] [PubMed]

96. Higashi, Y.; Noma, K.; Yoshizumi, M.; Kihara, Y. Endothelial function and oxidative stress in cardiovascular diseases. *Circ. J.* **2009**, *73*, 411–448. [CrossRef] [PubMed]

97. Oboh, G.; Agunloye, O.M.; Adefegha, S.A.; Akinyemi, A.J.; Ademiluyi, A.O. Caffeic and chlorogenic acids inhibit key enzymes linked to type 2 diabetes (in vitro): A comparative study. *J. Basic Clin. Physiol. Pharmacol.* **2015**, *26*, 165–170. [CrossRef] [PubMed]

98. Akerele, O.A.; Cheema, S.K. Fatty acyl composition of lysophosphatidylcholine is important in atherosclerosis. *Med. Hypotheses* **2015**, *85*, 754–760. [CrossRef] [PubMed]

99. Jung, H.J.; Im, S.S.; Song, D.K.; Bae, J.H. Effects of chlorogenic acid on intracellular calcium regulation in lysophosphatidylcholine-treated endothelial cells. *BMB Rep.* **2017**, *50*, 323–328. [CrossRef] [PubMed]
100. Griendling, K.K.; Sorescu, D.; Ushio-Fukai, M. NAD(P)H oxidase: Role in cardiovascular biology and disease. *Circ. Res.* **2000**, *86*, 494–501. [CrossRef] [PubMed]

![antioxidants logo] *antioxidants*

MDPI

Review

Effects of Mulberry Fruit (*Morus alba* L.) Consumption on Health Outcomes: A Mini-Review

Hongxia Zhang [1,†]**, Zheng Feei Ma** [2,3,*,†]**, Xiaoqin Luo** [4] **and Xinli Li** [5]

1 Department of Food Science, University of Otago, Dunedin 9016, New Zealand; zhanghongxia326@hotmail.com
2 Department of Public Health, Xi'an Jiaotong-Liverpool University, Suzhou 215123, China
3 School of Medical Sciences, Universiti Sains Malaysia, Kota Bharu 15200, Malaysia
4 Department of Nutrition and Food Safety, School of Public Health, Xi'an Jiaotong University Health Science Center, Xi'an 710061, China; luoxiaoqin2012@mail.xjtu.edu.cn
5 Department of Nutrition and Food Hygiene, School of Public Health, Medical College of Soochow University, Suzhou 215123, China; lixinli@suda.edu.cn
* Correspondence: Zhengfeei.Ma@xjtlu.edu.cn; Tel.: +86-512-8188-4938
† These authors contributed equally to this work.

Received: 22 March 2018; Accepted: 18 May 2018; Published: 21 May 2018

Abstract: Mulberry (*Morus alba* L.) belongs to the Moraceae family and is widely planted in Asia. Mulberry fruits are generally consumed as fresh fruits, jams and juices. They contain considerable amounts of biologically active ingredients that might be associated with some potential pharmacological activities that are beneficial for health. Therefore, they have been traditionally used in traditional medicine. Studies have reported that the presence of bioactive components in mulberry fruits, including alkaloids and flavonoid, are associated with bioactivities such as antioxidant. One of the most important compounds in mulberry fruits is anthocyanins which are water-soluble bioactive ingredients of the polyphenol class. Studies have shown that mulberry fruits possess several potential pharmacological health benefits including anti-cholesterol, anti-obesity and hepatoprotective effects which might be associated with the presence of some of these bioactive compounds. However, human intervention studies on the pharmacological activities of mulberry fruits are limited. Therefore, future studies should explore the effect of mulberry fruit consumption on human health and elucidate the detailed compounds. This paper provides an overview of the pharmacological activities of mulberry fruits.

Keywords: mulberry; polyphenols; anthocyanins; health; nutrition

---

1. Introduction

Natural products have always been a rich source of biologically active compounds [1–3]. These substances present in fruits and vegetables have received increasing attention because of their antioxidant properties and potential strategy in reducing the risk of certain types of diseases such as metabolic syndrome [1,4,5]. About 50% of the drugs approved are natural products [5]. About 80% of the populations living in many countries rely on the phytomedicines and the plant-derived drug market is estimated to reach approximately $35 billion in 2020 [5,6].

Mulberry (*Morus alba* L.) belongs to the Morus genus of the Moraceae family [7]. Mulberry is also known as *Ramulus Mori* or Sangzhi [8]. To date, this genus has 24 species and 100 varieties that have been known [7]. Mulberry is a species native to China and has been widely cultivated in many regions including Asia, Africa, America, Europe and India [9]. China has planted mulberry for more than 5000 years and mulberry is a traditional Chinese edible fruit that can be eaten fresh [10]. According to traditional Chinese Medicine, mulberry fruits are used to improve eyesight and protect against liver damage [11]. They are grown to feed silkworms [12,13] The season of fresh mulberry fruit in China is usually less than 1 month. Mulberry fruits are difficult to preserve because they have high water content (i.e., ~80%) [11]. Mulberry has been used in traditional Oriental medicine to treat diabetes and premature white hair [14].

Mulberry fruits are appetising and low in calories [15]. Mulberry fruits have a sour taste with a pH < 3.5, providing a more concentrated flavour for fruit production and fresh-eating [16]. Mulberry fruits possess several potential pharmacological properties including anti-cholesterol, anti-diabetic, antioxidative and anti-obesity effects [8,17–19]. These pharmacological properties are due to the presence of polyphenol compounds including anthocyanins, however, different colours of mulberry fruits even from the same species may have different amounts of anthocyanins [20]. Cyanidin-3-rutinoside and cyanidin-3-glucoside are the major anthocyanins isolated from mulberry fruits [21,22].

Although different mulberry varieties with the same genotype are likely to have differences in nutritional values and pharmacological properties [23], the aim of this work was to review some potential roles of mulberry fruits (*Morus alba* L.) and their bioactive compounds in health. Also, some of the potential mechanisms of their actions will be discussed briefly. We hope that this work would provide a valuable reference resource for future studies in this area.

Search Strategy

An electronic literature search was conducted using Google Scholar, Medline (OvidSP) and PubMed until February 2018. Additional articles were identified and obtained from references in the retrieved articles. Search terms included combinations of the following: mulberry, fruits, hypertension, diabetes, anti-tumour, hepatoprotective, anti-obesity, anti-oxidative stress and phytochemicals. For the purpose of this mini-review, the search was restricted to experimental, epidemiological and clinical studies published in English that address the phytochemical constituents and pharmacological properties of mulberry fruits (*Morus alba* L.).

2. Phytochemical Compounds

Compared with mulberry leaves and barks, mulberry fruits are less commonly used in traditional Chinese Medicine. The possible reasons might be due to the lack of awareness of their health benefits and limited production [24]. However, there is increasing interest in isolating and quantifying the phytochemical compounds from mulberry fruits. This is because mulberry fruits can be also consumed as foods [24]. Mulberry fruits have strong antioxidant property which is due primarily to the presence of polyphenols [25]. Figure 1 shows the major polyphenol composition found in mulberry fruits.

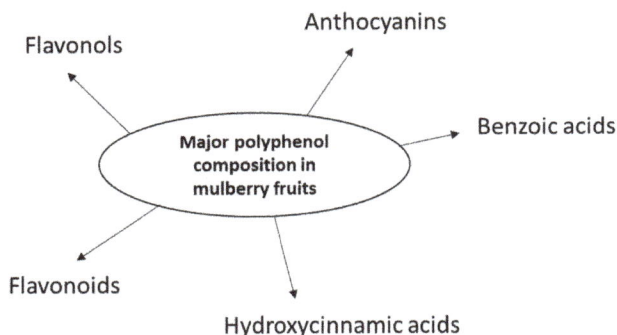

Figure 1. Major polyphenol composition in mulberry fruits.

Phytochemical compounds of mulberry fruits (*Morus nigra*, *Morus indica* and *Morus rubra*) have been reported in several studies [26–28]. Kang, Hur, Kim, Ryu and Kim [17] isolated cyanidin-3-*O*-β-D-glucopyranoside (C3G) from 1% HCI-MeOH mulberry fruit extracts using Amberlite IRC-50 ion exchange chromatography. C3G was identified and quantified by liquid chromatography-mass spectroscopy (LC-MS) and High-Performance Liquid Chromatography (HPLC) [17]. C3G is an aglycon of anthocyanin that has inflammation-suppressing and free radical scavenging activity, which might protect against endothelial dysfunction [17].

In a study assessing the polyphenolic composition of five major mulberry fruit varieties (i.e., Pachungsipyung, Whazosipmunja, Suwonnosang, Jasan and Mocksang) cultivated in Korea using spectrophotometric methods, Bae and Suh [29] reported that the total phenols, total anthocyanins, coloured (ionised) anthocyanins and total flavanols ranged from 960 to 2570 μg/g gallic acid equivalents, 137 to 2057 μg/g malvidin-3-glucoside equivalents, 10 to 190 μg/g malvidin-3-glucoside equivalents and 6 to 65 μg/g catechin equivalents.

Kusano, Orihara, Tsukamoto, Shibano, Coskun, Guvenc and Erdurak [24] isolated five new nortropane alkaloids (i.e., 2α,3β-dihydroxynortropane, 2β,3β-dihydroxynortropane, 2α,3β-6exo-trihydroxynortropane, 2α,3β,4α-trihydroxynortropane, 3β,6exo-dihydroxynortropane) along with nor-ψ-tropine from ripened mulberry fruits grown in Turkey. In addition, Kusano, Orihara, Tsukamoto, Shibano, Coskun, Guvenc and Erdurak [24] also isolated and determined the new structures of six amino acids, which were morusimic acid A, morusimic acid B, morusimic acid C, morusimic acid D, morusimic acid E and morusimic acid F using spectroscopic data.

Kim, et al. [30] identified five pyrrole alkaloids in mulberry fruits, which were morrole B, morrole C, morrole D, morrole E and morrole F based on spectroscopic data. In addition, the authors [30] also isolated 11 pyrrole alkaloids, which were 4-[formyl-5-(hydroxymethyl)-1H-pyrrol-1-yl]butanoate, 2-(5-hydroxymethyl-2′,5′-dioxo-2′,3′,4′,5′-tetrahydro-1′H-1,3′-bipyrrole)carbaldehyde, 4-[formyl-5-(hydroxymethyl)-1H-pyrrol-1-yl]butanoate, 4-[formyl-5-(methoxymethyl)-1H-pyrrol-1-yl]butanoic acid, methyl 2-[2-formyl-5-(methoxymethyl)-1H-pyrrole-1-yl]propanoate, 2-(5′-hydroxymethyl-2′-formylpyrrol-1′-yl)-3-phenyl-propionic acid lactone, methyl 2-[2-formyl-5-(methoxymethyl)-1H-pyrrol-1-yl]-3-(4-hydroxyphenyl)propanoate, 2-(5′- hydroxymethyl-2′-formylpyrrol-1′-yl)-3-(4-hydroxyphenyl)-propionic acid lactone, 2-(5-hydroxymethyl- 2-formylpyrrole-1-yl)propionic acid lactone, 2-(5-hydroxymethyl-2-formylpyrrol-1-yl)isovaleric acid lactone, 2-(5-hydroxymethyl-2-formylpyrrole-1-yl)isocaproic acid lactone and 2-[2-formyl-5-(hydroxymethyl)-1-pyrrolyl-]3-methylpentanoic acid lactone.

Natić, et al. [31] isolated epigallocatechin, epigallocatechin gallate, gallocatechin, gallocatechin gallate, isorhamnetin glucuronide, isorhamnetin hexoside, isorhamnetin hexosylhexoside, kaempferol glucuronide, kaempferol hexoside, kaempferol hexosylhexoside, kaempferol rhamnosylhexoside, morin and naringin from mulberry fruits grown in Vojvodina, North Serbia. Quercetin glucoronide, quercetin hexoside, quercetin hexosylhexoside, quercetrin from mulberry fruits were also isolated using ultra HPLC (UHPLC) system coupled to a high resolution mass spectrophotometer [31]. In addition, the authors [31] also reported the presence of cyanidin galloylhexoside, cyanidin hexoside, cyanidin hexosylhexoside, cyanidin pentoside, cyanidin rhamnosylhexoside, delphinidin acetylhexoside, delphinidin hexoside, delphinidin rhamnosylhexoside, pelargonidin hexoside, pelargonidin rhamnosylhexoside and petunidin rhamnosylhexoside from mulberry fruits.

Qin, et al. [32] isolated cyanidin 3-*O*-glucoside, cyanidin 3-*O*-rutinoside, pelargonidin 3-*O*-glucoside and pelargonidin 3-*O*-rutinoside ultraviolet-visible from mulberry fruits grown in Shaanxi, China using UV-Visible spectroscopy, HPLC-pulsed amperometric detector (PAD), LC-MS and proton nuclear magnetic resonance (1HNMR). Du, et al. [33] isolated cyanidin 3-*O*-β-D-galactopyranoside, cyanidin 3-*O*-β-D-glucopyranoside and cyanidin 7-*O*-β-D-glucopyranoside from mulberry fruits bought from local stores in Hangzhou, China. In addition, the authors [33] also isolated cyanidin 3-*O*-(6″-*O*-α-rhamnopyranosyl-β-D-galactopyranoside) and cyanidin

3-*O*-(6''-*O*-α-rhamnopyranosyl-β-D-glucopyranoside) from mulberry fruits. While Memon, et al. [34] isolated gallic acid, protocatechuic acid, protocatechuic aldehyde, *p*-hydroxybenzoic acid, vanillic acid, chlorogenic acid, syringic acid, syringealdehyde and m-coumaric acid from mulberry fruits grown in Pakistan. A study by Peng, et al. [35] identified eight major compounds which were gallic acid, chlorogenic acid, protocatechuic acid, rutin, caffeic acid, 3-caffeoyl quinic acid, 4-caffeoyl quinic acid and quercetin-3-*O*-glucoside in mulberry fruit water extract.

Another study by Kim, et al. [36] identified four pyrrole alkaloids from mulberry fruits planted in Chonbuk, Korea which were 2-formyl-5-(hydroxymethyl)-1H-pyrrole-1-butanoic acid, 5-(hydroxymethyl)-1H-pyrrole-2-carboxaldehyde, 2-formyl-1H-pyrrole-1-butanoic acid and 2-formyl-5-(methoxymethyl)-1H-pyrrole-1-butanoic acid. In addition, the authors [36] also isolated a new pyrrole alkaloid, which was morrole A. All the structures of isolated pyrrole alkaloids were determined using 1D and 2D nuclear magnetic resonance (NMR) analyses [36].

Isabelle, et al. [37] reported the presence of 3-caffeoyl quinic acid, 5-caffeoyl quinic acid, cyanidin-3-glucoside, 4-caffeoyl quinic acid, cyanidin-3-rutinoside, pelargonidin-3-glucoside, rutin, quercetin and kaempferol-3-rutinoside in the Chinese mulberry fruit cultivar Guo-2. In addition, the authors [37] also found the presence of α-tocopherol, α-tocotrienol, δ-tocopherol, γ-tocopherol, β-carotene, lutein, neoxanthin and violaxanthin in the Chinese mulberry fruit cultivar Bei-2-5, Guiyou-154, Heipisang, Xuan-27 and Tang-10. Rutin, 1-deoxynojirimycin (DJN), cyanidin-3-*O*-β-glucoside, cyanidin-3-*O*-β-rutinoside, resveratrol and oxyresveratrol were also present in the Chinese mulberry fruits [38,39].

Wang, Xiang, Wang, Tang and He [15] isolated quercetin-3-*O*-β-D-glucopyranoside, quercetin 3-*O*-(6''-*O*-acetyl)-β-D-glucopyranoside, quercetin 3-*O*-β-D-rutinoside, quercetin 7-*O*-β-D-glucopyranoside, quercetin 3,7-di-*O*-β-D-glucopyranoside, kaempferol 3-*O*-β-D-glucopyranoside, kaempferol 3-*O*-β-D-rutinoside, isobavachalcone, 2,4,2',4',-tetrahydroxy-3'-(3-methyl-2-butenyl)-chalcone (morachalcone), (2E)-1-[2,3-dihydro-4-hydroxy-2-(1-methylethenyl)-5-benzofuranyl]-3-(4-hydroxyphenyl)-1-propanone, 5,7,3'-trihydroxy-flavanone-49-*O*-β-D-glucopyranoside, 5,7,4'-trihydroxy-flavanone-3'-*O*-β-D-glucopyranoside, dihydrokaempferol 7-*O*-ß-D-glucopyranoside, 2-*O*-(3,4-dihydroxybenzoyl)-2,4,6-trihydroxyphenylacetic acid, 2-*O*-(3,4-dihydroxybenzoyl)-2,4,6-trihydroxyphenylmethylacetate (jaboticabin), *p*-hydroxybenzoic acid, protocatechuic acid, 3-methoxy-4-hydroxybenzoic acid (vanillic acid), protocatechuic acid methyl ester, protocatechuic acid ethyl ester, 4-hydroxyphenylacetic acid methyl ester, 5,7-dihydroxychromone, 2-(4-hydroxyphenyl)ethanol (tyrosol) and pyrocatecholin in ethyl acetate-soluble extract of mulberry fruits. The authors [15] determined the structures of isolated compounds based on MS and NMR analysis.

Jiang and Nie [40] reported that mulberry fruit cultivar Hetianbaisang contains many types of essential amino acids (i.e., isoleucine, leucine, threonine, lysine, valine, phenylalanine, tyrosine, tryptophan, histidine, methionine and cysteine) and seven non-essential amino acids (i.e., arginine, alanine, proline, glutamic acid, glycine, serine and aspartic acid). In addition, the authors [40] also found the presence of minerals including potassium, calcium, magnesium, iron, sodium, zinc, copper, selenium and manganese in mulberry fruit cultivar Hetianbaisang. Mulberry fruit cultivar Hetianbaisang also contains organic acids including malic acid, succinic acid, citric acid, tartaric acid, acetic acid [40]. In addition, linoleic acid, myristic acid, stearic acid, palmitic acid and α-linoleic acid were also detected in mulberry fruit cultivar Hetianbaisang [40].

Yang, Yang and Zheng [11] reported that the total phenolics, total flavonoids and anthocyanins in the freeze-dried powder of mulberry fruits were 23.0 mg/g gallic acid equivalents, 3.9 mg/g rutin equivalents, 0.87 mg/g cyanidin-3-glucoside equivalents, respectively. The major flavonol in mulberry fruit powder was rutin (0.43 mg/g), followed by morin (0.16 mg/g), quercetin (0.01 mg/g) and myricetin (0.01 mg/g) [11]. HPLC was used to determine the flavonols in mulberry fruit powder [11]. In addition, the freeze-dried powder of mulberry fruits also contained 1.20 mg/g ascorbic acid, 0.32 mg/g vitamin E and 243.0 mg/g dietary fibre [11].

Fatty acid content and composition of mulberry can vary according to different ecological conditions. For example, Yang, Yang and Zheng [11] found that Chinese mulberry fruits had 7.55% total lipids, with 87.5% of unsaturated fatty acids. The highest fatty acid content in Chinese mulberry fruits was linoleic acid C18:2 (79.4%), followed by palmitic acid C16:2 (8.6%) and oleic acid C18:1 (7.5%) [11]. In addition, Chinese mulberry also contained 0.6% α-linolenic acid C18:3 [11]. Although the highest fatty acid content in Turkish mulberry was linoleic acid C18:2 (57.3%) followed by palmitic acid C16:0 (22.4%); no presence of linolenic acid C18:3 was reported [7].

Different colours of mulberry fruits (*M. alba* L) such as red, purple and purple-red have been reported [41]. Aramwit, Bang and Srichana [41] reported that purple mulberry fruit extract had higher contents of total sugars and anthocyanins than red and purple-red mulberry fruit extracts. This is because sugars are needed as the precursors to synthesis anthocyanins [41]. However, red mulberry fruit had a higher ascorbic acid and ß-carotene than purple and purple-red mulberry fruit extracts [41].

Many volatile compounds have also been found in mulberry fruits [42]. Calin-Sanchez, Martinez-Nicolas, Munera-Picazo, Carbonell-Barrachina, Legua and Hernandez [42] reported that volatile compounds found in mulberry fruits grown in Spain included acetic acid, 3-hydroxyl-2-butanone, ethyl butyrate, ethyl acetate, 3-methylbutanal, 2-methybutanal, heptanal, methional, hexanal, trans-2-hexanal, 2-octenone,hexanoic acid, benzaldehyde, methyl hexanoate, 2-ethylhexanal, octanal, limonene, 6-methyl-5-hepten-2on, ethyl hexanoate, 2,4-nonanadienal, phenylacetaldehyde, trans-2-octenal, cis-α-ocimene, terpinonene, 2-nonanone, nonanal, octanoic acid, cis-2-nonenal, dodecanoic acid, terpinen-4-ol, ethyl octanoate, ethyl dodecanoate, decanal, decanoic acid and ethyl decanoate. The authors [42] suggested that these volatile compounds in mulberry fruits might present better sensory profiles for the market demands from consumers.

Chen, et al. [43] reported that the levels of phenolic compounds in mulberry fruits are higher than blackberry, blueberry, raspberry and strawberry, suggesting that mulberry fruits can be used as good sources of phenolic compounds. Therefore, mulberry fruits are rich in diverse phenolic compounds including polyhenols, anthocyanins and flavonoids.

3. Pharmacological Properties

As mentioned previously, mulberry fruits are rich in anthocyanins [44], which have attracted attention of researchers and consumers because of their potential pharmacological activities on health [45–49]. Anthocyanins from mulberry fruits can inhibit the oxidation of low-density lipoprotein (LDL) and scavenge free radicals [33,50]. Many studies have showed that mulberry leaves exhibit a wide range of pharmacological activities [51–57]. However, there are limited studies that have been conducted on the pharmacological properties of mulberry fruits [15,58,59]. Also, most studies have been conducted in animal models using mulberry fruits as a dietary supplement [15,58,59]. Although existing literature shows that there is relationship between mulberry fruit consumption and improved health outcomes, these studies often infer a causal correlation between a bioactive substance of mulberry fruits and the observed health outcomes. This approach is more likely to oversimplify the complicated body mechanisms that will eventually lead to the observed health outcomes. Therefore, the conclusions based on such studies should always be interpreted with caution [60] because the observed health outcomes may not be attributed to the action of a single bioactive compound of mulberry fruits.

3.1. Hypolipidemic

Cardiovascular disease (CVD) is one of the most common causes of deaths, with about 17 million people die of CVD (including stroke and coronary heart disease) every year worldwide [61,62]. It is estimated that CVD will continue to be the largest contributor to global mortality in the future [63] Hyperlipidemia is one of the major risk factors for CVD [64]. Therefore, an increasing focus has been reported in research studies that determine the effectiveness of natural alternative medicine in reducing blood lipid levels [11]. This is because majority of the hypolipidemic drugs can potentially cause side effects and they are expensive [11].

Yang, Yang and Zheng [11] reported that rat fed with high fat diet supplemented with 5% or 10% mulberry fruit powder had a significant decrease in the concentration of serum and liver triglyceride, total cholesterol and serum LDL cholesterol. An increase in the serum high-density lipoprotein (HDL) cholesterol was reported in rat fed with high fat diet supplemented with 5% or 10% mulberry fruit powder [11]. It is suggested that the presence of dietary fiber in mulberry fruits inhibits the hepatic lipogenesis and increases LDL-receptor activity [65]. In addition, the authors suggested that mulberry fruits might have a hypolipidemic effect because mulberry fruits have high content of dietary fiber and linoleic acid [11].

Chen, Liu, Hsu, Huang, Yang and Wang [50] reported that New Zealand white rabbits fed with high cholesterol diet (HCD) (containing 95.7% standard Purina chow, 3% lard oil and 1.3% cholesterol) plus 0.5% or 1.0% water extract of mulberry fruits for 10 weeks had lower levels of total cholesterol, LDL cholesterol, and triglycerides than those fed with only lard oil diet. The authors [50] also showed that rabbits fed with HCD plus 0.5% or 1.0% water extract of mulberry fruits had significantly reduced severe atherosclerosis in the aorta by 42–63% and these findings were supported by histopathological examination of blood vessel of rabbits. The effect of water extract of mulberry fruits on the levels of total cholesterol and LDL cholesterol was reported to be dose-dependent [50]. No adverse effects on the changes of liver or renal functions in rabbits fed with HCD plus 0.5% or 1.0% water extract of mulberry fruits were reported [50].

In a randomised controlled study of 58 hypercholesterolemic adults aged 30–60 years, Sirikanchanarod, et al. [66] reported that after 6 weeks of 45 g freeze-dried mulberry fruit consumption (325 mg anthocyanins), the intervention group had a significantly lower level of total cholesterol and LDL (both *p*-values < 0.001) than the control group. The authors [66] suggested that mulberry fruits might be used as an alternative treatment for hypercholesterolemic patients. Therefore, the consumption of mulberry fruits might reduce the risk of atherosclerosis because mulberry fruits possess anti-hyperlipidemic and anti-oxidative abilities to prevent the oxidation of LDL [50].

3.2. Anti-Diabetic

Diabetes is characterised by hyperglycemia which results from the defects of secretion of insulin [67]. It is associated with a series of health complications including CVD and failure of various organs [67]. Jiao, Wang, Jiang, Kong, Wang and Yan [59] reported that diabetic rats fed with two different fractions of mulberry fruit polysaccharides (MFP50 and MFP90) for seven weeks had a significant decrease in the levels of fasting glucose, fasting serum insulin, homeostasis model of assessment-insulin resistance, triglyceride and oral glucose tolerance test-area under the curve. The MFP50 and MFP90 had a final ethanol concentration of 50% and 90%, respectively [59]. When compared with diabetic rats fed with pure water, diabetic rats fed with MFP50 and MFP90 had a lower serum insulin level at a rate of 26.5% and 32.5%, respectively [59]. The MFP50 group had a significant increase in the level of HDL cholesterol and the proportion of HDL cholesterol to total cholesterol [59]. The authors [59] also found that both MFP50 and MFP90 reduced the levels of serum alanine transaminase (ALT), suggesting that they have potential hepatoprotective effects. Although MFP50 had a more stable hypoglycemic effect than MFP90, MFP90 had a better hypolipidemic effect than MFP50 [59].

Similar findings were also reported by Guo, Li, Zheng, Xu, Liang and He [58] who found that diabetic rats fed with mulberry fruit polysaccharides for 2 weeks had a decrease in fasting blood glucose. Another study by Wang, Xiang, Wang, Tang and He [15] reported that diabetic rats fed with ethyl acetate-soluble extract of mulberry fruits for 2 weeks had a significant decrease in the levels of fasting blood glucose and glycosylated serum protein. The authors [15] also found that ethyl acetate-soluble extract of mulberry fruits had significantly increased the antioxidant activities of catalase (CAT), glutathione peroxidase (GSH-Px) and superoxide dismutase (SOD) in diabetic rats. Ethyl acetate-soluble extract of mulberry fruits also possesses strong α-glucoside inhibitory activity and radical-scavenging activities against 2,2-diphenyl-1-picrylhdrazyl (DPPH) and superoxide anion radicals [15]. A study by Xu, et al. [68] reported that diabetic mice fed with mulberry fruit polysaccharides had a lower level of haemoglobin A1c (HbA1c) and a reduction in streptozotocin (STZ)-lesioned pancreatic cells. In addition, diabetic mice fed with mulberry fruit polysaccharides also had an increase in insulin level and B-cell lymphoma 2 (bcl-2) expression [68].

Yan, et al. [69] reported that male C57BL6/J genetic background (db/db) mice fed with anthocyanin extract of mulberry fruit in the doses of 50 and 125 mg/kg body weight per day for 8 weeks had a significant decrease in the levels of cholesterol, fasting blood glucose, leptin, serum insulin and triglyceride as well as an increase in adiponectin level. Therefore, the authors [69] suggested that anthocyanin extract of mulberry fruit can be used to improve the resistance of insulin and leptin. Taken together, these results [15,58,59] suggest that mulberry fruits might play an important role in the treatment of diabetes because of their anti-hyperglycemic and anti-hyperlipidemic effects.

3.3. Anti-Obesity

Several studies have shown that obesity plays a major role in contributing to dyslipidemia [70–73]. Lim, et al. [74] reported that high fat diet-induced obese mice fed with a combination of mulberry leaf extract and mulberry fruit extract at low and high doses had a significant decrease in body weight gain, fasting plasma glucose, insulin and homeostasis model assessment of insulin resistance. The low dose of combination of mulberry leaf extract and mulberry fruit extract was 133 mg mulberry leaf extract and 67 mg mulberry fruit extract/kg/day, while the high dose of combination of mulberry leaf extract and mulberry fruit extract was 333 mg mulberry leaf extract and 167 mg mulberry fruit extract/kg/day [74]. The high dose of combination of mulberry leaf extract and mulberry fruit extract had significantly improved the glucose control [74]. In addition, the high dose of combination of mulberry leaf extract and mulberry fruit extract also decreased the protein levels of manganese superoxide dismutase, inducible nitric oxide synthase, monocyte chemoattractant protein-1, C-reactive protein (CRP), tumour necrosis factor-α and interleukin-1 [74]. Therefore, it is suggested that the combination of mulberry leaf extract and mulberry fruit extract possess the anti-obesity and anti-diabetic properties by modulating oxidative stress and inflammation induced by obesity [74].

Peng, Liu, Chuang, Chyau, Huang and Wang [35] reported that male hamsters fed with mulberry fruit water extract for 12 weeks had a lower high fat diet-induced body weight and visceral fat, accompanied with a decrease in serum triacylglycerol, cholesterol, LDL/HDL ratio and free fatty acid. In addition, mulberry fruit water extract also reduced fatty acid synthase and 3-hydroxy-3-methylglutaryl-coenzyme A (HMG-CoA) reductase and elevated hepatic peroxisome proliferator-activated receptor α and carnitine palmitoyltransferase-1 [35]. No physiological burdens in terms of levels of serum blood urea nitrogen, creatinine, potassium and sodium ions were exerted by the administration of mulberry fruit extract [35]. The authors [35] suggested that mulberry fruit water extracts regulate lipolysis and lipogenesis, which can be used to reduce the body weight.

3.4. Anti-Tumour

Gastrointestinal tract cancers are also one of the most common types of cancers in the world [75,76] and *Helicobacter pylori* is one of the common suspects in triggering the gastric carcinogenesis [77,78]. Huang, et al. [79] reported that after male balb/c nude mice were fed with anthocyanin-rich mulberry fruit extract for 7 weeks, atypical glandular cells (AGS) tumour xenograft growth in mice was inhibited, suggesting that anthocyanins from mulberry fruits might be used to prevent gastric carcinoma formation.

3.5. Hepatoprotective

In a study investigating the protective effect of mulberry fruit marc (the solid component after juicing) anthocyanins on carbon tetrachloride (CC14)-induced liver fibrosis in male Sprague Dawley rats, Li, et al. [80] reported that rats fed with mulberry fruit marc anthocyanins had a decrease in the levels of ALT, aspartate amino transferase, collagen type-III hyaluronidase acid and hydroxyproline. Another study by Chang, et al. [81] reported that mulberry fruit extracts suppressed the synthesis and enhanced the oxidation of fatty acids. Therefore, the mulberry fruits might prevent the non-alcoholic fatty liver disease.

3.6. Protective against Cytotoxicity and Oxidative Stress

In a study investigating the protective effect of mulberry fruit extract against ethyl carbamate (EC)-induced cytotoxicity in human liver HepG2 cells, Chen, Li, Bao and Gowd [43] reported no decrease in cell viability with the treatments of mulberry fruit extract (0.5 mg/mL, 1.0 mg/mL and 2.0 mg/mL). Therefore, the authors [43] suggested that mulberry fruits can be used to protect against EC-induced cytotoxicity and oxidative stress. Also, in a study investigating the effect of mulberry fruit consumption on the anti-fatigue activity in mice using a weight-loaded swimming test, Jiang, Guo, Xu, Huang, Yuan and Lv [44] reported that mice fed with mulberry juice purification and mulberry marc purification had an increase endurance capacity than the control group. The authors [44] suggested that the presence of anthocyanins in mulberry fruits might act as an antioxidant to reduce exercise-induced oxidative stress and physical fatigue.

3.7. Protective against Brain Damage

Kang, Hur, Kim, Ryu and Kim [17] reported that that C3G isolated from mulberry fruit extracts had shown a cytoprotective effect on PC12 cells exposed to hydrogen peroxide in vitro and a neuroprotective effect on cerebral ischemic damage caused by oxygen glucose deprivation (OGD) in vivo. Therefore, it is suggested that mulberry fruits possess neuroprotective effects in vivo and vitro ischemic oxidative stress [17]. Table 1 shows an overview of animal studies investigating the pharmacological properties of mulberry fruits.

Table 1. An overview of animal studies investigating the pharmacological properties of mulberry fruits.

Pharmacological Properties	References
Hypolipidemic	Yang et al. [11]; Chen et al. [50]; Sirikanchanarod et al. [66]
Anti-diabetic	Wang et al. [15]; Jiao et al. [59]; Guo et al. [58]; Xu et al. [68]; Yan et al. [69]
Anti-obesity	Peng et al. [35]; Lim et al. [74]
Anti-tumour	Huang et al. [79]
Hepatoprotective	Li et al. [80]; Chang et al. [81]
Protective against cytotoxicity and oxidative stress	Jiang et al. [44]; Chang et al. [81]
Protective against brain damage	Kang et al.[17]

3.8. Adverse Effects

Due to a limited number of human studies, it is difficult to assess the safety of mulberry fruit consumption. Moreover, there is insufficient evidence regarding the recommended consumption of mulberry fruits (i.e., dosage) and its treatment duration. It is necessary that all future clinical studies that investigate the effects of mulberry fruit consumption on health should follow the Consolidated Standards of Reporting Trials (CONSORT) guidelines for generating scientifically rigorous evidence [82–84].

4. Conclusions and Future Research

Literature reviews have highlighted that mulberry fruits contain high content of polyphenolic compounds and antioxidants [85]. This suggests that there are many opportunities for the food and healthcare industry to explore the health benefits of mulberry fruits because there is a potential growing market for mulberry fruits. However, the contents of bioactive compounds such as anthocyanins, alkaloids, flavonoids and polyphenols are dependent on the cultivars. Although the bioactive compounds may work synergistically to promote health, such claims still require further investigation in order to establish the causative relationship between mulberry fruit consumption and health.

There are limited studies with sufficient data to support whether mulberry fruits are beneficial to human health especially in terms of the management and prevention of chronic diseases such as diabetes and CVD. The majority of the studies that reported beneficial effects of mulberry fruits on health are animal-based studies. Moreover, these studies used different varieties of mulberry fruits, types of solvents and methods of preparation, which cause the evaluation of activity of mulberry fruits to be difficult and these studies involve quite heterogeneous data. Therefore, larger well-designed, randomised controlled trials are needed to examine the effects of mulberry fruit consumption on human health. Similar to other plants and food products [1,86], the fate of polyphenol compounds in the body, especially after undergoing intestinal transformations by enzymes produced by gut microbiota should also be addressed. The elucidation of some active ingredient structures in mulberry fruits and their mechanisms in promoting pharmacological properties are also worthy of further research.

Author Contributions: The project idea was developed by Z.F.M. Z.F.M. wrote the first draft of the manuscript. Z.F.M., H.X., X.L. and X.L. conducted the literature search and revised the manuscript.

Acknowledgments: The authors received no specific funding for this work.

Conflicts of Interest: The authors declare no conflict of interest.

References

1. Ma, Z.F.; Zhang, H. Phytochemical constituents, health benefits, and industrial applications of grape seeds: A mini-review. *Antioxidants* **2017**, *6*, 71. [CrossRef] [PubMed]
2. Ji, H.-F.; Li, X.-J.; Zhang, H.-Y. Natural products and drug discovery. Can thousands of years of ancient medical knowledge lead us to new and powerful drug combinations in the fight against cancer and dementia? *EMBO Rep.* **2009**, *10*, 194–200. [CrossRef] [PubMed]
3. Zhang, H.; Ma, Z.F. Phytochemical and pharmacological properties of *Capparis spinosa* as a medicinal plant. *Nutrients* **2018**, *10*, 116. [CrossRef] [PubMed]
4. Cao, Y.; Ma, Z.F.; Zhang, H.; Jin, Y.; Zhang, Y.; Hayford, F. Phytochemical properties and nutrigenomic implications of yacon as a potential source of prebiotic: Current evidence and future directions. *Foods* **2018**, *7*, 59. [CrossRef] [PubMed]
5. Veeresham, C. Natural products derived from plants as a source of drugs. *J. Adv. Pharm. Technol. Res.* **2012**, *3*, 200–201. [CrossRef] [PubMed]
6. Gryn-Rynko, A.; Bazylak, G.; Olszewska-Slonina, D. New potential phytotherapeutics obtained from white mulberry (*Morus alba* L.) leaves. *Biomed. Pharmacother.* **2016**, *84*, 628–636. [CrossRef] [PubMed]
7. Ercisli, S.; Orhan, E. Chemical composition of white (*Morus alba*), red (*Morus rubra*) and black (*Morus nigra*) mulberry fruits. *Food Chem.* **2007**, *103*, 1380–1384. [CrossRef]
8. Ye, F.; Shen, Z.; Xie, M. Alpha-glucosidase inhibition from a Chinese medical herb (*Ramulus mori*) in normal and diabetic rats and mice. *Phytomedicine* **2002**, *9*, 161–166. [CrossRef] [PubMed]
9. Khan, M.A.; Rahman, A.A.; Islam, S.; Khandokhar, P.; Parvin, S.; Islam, M.B.; Hossain, M.; Rashid, M.; Sadik, G.; Nasrin, S.; et al. A comparative study on the antioxidant activity of methanolic extracts from different parts of *Morus alba* L. (moraceae). *BMC Res. Notes* **2013**, *6*, 24. [CrossRef] [PubMed]
10. Ning, D.; Lu, B.; Zhang, Y. The processing technology of mulberry series product. *China Fruit Veg. Proc.* **2005**, *5*, 38–40.
11. Yang, X.; Yang, L.; Zheng, H. Hypolipidemic and antioxidant effects of mulberry (*Morus alba* L.) fruit in hyperlipidaemia rats. *Food Chem. Toxicol.* **2010**, *48*, 2374–2379. [CrossRef] [PubMed]
12. Arabshahi-Delouee, S.; Urooj, A. Antioxidant properties of various solvent extracts of mulberry (*Morus Indica* L.) leaves. *Food Chem.* **2007**, *102*, 1233–1240. [CrossRef]
13. Sohn, B.-H.; Park, J.-H.; Lee, D.-Y.; Cho, J.-G.; Kim, Y.-S.; Jung, I.-S.; Kang, P.-D.; Baek, N.-I. Isolation and identification of lipids from the silkworm (*Bombyx mori*) droppings. *J. Korean Soc. Appl. Biol. Chem.* **2009**, *52*, 336–341. [CrossRef]
14. Liu, H.; Qiu, N.; Ding, H.; Yao, R. Polyphenols contents and antioxidant capacity of 68 Chinese herbals suitable for medical or food uses. *Food Res. Int.* **2008**, *41*, 363–370. [CrossRef]
15. Wang, Y.; Xiang, L.; Wang, C.; Tang, C.; He, X. Antidiabetic and antioxidant effects and phytochemicals of mulberry fruit (*Morus alba* L.) polyphenol enhanced extract. *PLoS ONE* **2013**, *8*, e71144. [CrossRef] [PubMed]

16. Yang, Y.; Zhang, T.; Xiao, L.; Yang, L.; Chen, R. Two new chalcones from leaves of *Morus alba* L. *Fitoterapia* **2010**, *81*, 614–616. [CrossRef] [PubMed]

17. Kang, T.H.; Hur, J.Y.; Kim, H.B.; Ryu, J.H.; Kim, S.Y. Neuroprotective effects of the cyanidin-3-*O*-beta-D-glucopyranoside isolated from mulberry fruit against cerebral ischemia. *Neurosci. Lett.* **2006**, *391*, 122–126. [CrossRef] [PubMed]

18. Kim, A.J.; Park, S. Mulberry extract supplements ameliorate the inflammation-related hematological parameters in carrageenan-induced arthritic rats. *J. Med. Food* **2006**, *9*, 431–435. [CrossRef] [PubMed]

19. Zhang, Z.; Shi, L. Anti-inflammatory and analgesic properties of *cis*-mulberroside a from *Ramulus mori*. *Fitoterapia* **2010**, *81*, 214–218. [CrossRef] [PubMed]

20. Gerasopoulos, D.; Stavroulakis, G. Quality characteristics of four mulberry (*Morus* sp) cultivars in the area of Chania, Greece. *J. Sci. Food Agric.* **1997**, *73*, 261–264. [CrossRef]

21. Suhl, H.J.; Noh, D.O.; Kang, C.S.; Kim, J.M.; Lee, S.W. Thermal kinetics of color degradation of mulberry fruit extract. *Die Nahr.* **2003**, *47*, 132–135.

22. Liu, X.; Xiao, G.; Chen, W.; Xu, Y.; Wu, J. Quantification and purification of mulberry anthocyanins with macroporous resins. *J. Biomed. Biotechnol.* **2004**, 326–331. [CrossRef] [PubMed]

23. Bao, T.; Xu, Y.; Gowd, V.; Zhao, J.; Xie, J.; Liang, W.; Chen, W. Systematic study on phytochemicals and antioxidant activity of some new and common mulberry cultivars in China. *J. Funct. Foods* **2016**, *25*, 537–547. [CrossRef]

24. Kusano, G.; Orihara, S.; Tsukamoto, D.; Shibano, M.; Coskun, M.; Guvenc, A.; Erdurak, C.S. Five new nortropane alkaloids and six new amino acids from the fruit of *Morus alba* Linne growing in Turkey. *Chem. Pharm. Bull.* **2002**, *50*, 185–192. [CrossRef] [PubMed]

25. Yang, J.; Liu, X.; Zhang, X.; Jin, Q.; Li, J. Phenolic profiles, antioxidant activities, and neuroprotective properties of mulberry (*Morus atropurpurea* Roxb.) fruit extracts from different ripening stages. *J. Food Sci.* **2016**, *81*, C2439–C2446. [CrossRef] [PubMed]

26. Chan, E.W.; Lye, P.Y.; Wong, S.K. Phytochemistry, pharmacology, and clinical trials of *Morus alba*. *Chin. J. Nat. Med.* **2016**, *14*, 17–30. [PubMed]

27. Arfan, M.; Khan, R.; Rybarczyk, A.; Amarowicz, R. Antioxidant activity of mulberry fruit extracts. *Int. J. Mol. Sci.* **2012**, *13*, 2472–2480. [CrossRef] [PubMed]

28. Imran, M.; Khan, H.; Shah, M.; Khan, R.; Khan, F. Chemical composition and antioxidant activity of certain *Morus* species. *J. Zhejiang Univ. Sci. B* **2010**, *11*, 973–980. [CrossRef] [PubMed]

29. Bae, S.-H.; Suh, H.-J. Antioxidant activities of five different mulberry cultivars in Korea. *Food Sci. Technol.* **2007**, *40*, 955–962. [CrossRef]

30. Kim, S.B.; Chang, B.Y.; Hwang, B.Y.; Kim, S.Y.; Lee, M.K. Pyrrole alkaloids from the fruits of *Morus alba*. *Bioorgan. Med. Chem. Lett.* **2014**, *24*, 5656–5659. [CrossRef] [PubMed]

31. Natić, M.M.; Dabić, D.Č.; Papetti, A.; Fotirić Akšić, M.M.; Ognjanov, V.; Ljubojević, M.; Tešić, Ž.L. Analysis and characterisation of phytochemicals in mulberry (*Morus alba* L.) fruits grown in Vojvodina, North Serbia. *Food Chem.* **2015**, *171*, 128–136. [CrossRef] [PubMed]

32. Qin, C.; Li, Y.; Niu, W.; Ding, Y.; Zhang, R.; Shang, X. Analysis and characterisation of anthocyanins in mulberry fruit. *Czech J. Food Sci.* **2010**, *28*, 117–126. [CrossRef]

33. Du, Q.; Zheng, J.; Xu, Y. Composition of anthocyanins in mulberry and their antioxidant activity. *J. Food Comp. Anal.* **2008**, *21*, 390–395. [CrossRef]

34. Memon, A.A.; Memon, N.; Luthria, D.L.; Bhanger, M.I.; Pitafi, A.A. Phenolic acids profiling and antioxidant potential of mulberry (*Morus laevigata* W., *Morus nigra* L., *Morus alba* L.) leaves and fruits grown in Pakistan. *Pol. J. Food Nutr. Sci.* **2010**, *60*, 25–32.

35. Peng, C.-H.; Liu, L.-K.; Chuang, C.-M.; Chyau, C.-C.; Huang, C.-N.; Wang, C.-J. Mulberry water extracts possess an anti-obesity effect and ability to inhibit hepatic lipogenesis and promote lipolysis. *J. Agric. Food Chem.* **2011**, *59*, 2663–2671. [CrossRef] [PubMed]

36. Kim, S.B.; Chang, B.Y.; Jo, Y.H.; Lee, S.H.; Han, S.-B.; Hwang, B.Y.; Kim, S.Y.; Lee, M.K. Macrophage activating activity of pyrrole alkaloids from *Morus alba* fruits. *J. Ethnopharmacol.* **2013**, *145*, 393–396. [CrossRef] [PubMed]

37. Isabelle, M.; Lee, B.L.; Ong, C.N.; Liu, X.; Huang, D. Peroxyl radical scavenging capacity, polyphenolics, and lipophilic antioxidant profiles of mulberry fruits cultivated in southern China. *J. Agric. Food Chem.* **2008**, *56*, 9410–9416. [CrossRef] [PubMed]

38. Liu, C.; Xiang, W.; Yu, Y.; Shi, Z.-Q.; Huang, X.-Z.; Xu, L. Comparative analysis of 1-deoxynojirimycin contribution degree to α-glucosidase inhibitory activity and physiological distribution in *Morus alba* L. *Ind. Crops Prod.* **2015**, *70*, 309–315. [CrossRef]

39. Song, W.; Wang, H.J.; Bucheli, P.; Zhang, P.F.; Wei, D.Z.; Lu, Y.H. Phytochemical profiles of different mulberry (*Morus* sp.) species from China. *J. Agric. Food Chem.* **2009**, *57*, 9133–9140. [CrossRef] [PubMed]

40. Jiang, Y.; Nie, W.-J. Chemical properties in fruits of mulberry species from the Xinjiang province of China. *Food Chem.* **2015**, *174*, 460–466. [CrossRef] [PubMed]

41. Aramwit, P.; Bang, N.; Srichana, T. The properties and stability of anthocyanins in mulberry fruits. *Food Res. Int.* **2010**, *43*, 1093–1097. [CrossRef]

42. Calin-Sanchez, A.; Martinez-Nicolas, J.J.; Munera-Picazo, S.; Carbonell-Barrachina, A.A.; Legua, P.; Hernandez, F. Bioactive compounds and sensory quality of black and white mulberries grown in Spain. *Plant Foods Hum. Nutr.* **2013**, *68*, 370–377. [CrossRef] [PubMed]

43. Chen, W.; Li, Y.; Bao, T.; Gowd, V. Mulberry fruit extract affords protection against ethyl carbamate-induced cytotoxicity and oxidative stress. *Oxid. Med. Cell. Longev.* **2017**, *2017*, 1594963. [CrossRef] [PubMed]

44. Jiang, D.Q.; Guo, Y.; Xu, D.H.; Huang, Y.S.; Yuan, K.; Lv, Z.Q. Antioxidant and anti-fatigue effects of anthocyanins of mulberry juice purification (MJP) and mulberry marc purification (MMP) from different varieties mulberry fruit in China. *Food Chem. Toxicol.* **2013**, *59*, 1–7. [CrossRef] [PubMed]

45. Carvalho, J.C.T.; Perazzo, F.F.; Machado, L.; Bereau, D. Biologic activity and biotechnological development of natural products. *Biomed. Res. Int.* **2013**, 971745. [CrossRef] [PubMed]

46. Lila, M.A. Anthocyanins and human health: An in vitro investigative approach. *J. Biomed. Biotechnol.* **2004**, *2004*, 306–313. [CrossRef] [PubMed]

47. Lee, Y.M.; Yoon, Y.; Yoon, H.; Park, H.M.; Song, S.; Yeum, K.J. Dietary anthocyanins against obesity and inflammation. *Nutrients* **2017**, *9*. [CrossRef] [PubMed]

48. Yang, S.; Wang, B.L.; Li, Y. Advances in the pharmacological study of *Morus alba* L. *Acta Pharm. Sin.* **2014**, *49*, 824–831.

49. Huang, H.P.; Ou, T.T.; Wang, C.J. Mulberry (sang shen zi) and its bioactive compounds, the chemoprevention effects and molecular mechanisms in vitro and in vivo. *J. Tradit. Complement. Med.* **2013**, *3*, 7–15. [CrossRef] [PubMed]

50. Chen, C.-C.; Liu, L.-K.; Hsu, J.-D.; Huang, H.-P.; Yang, M.-Y.; Wang, C.-J. Mulberry extract inhibits the development of atherosclerosis in cholesterol-fed rabbits. *Food Chem.* **2005**, *91*, 601–607. [CrossRef]

51. Adisakwattana, S.; Ruengsamran, T.; Kampa, P.; Sompong, W. In vitro inhibitory effects of plant-based foods and their combinations on intestinal α-glucosidase and pancreatic α-amylase. *BMC Complement. Altern. Med.* **2012**, *12*, 110. [CrossRef] [PubMed]

52. Chang, Y.-C.; Yang, M.-Y.; Chen, S.-C.; Wang, C.-J. Mulberry leaf polyphenol extract improves obesity by inducing adipocyte apoptosis and inhibiting preadipocyte differentiation and hepatic lipogenesis. *J. Funct. Foods* **2016**, *21*, 249–262. [CrossRef]

53. Kwon, H.J.; Chung, J.Y.; Kim, J.Y.; Kwon, O. Comparison of 1-deoxynojirimycin and aqueous mulberry leaf extract with emphasis on postprandial hypoglycemic effects: In vivo and in vitro studies. *J. Agric. Food. Chem.* **2011**, *59*, 3014–3019. [CrossRef] [PubMed]

54. Li, Y.-G.; Ji, D.-F.; Zhong, S.; Lv, Z.-Q.; Lin, T.-B.; Chen, S.; Hu, G.-Y. Hybrid of 1-deoxynojirimycin and polysaccharide from mulberry leaves treat diabetes mellitus by activating PDX-1/insulin-1 signaling pathway and regulating the expression of glucokinase, phosphoenolpyruvate carboxykinase and glucose-6-phosphatase in alloxan-induced diabetic mice. *J. Ethnopharmacol.* **2011**, *134*, 961–970. [PubMed]

55. Naowaratwattana, W.; De-Eknamkul, W.; De Mejia, E.G. Phenolic-containing organic extracts of mulberry (*Morus alba* L.) leaves inhibit HepG2 hepatoma cells through G2/M phase arrest, induction of apoptosis, and inhibition of topoisomerase II alpha activity. *J. Med. Food.* **2010**, *13*, 1045–1056. [CrossRef] [PubMed]

56. Naowaboot, J.; Pannangpetch, P.; Kukongviriyapan, V.; Kukongviriyapan, U.; Nakmareong, S.; Itharat, A. Mulberry leaf extract restores arterial pressure in streptozotocin-induced chronic diabetic rats. *Nutr. Res.* **2009**, *29*, 602–608. [CrossRef] [PubMed]

57. De Oliveira, A.M.; do Nascimento, M.F.; Ferreira, M.R.A.; de Moura, D.F.; dos Santos Souza, T.G.; da Silva, G.C.; da Silva Ramos, E.H.; Paiva, P.M.G.; de Medeiros, P.L.; da Silva, T.G.; et al. Evaluation of acute toxicity, genotoxicity and inhibitory effect on acute inflammation of an ethanol extract of *Morus alba* L. (moraceae) in mice. *J. Ethnopharmacol.* **2016**, *194*, 162–168. [CrossRef] [PubMed]

58. Guo, C.; Li, R.; Zheng, N.; Xu, L.; Liang, T.; He, Q. Anti-diabetic effect of *Ramulus mori* polysaccharides, isolated from *Morus alba* L., on STZ-diabetic mice through blocking inflammatory response and attenuating oxidative stress. *Int. Immunopharmacol.* **2013**, *16*, 93–99. [CrossRef] [PubMed]

59. Jiao, Y.; Wang, X.; Jiang, X.; Kong, F.; Wang, S.; Yan, C. Antidiabetic effects of *Morus alba* fruit polysaccharides on high-fat diet- and streptozotocin-induced type 2 diabetes in rats. *J. Ethnopharmacol.* **2017**, *199*, 119–127. [CrossRef] [PubMed]

60. Willett, W.C. Dietary fats and coronary heart disease. *J. Intern. Med.* **2012**, *272*, 13–24. [CrossRef] [PubMed]

61. Townsend, N.; Wilson, L.; Bhatnagar, P.; Wickramasinghe, K.; Rayner, M.; Nichols, M. Cardiovascular disease in Europe: Epidemiological update 2016. *Eur. Heart J.* **2016**, *37*, 3232–3245. [CrossRef] [PubMed]

62. Ma, Z.F.; Lee, Y.Y. Virgin coconut oil and its cardiovascular health benefits. *Nat. Prod. Commun.* **2016**, *11*, 1151–1152.

63. Lu, H.; Pan, W.-Z.; Wan, Q.; Cheng, L.-L.; Shu, X.-H.; Pan, C.-Z.; Qian, J.-Y.; Ge, J.-B. Trends in the prevalence of heart diseases over a ten-year period from single-center observations based on a large echocardiographic database. *J. Zhejiang Univ. Sci. B* **2016**, *17*, 54–59. [CrossRef] [PubMed]

64. Chobanian, A.V. Single risk factor intervention may be inadequate to inhibit atherosclerosis progression when hypertension and hypercholesterolemia coexist. *Hypertension* **1991**, *18*, 130–131. [CrossRef] [PubMed]

65. Venkatesan, N.; Devaraj, S.N.; Devaraj, H. Increased binding of LDL and VLDL to apo B,E receptors of hepatic plasma membrane of rats treated with Fibernat. *Eur. J. Nutr.* **2003**, *42*, 262–271. [CrossRef] [PubMed]

66. Sirikanchanarod, A.; Bumrungpert, A.; Kaewruang, W.; Senawong, T.; Pavadhgul, P. The effect of mulberry fruits consumption on lipid profiles in hypercholesterolemic subjects: A randomized controlled trial. *J. Pharm. Nutr. Sci.* **2016**, *60*, 7–14.

67. Kalofoutis, C.; Piperi, C.; Kalofoutis, A.; Harris, F.; Phoenix, D.; Singh, J. Type II diabetes mellitus and cardiovascular risk factors: Current therapeutic approaches. *Exp. Clin. Cardiol.* **2007**, *12*, 17–28. [PubMed]

68. Xu, L.; Yang, F.; Wang, J.; Huang, H.; Huang, Y. Anti-diabetic effect mediated by *Ramulus mori* polysaccharides. *Carbohydr. Polym.* **2015**, *117*, 63–69. [CrossRef] [PubMed]

69. Yan, F.; Dai, G.; Zheng, X. Mulberry anthocyanin extract ameliorates insulin resistance by regulating PI3K/AKT pathway in HepG2 cells and db/db mice. *J. Nutr. Biochem.* **2016**, *36*, 68–80. [CrossRef] [PubMed]

70. Ebbert, J.O.; Jensen, M.D. Fat depots, free fatty acids, and dyslipidemia. *Nutrients* **2013**, *5*, 498–508. [CrossRef] [PubMed]

71. Jung, U.J.; Choi, M.-S. Obesity and its metabolic complications: The role of adipokines and the relationship between obesity, inflammation, insulin resistance, dyslipidemia and nonalcoholic fatty liver disease. *Int. J. Mol. Sci.* **2014**, *15*, 6184–6223. [CrossRef] [PubMed]

72. Klop, B.; Elte, J.W.F.; Castro Cabezas, M. Dyslipidemia in obesity: Mechanisms and potential targets. *Nutrients* **2013**, *5*, 1218–1240. [CrossRef] [PubMed]

73. DeFronzo, R.A.; Ferrannini, E. Insulin resistance. A multifaceted syndrome responsible for NIDDM, obesity, hypertension, dyslipidemia, and atherosclerotic cardiovascular disease. *Diabetes Care* **1991**, *14*, 173–194. [CrossRef] [PubMed]

74. Lim, H.H.; Lee, S.O.; Kim, S.Y.; Yang, S.J.; Lim, Y. Anti-inflammatory and antiobesity effects of mulberry leaf and fruit extract on high fat diet-induced obesity. *Exp. Biol. Med.* **2013**, *238*, 1160–1169. [CrossRef] [PubMed]

75. Torre, L.A.; Bray, F.; Siegel, R.L.; Ferlay, J.; Lortet-Tieulent, J.; Jemal, A. Global cancer statistics, 2012. *CA Cancer J.* **2015**, *65*, 87–108. [CrossRef] [PubMed]

76. Pourhoseingholi, M.A.; Vahedi, M.; Baghestani, A.R. Burden of gastrointestinal cancer in Asia: An overview. *Gastroenterol. Hepatol. Bed Bench* **2015**, *8*, 19–27. [PubMed]

77. Ma, Z.F.; Majid, N.A.; Yamaoka, Y.; Lee, Y.Y. Food allergy and helicobacter pylori infection: A systematic review. *Front. Microbiol.* **2016**, *7*, 368. [CrossRef] [PubMed]

78. Nishizawa, T.; Suzuki, H. Gastric carcinogenesis and underlying molecular mechanisms: Helicobacter pylori and novel targeted therapy. *BioMed Res. Int.* **2015**, *2015*, 794378. [CrossRef] [PubMed]

79. Huang, H.-P.; Chang, Y.-C.; Wu, C.-H.; Hung, C.-N.; Wang, C.-J. Anthocyanin-rich *Mulberry* extract inhibit the gastric cancer cell growth in vitro and xenograft mice by inducing signals of p38/p53 and c-jun. *Food Chem.* **2011**, *129*, 1703–1709. [CrossRef]

80. Li, Y.; Yang, Z.; Jia, S.; Yuan, K. Protective effect and mechanism of action of mulberry marc anthocyanins on carbon tetrachloride-induced liver fibrosis in rats. *J. Funct. Foods* **2016**, *24*, 595–601. [CrossRef]

81. Chang, J.-J.; Hsu, M.-J.; Huang, H.-P.; Chung, D.-J.; Chang, Y.-C.; Wang, C.-J. Mulberry anthocyanins inhibit oleic acid induced lipid accumulation by reduction of lipogenesis and promotion of hepatic lipid clearance. *J. Agric. Food Chem.* **2013**, *61*, 6069–6076. [CrossRef] [PubMed]

82. Schulz, K.F.; Altman, D.G.; Moher, D. Consort 2010 statement: Updated guidelines for reporting parallel group randomised trials. *BMJ* **2010**, *340*, c332. [CrossRef] [PubMed]

83. Moher, D.; Hopewell, S.; Schulz, K.F.; Montori, V.; Gøtzsche, P.C.; Devereaux, P.J.; Elbourne, D.; Egger, M.; Altman, D.G. Consort 2010 explanation and elaboration: Updated guidelines for reporting parallel group randomised trials. *BMJ* **2010**, *340*, c869. [CrossRef] [PubMed]

84. Pandis, N.; Fleming, P.S.; Hopewell, S.; Altman, D.G. The consort statement: Application within and adaptations for orthodontic trials. *Am. J. Orthod. Dentofac. Orthop.* **2015**, *147*, 663–679. [CrossRef] [PubMed]

85. Yuan, Q.; Zhao, L. The mulberry (*Morus alba* L.) fruit—A review of characteristic components and health benefits. *J. Agric. Food Chem.* **2017**, *65*, 10383–10394. [CrossRef] [PubMed]

86. Ravichanthiran, K.; Ma, Z.F.; Zhang, H.; Cao, Y.; Wang, C.W.; Muhammad, S.; Aglago, E.K.; Zhang, Y.; Jin, Y.; Pan, B. Phytochemical profile of brown rice and its nutrigenomic implication. *Antioxidants* **2018**, accepted for publication.

antioxidants

MDPI

Review

Polyphenols as Promising Drugs against Main Breast Cancer Signatures

María Losada-Echeberría [1], María Herranz-López [1], Vicente Micol [1,2] and Enrique Barrajón-Catalán [1,*]

[1] Institute of Molecular and Cell Biology (IBMC), Miguel Hernández University (UMH), Avda. Universidad s/n, Elche 03202, Spain; mlosada@umh.es (M.L.-E.); mherranz@umh.es (M.H.-L.); vmicol@umh.es (V.M.)

[2] CIBER, Fisiopatología de la Obesidad y la Nutrición, CIBERobn, Instituto de Salud Carlos III (CB12/03/30038), Palma de Mallorca 07122, Spain

* Correspondence: e.barrajon@umh.es; Tel.: +34-965-222-586

Received: 4 October 2017; Accepted: 3 November 2017; Published: 7 November 2017

Abstract: Breast cancer is one of the most common neoplasms worldwide, and in spite of clinical and pharmacological advances, it is still a clinical problem, causing morbidity and mortality. On the one hand, breast cancer shares with other neoplasms some molecular signatures such as an imbalanced redox state, cell cycle alterations, increased proliferation and an inflammatory status. On the other hand, breast cancer shows differential molecular subtypes that determine its prognosis and treatment. These are characterized mainly by hormone receptors especially estrogen receptors (ERs) and epidermal growth factor receptor 2 (HER2). Tumors with none of these receptors are classified as triple negative breast cancer (TNBC) and are associated with a worse prognosis. The success of treatments partially depends on their specificity and the adequate molecular classification of tumors. New advances in anticancer drug discovery using natural compounds have been made in the last few decades, and polyphenols have emerged as promising molecules. They may act on various molecular targets because of their promiscuous behavior, presenting several physiological effects, some of which confer antitumor activity. This review analyzes the accumulated evidence of the antitumor effects of plant polyphenols on breast cancer, with special attention to their activity on ERs and HER2 targets and also covering different aspects such as redox balance, uncontrolled proliferation and chronic inflammation.

Keywords: breast cancer; polyphenols; luminal; TNBC; redox balance; apoptosis; autophagy; inflammation; ER; HER2

1. Introduction

Nowadays, cancer is one of the main causes of mortality worldwide. In 2012, 14 million new cases were diagnosed, and there were 8.2 million cancer-related deaths [1]. Breast cancer is the most common tumor in occidental women; one in eight women will have a breast tumor during their lifetime, and every year, up to 1.4 million new cases are diagnosed worldwide [1]. An annual mortality rate of about 450,000 is estimated, which accounts for 20–30% of all tumors.

Currently, treatments are based mainly on two molecular markers: hormone receptors and epidermal growth factor receptor 2 (HER2). The expression of these molecular markers determines both prognosis and treatment. Although new therapies against breast cancer have been able to reduce mortality, the prognosis, especially in the more advanced stages, remains unpromising, and therefore, further research in this field is needed [2]. The main advances have been obtained for HER2 positive tumors where monoclonal antibodies such as trastuzumab (Herceptin®) have improved prognosis in HER2 overexpressing tumors [3]. Unfortunately, even in these cases, resistance is also frequent, leading to nonspecific therapeutic options. In addition, approximately 10–20% of breast tumors are considered triple negative, which implies that no specific therapy is available, and only classic chemotherapy may be applied.

In this clinical scenario, new efforts have been made to obtain new drugs for breast cancer treatment leading to several promising molecules [4]. However, new molecules are still required. Natural compounds from different origins such as vegetal [5], microbial [6] and marine [7] species are a source of new molecules demonstrating activity against cancer and other diseases. These compounds derive from the secondary metabolism of these organisms and have been selected by nature through evolution. Between natural compounds, polyphenols have emerged as one of the main families of compounds with potential biological activity in many diseases such as cancer [8–13], diabetes [14,15], inflammation [16–19], obesity-related diseases [20], neurodegenerative disorders [21–23], bacterial [24–27] and viral infections [28,29] or cardiovascular diseases [30]. In addition, they possess a relevant antioxidant activity [24,31–35], which is the basis of part of their biological activity.

Polyphenols are widely distributed in fruits, vegetables, tea, essential oils and cereals; their molecular structure is characterized by the presence of one or more phenolic rings substituted with at least one hydroxyl group. Different classes and subclasses of polyphenols generate a large structural variability that is characterized by the number of phenolic rings they possess and the moieties that substitute their aromatic rings (see http://phenol-explorer.eu/compounds/classification for an updated classification). The main groups of polyphenols are: phenolic acids, flavonoids, stilbenes and lignans (Figure 1).

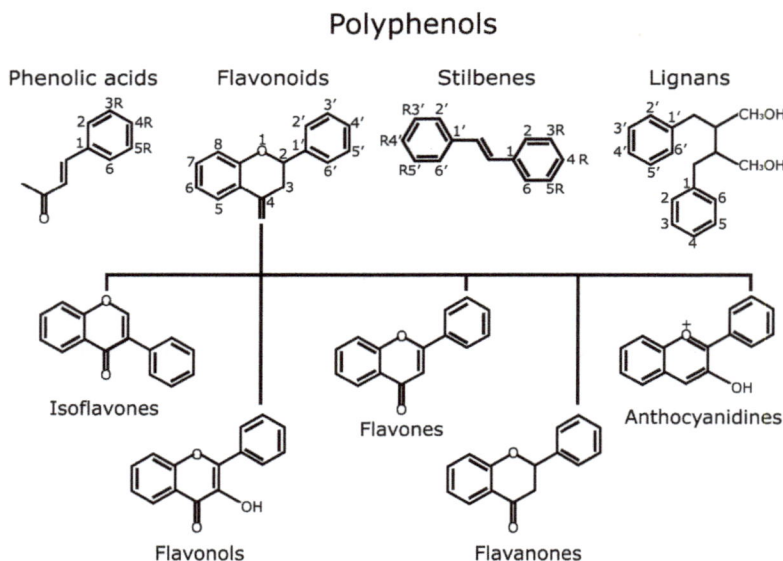

Figure 1. Polyphenols structure and classification.

As mentioned above, some polyphenols have demonstrated anticancer activity, showing biological activity against most of the main cancer molecular targets such as kinases, pre- and anti-apoptotic proteins, enzymes that regulate energy metabolism and regulatory proteins linked to proliferation and signaling pathways [36]. Their broad activity may be attributed to several mechanisms, including the interaction and modulation of a wide range of proteins, enzymes and membrane receptors, regulation of gene expression, apoptosis induction, vasodilation and modulation of cell pathways [37–42]. In addition, polyphenols have preventive effects against tumor initiation through numerous mechanisms, such as the avoidance of genotoxic molecule formation, the blockade of mutagenic transforming enzyme activity [43], the regulation of Phase I and II enzymes, such as cytochrome P450s (CYP) [44] and S-transferase (GST) [45], as well as preventing DNA damage [46,47].

For all these reasons, new treatments based on polyphenolic compounds are being studied as an alternative and/or adjuvant therapies in these pathologies using different models [21]. Potential benefits of their dietary intake on human health and, more specifically, on cancer risk (including breast cancer) have been also reviewed [48,49]. Specifically, for breast cancer, interesting results have been obtained with a mixture of tea extract and quercetin [50], with *Pinus radiata* [51], Indian lotus [52], *Hypogymnia physodes* lichen [53], *Morinda citrifolia* [54] or with olive leaf extracts [55–58], among others.

This review describes the different breast cancer types, molecular biomarkers and their main treatments. A compilation of the main molecular breast cancer targets and the use of polyphenols to address them is reviewed, covering different aspects such as redox balance, uncontrolled proliferation and chronic inflammation, with particular interest in ER (estrogen receptor) and HER2 and the use of polyphenols to modulate their pathways.

2. Breast Cancer Biomarkers Determine Both Prognosis and Treatment

The identification of molecular biomarkers plays a significant role in the diagnosis and prognosis of breast cancer. They represent therapeutic targets, and their expression is used to classify cancers according to the different molecular subtypes (Table 1). The major biomarkers of breast cancer include the hormonal estrogen and progesterone receptors (ER and PR) and HER2/ERBB2 and, with less relevance, Ki-67 protein [59]. These markers have been extensively studied, and their expression correlates with differences in tumor behavior and patient response to treatments [60,61]:

- Hormone receptors, ERs and PR, are the main factors responsible for hormone response. Breast cancer is a hormone-dependent tissue, and this response is controlled by these receptors [62,63]. ER and PR expression confer a better prognosis and are the basis of hormonal therapy.
- HER2 is a membrane receptor involved in cell proliferation signal transduction. It is present in normal cells and in most tumors, but in 5–15% of breast tumors is overexpressed, increasing tumor aggressiveness [59]. These tumors are very often sensitive to treatment with anti-HER2 treatments, such as humanized monoclonal antibodies or specific inhibitors [64].
- Ki-67: is a protein marker that can be only detected in proliferating cells and currently is used to rate tumor proliferation, particularly lymphomas, breast, endocrine and brain cancers [65]. Indeed, Ki-67 contributes greatly to the Oncotype score [66]. Tumors with high proliferation rates (>15%) have a poor prognosis [65].

Using these molecular markers, breast cancer can be divided into four major molecular subtypes: Luminal A, Luminal B, HER2 type and triple negative breast cancer (TNBC) [67–69]. This division determines treatment as shown in Table 1. Overexpression of HER2 is related to the lack of expression of hormone receptors in most cases. The same situation occurs with Ki67, which is usually elevated in cells that do not express these receptors. Between Ki67 and HER2, no relationship has been found.

Table 1. Breast cancer molecular subtypes and their main treatments. Representative cell lines for each subtype are also shown.

Subtype	ER/PR	HER2	Ki67	Treatment	Cell lines
Luminal A	+/+	−	<15%	Antihormonal	MCF7, T47D
Luminal B	+/+	−/+	>15%	Antihormonal	BT474
HER2-type	−/−	+		Anti-HER2	SkBr3, AU565
TNBC	−/−	−/−	>15%	Chemotherapy	MDA-MB231

Luminal A tumors are hormone dependent, with hormone receptors positive expression (ER/PR-positive). They are HER2 negative and present one or two tumor grades. They represent

30–70% of breast cancers [70–72] and have the best prognosis, with high survival and low recurrence rates [73–75].

Luminal B tumors tend to be ER/PR-positive. They can be HER2-negative or positive; in this last case, some authors consider it as a new sub-type called Luminal C [76], but this classification is not widespread. They are also characterized by a higher tumor grade, a larger tumor size and a positive lymph node dissemination. Patients with Luminal B tumors are usually diagnosed at more advanced ages than in cases with Luminal A [75,77]. Compared to Luminal A tumors, they also tend to have factors that lead to a poorer prognosis, mainly an increase in Ki67 protein of 15–20% [78,79]. The prevalence of Luminal B tumors is approximately 10–20% [70–72] and still shows high survival rates, although not as high as those of Luminal A tumors [75]. Treatments for both Luminal A and B are based on hormone therapy regimens [80], which is based on the use of SERMs (selective estrogen receptor modulators) such as tamoxifen or fulvestrant and aromatase inhibitors like anastrozole, exemestane and letrozole [81].

HER2-type tumors are characterized by being ER/PR-negative, overexpressing HER2, having lymph node positive implication and present poorer tumor grade [74,75]. Patients with HER2 tumors are usually diagnosed at an earlier age than Luminal A and Luminal B [75]. Approximately 5–15% of breast cancers are HER2-positive [71,72] and can be treated with specific anti-HER2 drugs. This group includes monoclonal antibodies like trastuzumab and pertuzumab and specific HER2 inhibitors like lapatinib [64]. Before these drugs were available, HER2-type tumors had a rather poor prognosis [71,82].

TNBC encompasses all tumors that are negative for ER, PR and HER2. It is considered the most metastatic type of breast cancer and has highly invasive properties, is larger, has a poorer prognosis with a high probability of relapse, no response to hormonal therapy and has nodal involvement. Around 10–20% of tumors correspond to triple negative tumors [70–72]. There are several approaches to counteract TNBC, but all based on classical chemotherapy using anthracyclines, taxanes, poly(ADP-ribose) polymerase protein inhibitors and platinum-containing chemotherapeutic agents [83].

3. Breast Cancer Signatures and Polyphenols

3.1. Redox Balance

Oxidative stress is caused by an imbalance between the production of reactive oxygen species (ROS) and the efficacy of the endogenous antioxidant system. Tissues are continuously exposed to free radicals derived from metabolism or due to external factors such as pollution or radiation [84–86]. In fact, ROS participate in physiological functions such as metabolism signaling and defense against infections [87,88]. However, uncontrolled ROS production or accumulation can induce lipid peroxidation, protein modifications and DNA damage. These events lead to membrane alterations, protein dysfunctions and genetic alterations, all of which are linked to carcinogenesis and tumor progression.

Balance between oxidant species and antioxidants (redox balance) is essential to maintain a healthy cell status. Breast cancer is characterized by a systemic prooxidant status [89], and an increased ROS presence is determinant for some relevant events such as tumor progression mediated by stromal cells [90]. However, ROS can play a dual role [91], not only in breast [92], but in all cancers. Cells have several mechanisms to transform and eliminate ROS and avoid their harmful effects, such as superoxide dismutase (SOD), catalase (CAT) or glutathione peroxidase (GTX) enzymes [93–95]. The synergistic action between these enzymes, vitamins and exogenous antioxidants such as polyphenols allows neutralizing free radicals and modulating cellular signaling [96].

Initially, increased ROS production leads to a pro-oncogenic situation as it provokes two crucial effects: mitochondrial dysfunction that conduces to protein oxidation, lipid peroxidation and DNA damage. On the other hand, once tumor cells have developed, an increase in ROS presence can lead to tumor cell death. This fact has been linked to some anticancer drugs such as doxorubicin and paclitaxel.

Polyphenols can participate in these two situations. First, it is generally admitted that polyphenols are antioxidants and therefore counteract ROS production and inhibit oxidative DNA damage and mitochondrial dysfunction by acting as chemo-preventive agents [97,98]. For this reason, the preventive character of polyphenols acquired through diet in diseases such as cancer [99,100], diabetes [101,102] or atherosclerosis [103,104] has been studied.

However, there is increasing evidence suggesting that under certain conditions, polyphenols can act as prooxidants, leading to tumor cell death [105,106]. For example, it has been shown that in systems containing active redox metals such as copper, some polyphenols show prooxidant activity, catalyzing their redox cycle and leading to ROS formation [107,108]. Since copper levels in cancer cells compared to healthy cells are increased [109], this prooxidant mechanism would present preferential cytotoxicity against cancer cells, leaving the normal cells undamaged. This effect has been demonstrated using polyphenols like luteolin, apigenin, epigallocatechin-3-gallate and resveratrol [110].

3.2. Uncontrolled Proliferation

One of the main characteristics of tumor cells, no matter their origin, is their ability to grow and proliferate in an uncontrolled way. Cellular proliferation is mainly linked to cell cycle progression. This cycle shows different checkpoints in which the cell examines the internal and external signals and decides whether to proceed with cell division or not. This uncontrolled cell division is caused by a malfunction of these checkpoints in the cell cycle [111,112]. The most important regulators of the cell cycle are proteins called cyclins, enzymes called cyclin-dependent kinases (CDKs) and anaphase-promoting enzymatic complex (APC/C) [113,114]. In addition, the active CDK complexes are regulated by binding to CDK inhibitors (p21 and p27) and by other kinases and phosphatases, which control the cell cycle by balancing CDK activity. Variations in the concentration of these inhibitors can alter the normal sequence of the cell cycle as occurs in some tumors or in aged cells.

Polyphenols can act by modulating cyclins, Cdks or APC/C causing cell cycle arrest [115,116]. This cytostatic activity has been also studied in breast cancer, where some polyphenolic compounds have demonstrated their cytostatic activity. For example, ginnalins A–C induce cell cycle arrest in the S and G2/M phases in colon cancer HCT-116 cells and breast cancer MCF-7 cells by decreasing cyclin A and D1 levels [117]. Green tea polyphenols induce cell cycle arrest at G1/G0 phase in breast cancer MCF-7 cells [118]. Polyphenolic extracts from hawthorn fruit have shown a cytostatic effect on MCF-7 breast cancer cells by blocking the cycle in S phase [119]. Ellagic acid induces cell cycle arrest at G0–G1 in human breast cancer MCF-7 cells mediated by a cyclin A2 and cyclin E2 downregulation and an upregulation of the CDK-inhibitors $p21^{Cip1}$, $p15$ and $p19$ [120].

In addition to cytostatic activity, however, uncontrolled proliferation can be also treated through cytotoxic mechanisms like apoptosis, autophagy, necrosis or necroptosis. Polyphenols can participate in all these mechanisms as described below.

3.2.1. Apoptosis

Apoptosis is the main mechanism of cell-programmed death. It is characterized by a series of molecular processes that result in cell membrane blebbing, nuclear and chromosomal DNA fragmentation, chromatin condensation, fragmentation and the translocation of phosphatidylserine to the outer face of the plasma membrane, which means that they are eliminated by macrophages [121–123].

There are two different apoptotic pathways, mainly differentiated by their starting stimuli. These two pathways may overlap and share some molecular targets as caspases [121]. On the one hand, the extrinsic pathway is mediated by ligands that bind to receptors on the cell surface. Alternatively, the intrinsic pathway is mediated by cellular stress or by DNA injury. There are several molecular mechanisms that tumor cells use to suppress apoptosis. For example, tumor cells may acquire resistance to apoptosis by downregulation of anti-apoptotic Bcl-2 expression or mutation of the pro-apoptotic BAX protein. Expression of both is regulated by the tumor suppressor gene p53 [124], which is mutated in a large number of cancer types [125].

As occurred in previous sections, polyphenols can act over different stages of apoptosis. For example, anti-apoptotic Bcl-2 expression is decreased in human breast cancer cell lines, MCF-7 and T47D with a silibinin treatment [126]. Besides, silibinin upregulated phosphatidylinositol-3,4,5-trisphosphate-3-phosphatase (PTEN) and caused a slight increase in p21, thus promoting apoptosis [127,128]. The treatment with Annurca apple polyphenolic extract increases the levels of p53, p21 and the pro-apoptotic ratio of Bax/Bcl-2 in parallel with caspase-9, -6 and -7, in MCF-7 breast cancer cells [129]. Fruit peel polyphenolic extract from different sources (red grape, blackberry, black cherry, black currant, elderberry, blackthorn and plum) induced caspase-dependent cell death associated with an increase in oxidative stress, causing the release of pro- and anti-apoptotic mitochondrial proteins from the Bcl-2 family in breast cancer MCF-7 cells [130]. Oleuropein induced apoptosis due to upregulation of both p53 and Bax gene expression levels and downregulation of Bcl2 in human breast cancer MCF-7 cells [131] and in the HepG2 human hepatoma cell line [132].

Tea polyphenols such as epigallocatechin gallate (EGCG) downregulate telomerase activity in breast cancer cells thereby increasing cellular apoptosis and inhibiting cellular proliferation of MCF-7 and MDA-MB-231 breast cancer cells [133]. They can also inhibit cell growth and induce apoptosis through downregulation of survivin expression, a member of the inhibitor apoptosis protein family (IAP) that inhibits caspases and blocks cell death [134]; this effect has been observed in MCF-7, SK-BR-3 cells and MDA-MB-231 breast cancer cell models [135,136].

Artichoke polyphenols modify Bcl-2 and BAX expression in human breast cancer cell line MDA-MB-231, leading to a pro-apoptotic situation [137], which was accompanied with the upregulation of p21 [138].

Finally, other widely-distributed polyphenols such as resveratrol [139], quercetin [140] and catechin [141], promote apoptosis by decreasing IAP1 and survivin expression and increasing FAS ligand and its receptor [142] expression.

3.2.2. Autophagy

Initially, autophagy or cellular autodigestion is a route involved in the degradation of proteins and organelles that may be important in the pathogenesis of some diseases. Dysfunctions in the autophagy process are associated with cancer [143–145], neurodegeneration [146,147], aging [147,148] and infections [149,150]. The major proteins involved in the regulation of autophagy are the mammalian target of the rapamycin (mTOR), phosphatidylinositol 3 kinase (PI3K), AKT kinase (AKT), Beclin-1 and p53 [151]. Activation of the PI3K/AKT pathway implies mTOR activation, leading to autophagy inhibition [152]; on the contrary, mTOR inhibition is related to autophagy activation. In addition to mTOR, there is a family of proteins called ATG (autophagy-related protein), such as LC3, ATG5 or ATG12, that are also involved in the regulation of this mechanism.

Some polyphenols can induce tumor cell death through autophagy activation; for example, *Solanum nigrum* L. extract decreased p-AKT levels causing mTOR inactivation and triggering autophagy in AU565 human breast cancer cells [153], as well as blueberry polyphenols in MDA-MB231 cells [154] and grape skin extracts in a murine model of breast cancer [155]. Mango polyphenols cause downregulation of mTOR in human breast ductal carcinoma in situ xenograft models [156].

Resveratrol increases levels of LC3 and its lipidic form, LC3-II, which induces autophagy [157] in human breast cancer MCF-7 cells [158], as well as polyphenol-enriched extract of *Pimenta dioica* berries does in human breast cancer MCF-7, MDA-MB231, SkBr3, BT474 and T47D cells [159] and carnosol in MDA-M231 cells [160]. In another study, it was observed that LC3-II, Beclin 1 and Atg 7 were significantly upregulated by resveratrol [161], inducing autophagy.

3.3. Chronic Inflammation and Pro-Inflammatory Factors

Chronic inflammation has been linked to the development of some tumors [162–164], and it is known that diseases such as pancreatitis [165], hepatic steatosis or Crohn's disease [166], which present chronic inflammation, significantly increase the risk of cancer. This direct relationship with cancer development is also observed in infectious diseases that produce inflammation, such as hepatitis [167,168] or stomach infection by *Helicobacter pylori* [169–171].

Cell signaling by inflammatory cytokines had been shown to promote the development of cancer [172,173]. However, it was not until 2008, when the direct link between inflammation and cancer was first established, when it was observed that chronic inflammation causing DNA damage led to cancer development [174].

The abilities to inhibit or block the activity of NF-kB [175,176], cyclooxygenase (COX-2) [176–178] and lipoxygenase (LOX) [178,179] are the main causes of the anti-inflammatory capacity observed for polyphenols; thus, the role of phytochemicals in these pathways and in cancer-related inflammation has been extensively studied [180]. In this regard, individual polyphenols such as curcumin, EGCG and resveratrol have been reported to show anti-inflammatory effects [180], most of them by reducing NF-κB activation or expression. In addition to individual compounds, cocoa polyphenols decreased the nuclear levels of NF-κB and the expression of pro-inflammatory enzymes such as COX-2 and inducible NO synthase. Additionally, cocoa polyphenols effectively downregulated the levels of inflammatory markers induced by tumor necrosis factor-alpha (TNF-α) by inhibiting NF-κB translocation and JNK phosphorylation [181].

Pomegranate polyphenols also demonstrated anti-inflammatory activity as they were able to decrease NF-κB [182] and Nrf-2 [183] pathways in breast cancer cells. Cranberry extract also showed the ability to modulate PI3K/AKT, MAPK/ERK and STAT3 pathways, central nodes in the inflammatory signaling of cancer stem cells [184].

3.4. ER and Estrogen Synthesis

Estrogens exert their biological action by binding to ER-α and ER-β, both of which are members of the nuclear receptor superfamily of transcription factors. Although both ERs can develop estrogen-related responses, ER-α is the one that plays an important role in breast cancer (in this review, the term "ER" is used for "ER-α").

The effects of polyphenols on ER and estrogen pathway is mainly based on three actions: anti-estrogenic mechanisms [185–187], changes in ER expression and aromatase modulation [188–190] (Figure 2). Anti-estrogenic mechanisms are based on the structural similarities between some groups of flavonoids, such as isoflavones and lignans, and estrogens. In fact, these compounds are considered as phytoestrogens [191–197] due to their natural origin and their ability to interact with ERs. Polyphenols may exert different effects according to the dose administered. Low dose treatments have an activating effect on the estrogen receptor, which triggers the proliferative response favoring tumor development. Conversely, high concentrations promote processes such as apoptosis or cell cycle blockage resulting in antitumor effects as mentioned above. In addition, ER expression is also regulated by some polyphenols, for example, EGCG, which downregulates ER-α protein, mRNA and gene promoter activity in MCF-7 [198] and ER and PR in T-47D [199] cell lines. There are so many additional examples of flavonoids, like apigenin, luteolin myricetin, anthocyanidins or quercetin, and other compounds, such as stilbenes, ellagitannins, sulforaphanes, curcuminoids and tocopherols, among others, that are extensively reviewed [197,200].

Aromatase enzyme, a member of the CYP superfamily of enzymes, is encoded by the CYP19 gene and supposed to be another target of breast cancer. Its function is to aromatize androgens, producing estrogens. One strategy to counteract the proliferative effect of estrogen in breast cancer is to use aromatase inhibitors. Some polyphenols have shown aromatase inhibiting properties like biochanin A, an isoflavone extracted from red clover, and genistein, which have been reported to inhibit the activity of the aromatase enzyme in SK-BR3 [201]. Resveratrol also inhibits aromatase by significantly reducing the CYP19-encoding mRNA abundance in SK-BR-3 cells [202]. Isoliquiritigenin inhibited aromatase mRNA expression and suppressed the activity of CYP19 promoters I.3 and II in MCF-7 cells [203].

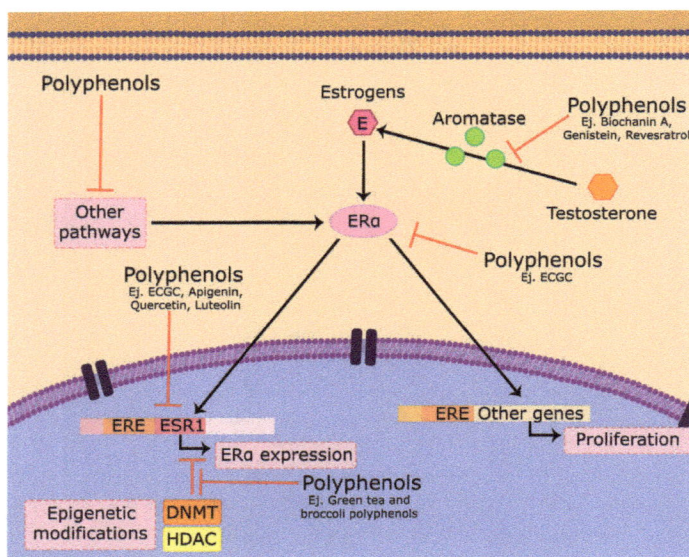

Figure 2. Estrogenic response can be modulated by polyphenol activity. ER activation leads to cell proliferation by activating estrogen response element (ERE)-controlled genes, including their own ER gene (ESR1). Some polyphenols are able both to inhibit ER activity (antiestrogenic action) or ER expression and consequently reduce cell proliferation. The ER activity can also be regulated by some transactivation pathways including the Ras/Raf, PI3K/Akt, AMPK or PKC pathways; all of them can be also modified by polyphenols. ESR1 gene expression can be also regulated by epigenetic modifications through chromatin modifying enzymes such as DNMT (DNA methyltransferases) and HDAC (histone deacetylases). Finally, estrogen synthesis from androgens by aromatase can be also blocked by some polyphenols.

Besides the above-mentioned mechanisms, some polyphenols are able to transform ER-negative phenotypes into ER-positive, thereby improving prognosis and allowing treatment with SERMs such as tamoxifen. This transformation may be due to epigenetic modifications mediated by chromatin modifying enzymes such as DNMT (DNA methyltransferases) and HDAC (histone deacetylases) [204–206]. Methylation of the ESR1 gene encoding for ER is associated with the ER-negative phenotype and therefore supports the hypothesis that treatment with DNMT inhibitors and HDAC activators can promote ER expression by reversing the ER-negative phenotype to ER-positive [207]. Green tea polyphenols have been shown to modify chromatin by inhibiting DNMTs and eliciting ER expression in the TNBC MDA-MB231 cell line [208]. It has also been observed that with a combined treatment of green tea extract and broccoli shoots, the epigenetic reactivation of ER is triggered, which in turn increases the sensitivity to tamoxifen in ER-negative breast cancer MDA-MB-231 and MDA-MB-157 cells.

3.5. HER2/ErRB Overexpression

HER2, also termed ERBB2 or EGFR2, is overexpressed in more than 30% of breast cancers and has been shown to play an important role in the progression of certain aggressive breast cancers. HER2 is a trans-membrane receptor tyrosine kinase that activates multiple proliferative signaling pathways, including PI3K/Akt and Ras/MAPK. Upon activation, HER2 can both homo- or hetero-dimerize with other HER family receptors, leading to signal transduction and cell proliferation (Figure 3). This is the main reason for which HER2 overexpression conduces to an abnormal proliferation and cancer spread. In addition to this main mechanism, however, it has also been shown that several apoptotic mechanisms are deregulated in cells overexpressing HER2 [209,210], and some anti-apoptotic proteins including Bcl-2, Bcl-xL and Mcl-1 are overexpressed [211]. These two additional mechanisms contribute to tumor cell survival and the spread of the cancer (Figure 3).

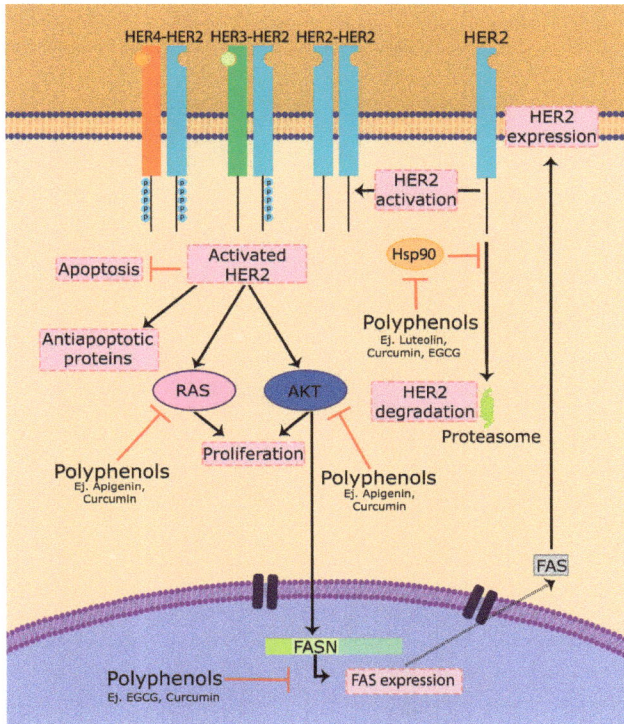

Figure 3. Polyphenols modulate the HER2 pathway. The HER2 membrane receptor, when activated, dimerizes with other HER2 (homodimerization) or other HER family receptor (heterodimerization) and transduces proliferation signals to downstream pathways such as the Ras/Raf or PI3K/Akt routes. This activation also inhibits apoptotic signals and promote antiapoptotic mechanisms. In addition, HER2 expression can be modulated by FASN, and some polyphenols are able to modify this action by actuating on FASN activity and/or expression. Finally, Hsp90 chaperone is required for HER2 to be functional, and some polyphenols modify this interaction leading to HER2 degradation by proteasome.

Nowadays, the most commonly-used therapy against HER2-positive tumors is the monoclonal antibody trastuzumab [212]. This treatment has positively revolutionized the prognosis of this kind of tumor; however, the main problem presented by these patients is the development of resistance to trastuzumab. Polyphenols can sensitize HER2-positive cells through different mechanisms: by decreasing its activation or downloading its expression. For example, some secoiridoids such as oleuropein promote a downregulation of HER2 by blocking ATP binding to the tyrosine kinase domain of the protein [213,214]. Silybin and silybin-phosphatidylcholine exhibit dose-dependent cell growth inhibitory effects and downregulation of HER2 in SkBr3 cells [215]. Apigenin induces apoptosis by depletion of HER2 protein and, in turn, suppression of signaling of the HER2/HER3-PI3K/Akt pathway [216]. HER2, p-Akt, p-MAPK and NF-κB oncoprotein levels decreased in a dose- and time-dependent way in BT-474, MCF- 7, MDA-MB-231 and SkBr3-hr (a herceptin-resistant strain from SkBr3 breast cancer cells) when treated with curcumin [217].

On the other hand, HER2 needs to interact with Hsp90 and its chaperone to acquire its function [218,219], so another possible therapeutic approach is the use of inhibitors of the HER2-Hsp90 binding, causing the dissociation of HER2 from its chaperone and, consequently, leading to HER2 degradation by the proteasome [220,221]. It has been observed that some polyphenols such as luteolin [222], geraniin [223], curcumin [224] or EGCG [225] are inhibitors of Hsp90 and therefore affect the HER2-Hsp90 binding.

Fatty acid synthase (FASN) has been linked with several types of cancer [226]. Overexpression of HER2 has been shown to increase FASN translation, which alters the activity of the mTOR and the PI3K/AKT signaling pathway in breast cancer cells [227]. FASN inhibition induces apoptosis and creates cytotoxicity [228–230], but also causes a marked decrease in the active forms of the HER2 protein, indicating that the HER2 oncogene presents positive feedback with FASN to ensure de novo overactive biogenesis of fatty acids. Consequently, FASN inhibition correlates with an inhibition of HER2 (Figure 3). Some polyphenols, such as curcumin [231], some olive oil lignans [232] and EGCG [233] have shown anticancer activity on HER-positive models through this FASN-linked mechanism.

4. Conclusions

Breast cancer is a heterogenic oncological disease in which, in addition to common cancer signatures, estrogenic response and HER2 expression constitute two clinical and molecular targets that must be taken into account.

As stated in this review, polyphenols are natural compounds that have demonstrated antitumor activity reaching different molecular targets. Some of them, especially EGCG, curcumin and flavonoids, such as luteolin, apigenin or quercetin, seem to be the most promising ones according to the current studies. Their promiscuous activity against different molecular targets is quite relevant and suggests that multitargeting therapy can be a future strategy for breast cancer treatment. However, additional aspects such as polyphenols' bioavailability and toxicity must be addressed. Polyphenols' bioavailability has been thoroughly studied for many polyphenols, especially those with the most relevant biological activity [234]. In general, polyphenols' bioavailability is low, and most of them are classified as Class IV by the Biopharmaceutical Classification System (BCS) (low permeability, low solubility) [235]. However, numerous studies have explored strategies to increase polyphenols' bioavailability, for example using different encapsulation techniques [236–239] or using pharmaceutical dispersions [240]. Intravenous administration represents an option for many anticancer drugs with poor bioavailability, so this could be a solution too for those polyphenols showing low bioavailability.

Toxicity is, as mentioned above, another relevant aspect to be studied. As occurred with their bioavailability, polyphenols' toxicity has been deeply studied [241–243], and in most cases, only very high doses present relevant toxic consequences. Moreover, new approaches to predict toxicity using bioinformatic tools such as admetSAR and DataWarrior [244,245] are also available, and new alternative methods both reducing and/or replacing animal use have been introduced in preclinical research (see http://www.oecd.org/chemicalsafety/testing/animal-welfare.htm). However, toxicity should not be a major problem since polyphenols are less toxic than most anticancer drugs.

All the accumulated evidence on the in vitro anticancer activity of polyphenols, as well as bioavailability and toxicity studies may be used to develop new anticancer drugs. Nevertheless, much preclinical research needs to be conducted before administration to oncologic patients with the required guaranties. In vivo studies and, especially, human clinical trials must be developed before real clinical use. However, most of these polyphenols can be obtained from the diet, allowing preventive or nutritional strategies that, although they must be challenged with clinical trials before being accepted, can be implemented in the short term as dietary recommendations. Some studies have been already conducted [48,49], but there is still a need for more evidence.

Future research must be focused on providing new and strong evidence about polyphenols' activity on every breast cancer signature. As mentioned above, preclinical and clinical evidence is scarce. New studies using omic sciences and in silico approaches could provide some of the pending results as occurred in other disciplines. Genomics, transcriptomics, proteomics, metabolomics and fluxomics strategies are continuously increasing our knowledge about polyphenols' mechanism of action [246,247]. In silico techniques, such as molecular docking, are being used to identify new ligands for specific cancer molecular targets [248,249]. Epigenetic studies are also providing relevant information [250,251] on polyphenols' effects, but a global approach using all the updated knowledge is undoubtedly the main pending task.

Acknowledgments: Some of the investigations expressed in this review have been partially or fully supported by competitive public grants from the following institutions: AGL2011-29857-C03-03 and IDI-20120751 grants (Spanish Ministry of Science and Innovation), Projects AGL2015-67995-C3-1-R, AGL2015-67995-C3-2-R, AGL2015-67995-C3-3-Rfrom the Spanish Ministry of Economy and Competitiveness (MINECO); PROMETEO/2012/007, PROMETEO/2016/006, ACOMP/2013/093, ACIF/2010/162 and ACIF/2016/230 grants from Generalitat Valenciana and CIBER (CB12/03/30038, Fisiopatologia de la Obesidad y la Nutricion, CIBERobn, Instituto de Salud Carlos III).

Author Contributions: María Losada-Echeberría gathered all the information and wrote the first version of the manuscript. María Herranz-López reviewed and added new information about polyphenols to the manuscript. Vicente Micol reviewed the manuscript and the English language. Enrique Barrajón-Catalán coordinated all the authors and edited and reviewed the different versions of the manuscript.

Conflicts of Interest: The authors declare no conflict of interest.

Abbreviations

HER2/ERB2	human epidermal growth factor receptor 2
CYP	cytochrome P450s
GST	s-transferase
TNBC	triple negative breast cancer
ER	estrogen receptor
PR	progesterone receptor
ROS	reactive oxygen species
SOD	superoxide dismutase
CAT	catalase
GTX	glutathione peroxidase
CDKs	cyclin-dependent kinases
APC/C	anaphase-promoting complex
PTEN	phosphatidylinositolo-3,4,5-trisphosphate-3-phosphatase
EGCG	epigallocatechin gallate
IAP	inhibitor apoptosis protein
mTOR	mammalian target of the rapamycin
PI3K	phosphatidylinositol 3 kinase
AKT	AKT kinase
COX-2	cyclooxygenase
LOX	lipoxygenase
TNF-α	Tumoral Necrosis Factor-alpha
DNMTs	DNA methyltransferases
HDACs	histone deacetylases
FASN	fatty acid synthase

References

1. Ferlay, J.; Soerjomataram, I.; Dikshit, R.; Eser, S.; Mathers, C.; Rebelo, M.; Parkin, D.M.; Forman, D.; Bray, F. Cancer incidence and mortality worldwide: Sources, methods and major patterns in globocan 2012. *Int. J. Cancer* **2015**, *136*, E359–E386. [CrossRef] [PubMed]

2. Torre, L.A.; Bray, F.; Siegel, R.L.; Ferlay, J.; Lortet-Tieulent, J.; Jemal, A. Global cancer statistics, 2012. *CA Cancer J. Clin.* **2015**, *65*, 87–108. [CrossRef] [PubMed]

3. Balduzzi, S.; Mantarro, S.; Guarneri, V.; Tagliabue, L.; Pistotti, V.; Moja, L.; D'Amico, R. Trastuzumab-containing regimens for metastatic breast cancer. *Cochrane Database Syst. Rev.* **2014**, *12*, CD006242.

4. Li, X.X.; Oprea-Ilies, G.M.; Krishnamurti, U. New developments in breast cancer and their impact on daily practice in pathology. *Arch. Pathol. Lab. Med.* **2017**, *141*, 490–498. [CrossRef] [PubMed]

5. Singh, S.; Sharma, B.; Kanwar, S.S.; Kumar, A. Lead phytochemicals for anticancer drug development. *Front. Plant Sci.* **2016**, *7*, 1667. [CrossRef] [PubMed]

6. Kawada, M.; Atsumi, S.; Wada, S.I.; Sakamoto, S. Novel approaches for identification of anti-tumor drugs and new bioactive compounds. *J. Antibiot.* **2017**. [CrossRef] [PubMed]

7. Ruiz-Torres, V.; Encinar, J.; Herranz-López, M.; Pérez-Sánchez, A.; Galiano, V.; Barrajón-Catalán, E.; Micol, V. An updated review on marine anticancer compounds: The use of virtual screening for the discovery of small-molecule cancer drugs. *Molecules* **2017**, *22*, 1037. [CrossRef] [PubMed]

8. Garcia-Villalba, R.; Carrasco-Pancorbo, A.; Oliveras-Ferraros, C.; Vazquez-Martin, A.; Menendez, J.A.; Segura-Carretero, A.; Fernandez-Gutierrez, A. Characterization and quantification of phenolic compounds of extra-virgin olive oils with anticancer properties by a rapid and resolutive LC-ESI-TOF MS method. *J. Pharm. Biomed. Anal.* **2010**, *51*, 416–429. [CrossRef] [PubMed]

9. Kunnumakkara, A.B.; Bordoloi, D.; Harsha, C.; Banik, K.; Gupta, S.C.; Aggarwal, B.B. Curcumin mediates anticancer effects by modulating multiple cell signaling pathways. *Clin. Sci.* **2017**, *131*, 1781–1799. [CrossRef] [PubMed]

10. Bae, J.; Kumazoe, M.; Yamashita, S.; Tachibana, H. Hydrogen sulphide donors selectively potentiate a green tea polyphenol EGCG-induced apoptosis of multiple myeloma cells. *Sci. Rep.* **2017**, *7*. [CrossRef] [PubMed]

11. Jahanafrooz, Z.; Motamed, N.; Bakhshandeh, B. Effects of miR-21 downregulation and silibinin treatment in breast cancer cell lines. *Cytotechnology* **2017**, *69*, 667–680. [CrossRef] [PubMed]

12. Tyszka-Czochara, M.; Bukowska-Strakova, K.; Majka, M. Metformin and caffeic acid regulate metabolic reprogramming in human cervical carcinoma SiHa/HTB-35 cells and augment anticancer activity of Cisplatin via cell cycle regulation. *Food Chem. Toxicol.* **2017**, *106*, 260–272. [CrossRef] [PubMed]

13. Zenthoefer, M.; Geisen, U.; Hofmann-Peiker, K.; Fuhrmann, M.; Kerber, J.; Kirchhofer, R.; Hennig, S.; Peipp, M.; Geyer, R.; Piker, L.; et al. Isolation of polyphenols with anticancer activity from the baltic sea brown seaweed fucus vesiculosus using bioassay-guided fractionation. *J. Appl. Phycol.* **2017**, *29*, 2021–2037. [CrossRef]

14. Pereira, T.M.C.; Pimenta, F.S.; Porto, M.L.; Baldo, M.P.; Campagnaro, B.P.; Gava, A.L.; Meyrelles, S.S.; Vasquez, E.C. Coadjuvants in the diabetic complications: Nutraceuticals and drugs with pleiotropic effects. *Int. J. Mol. Sci.* **2016**, *17*, 1273. [CrossRef] [PubMed]

15. Gothai, S.; Ganesan, P.; Park, S.Y.; Fakurazi, S.; Choi, D.K.; Arulselvan, P. Natural phyto-bioactive compounds for the treatment of type 2 diabetes: Inflammation as a target. *Nutrients* **2016**, *8*, 461. [CrossRef] [PubMed]

16. Abiodun, O.O.; Rodriguez-Nogales, A.; Algieri, F.; Gomez-Caravaca, A.M.; Segura-Carretero, A.; Utrilla, M.P.; Rodriguez-Cabezas, M.E.; Galvez, J. Antiinflammatory and immunomodulatory activity of an ethanolic extract from the stem bark of *Terminalia catappa* L. (Combretaceae): In vitro and in vivo evidences. *J. Ethnopharmacol.* **2016**, *192*, 309–319. [CrossRef] [PubMed]

17. Choi, J.Y.; Desta, K.T.; Saralamma, V.V.G.; Lee, S.J.; Lee, S.J.; Kim, S.M.; Paramanantham, A.; Lee, H.J.; Kim, Y.-H.; Shin, H.-C.; et al. LC-MS/Ms characterization, anti-inflammatory effects and antioxidant activities of polyphenols from different tissues of Korean Petasites japonicus (Meowi). *Biomed. Chromatogr.* **2017**. [CrossRef] [PubMed]

18. Liu, K.; Pi, F.; Zhang, H.; Ji, J.; Xia, S.; Cui, F.; Sun, J.; Sun, X. Metabolomics analysis to evaluate the anti-inflammatory effects of polyphenols: Glabridin reversed metabolism change caused by LPS in RAW 264.7 cells. *J. Agric. Food Chem.* **2017**, *65*, 6070–6079. [CrossRef] [PubMed]

19. Nakajima, V.M.; Moala, T.; Caria, C.R.E.P.; Moura, C.S.; Amaya-Farfan, J.; Gambero, A.; Macedo, G.A.; Macedo, J.A. Biotransformed citrus extract as a source of anti-inflammatory polyphenols: Effects in macrophages and adipocytes. *Food Res. Int.* **2017**, *97*, 37–44. [CrossRef] [PubMed]

20. Herranz-López, M.; Olivares-Vicente, M.; Encinar, J.A.; Barrajón-Catalán, E.; Segura-Carretero, A.; Joven, J.; Micol, V. Multi-targeted molecular effects of *Hibiscus sabdariffa* polyphenols: An opportunity for a global approach to obesity. *Nutrients* **2017**, *9*, 907. [CrossRef] [PubMed]

21. Molino, S.; Dossena, M.; Buonocore, D.; Ferrari, F.; Venturini, L.; Ricevuti, G.; Verri, M. Polyphenols in dementia: From molecular basis to clinical trials. *Life Sci.* **2016**, *161*, 69–77. [CrossRef] [PubMed]

22. Isac, S.; Panaitescu, A.M.; Spataru, A.; Iesanu, M.; Totan, A.; Udriste, A.; Cucu, N.; Peltecu, G.; Zagrean, L.; Zagrean, A.M. Trans-resveratrol enriched maternal diet protects the immature hippocampus from perinatal asphyxia in rats. *Neurosci. Lett.* **2017**, *653*, 308–313. [CrossRef] [PubMed]

23. Reglodi, D.; Renaud, J.; Tamas, A.; Tizabi, Y.; Socias, S.B.; Del-Bel, E.; Raisman-Vozari, R. Novel tactics for neuroprotection in parkinson's disease: Role of antibiotics, polyphenols and neuropeptides. *Prog. Neurobiol.* **2017**, *155*, 120–148. [CrossRef] [PubMed]

24. De Camargo, A.C.; Regitano-d'Arce, M.A.B.; Rasera, G.B.; Canniatti-Brazaca, S.G.; do Prado-Silva, L.; Alvarenga, V.O.; Sant'Ana, A.S.; Shahidi, F. Phenolic acids and flavonoids of peanut by-products: Antioxidant capacity and antimicrobial effects. *Food Chem.* **2017**, *237*, 538–544. [CrossRef] [PubMed]

25. Paunovic, S.M.; Maskovic, P.; Nikolic, M.; Miletic, R. Bioactive compounds and antimicrobial activity of black currant (*Ribes nigrum* L.) berries and leaves extract obtained by different soil management system. *Sci. Hortic.* **2017**, *222*, 69–75. [CrossRef]

26. Chen, M.S.; Zhao, Z.G.; Meng, H.C.; Yu, S.J. Antibiotic activity and mechanisms of sugar beet (beta vulgaris) molasses polyphenols against selected food-borne pathogens. *LWT Food Sci. Technol.* **2017**, *82*, 354–360. [CrossRef]

27. Gomez-Estaca, J.; Balaguer, M.P.; Lopez-Carballo, G.; Gavara, R.; Hernandez-Munoz, P. Improving antioxidant and antimicrobial properties of curcumin by means of encapsulation in gelatin through electrohydrodynamic atomization. *Food Hydrocoll.* **2017**, *70*, 313–320. [CrossRef]

28. Sanchez-Roque, Y.; Ayora-Talavera, G.; Rincon-Rosales, R.; Gutierrez-Miceli, F.A.; Meza-Gordillo, R.; Winkler, R.; Gamboa-Becerra, R.; Ayora-Talavera, T.D.; Ruiz-Valdiviezo, V.M. The flavonoid fraction from rhoeo discolor leaves acting as antiviral against influenza a virus. *Rec. Nat. Prod.* **2017**, *11*, 532–546. [CrossRef]

29. Vazquez-Calvo, A.; de Oya, N.J.; Martin-Acebes, M.A.; Garcia-Moruno, E.; Saiz, J.C. Antiviral properties of the natural polyphenols delphinidin and epigallocatechin gallate against the flaviviruses west nile virus, zika virus, and dengue virus. *Front. Microbiol.* **2017**, *8*, 1314. [CrossRef] [PubMed]

30. Scalbert, A.; Johnson, I.T.; Saltmarsh, M. Polyphenols: Antioxidants and beyond. *Am. J. Clin. Nutr.* **2005**, *81*, 215S–217S. [PubMed]

31. Lindsay, D.G.; Astley, S.B. European research on the functional effects of dietary antioxidants—EUROFEDA. *Mol. Asp. Med.* **2002**, *23*, 1–38. [CrossRef]

32. Bhattarai, G.; Poudel, S.B.; Kook, S.H.; Lee, J.C. Anti-inflammatory, anti-osteoclastic, and antioxidant activities of genistein protect against alveolar bone loss and periodontal tissue degradation in a mouse model of periodontitis. *J. Biomed. Mater. Res. Part A* **2017**, *105*, 2510–2521. [CrossRef] [PubMed]

33. Dehkharghanian, M.; Lacroix, M.; Vijayalakshmi, M.A. Antioxidant properties of green tea polyphenols encapsulated in caseinate beads. *Dairy Sci. Technol.* **2009**, *89*, 485–499. [CrossRef]

34. Dairi, S.; Carbonneau, M.A.; Galeano-Diaz, T.; Remini, H.; Dahmoune, F.; Aoun, O.; Belbahi, A.; Lauret, C.; Cristol, J.P.; Madani, K. Antioxidant effects of extra virgin olive oil enriched by myrtle phenolic extracts on iron-mediated lipid peroxidation under intestinal conditions model. *Food Chem.* **2017**, *237*, 297–304. [CrossRef] [PubMed]

35. Ursache, F.M.; Ghinea, I.O.; Turturica, M.; Aprodu, I.; Rapeanu, G.; Stanciuc, N. Phytochemicals content and antioxidant properties of sea buckthorn (*Hippophae rhamnoides* L.) as affected by heat treatment—Quantitative spectroscopic and kinetic approaches. *Food Chem.* **2017**, *233*, 442–449. [CrossRef] [PubMed]

36. Lambert, J.D.; Hong, J.; Yang, G.Y.; Liao, J.; Yang, C.S. Inhibition of carcinogenesis by polyphenols: Evidence from laboratory investigations. *Am. J. Clin. Nutr.* **2005**, *81*, 284S–291S. [PubMed]

37. Duluc, L.; Jacques, C.; Soleti, R.; Iacobazzi, F.; Simard, G.; Andriantsitohaina, R. Modulation of mitochondrial capacity and angiogenesis by red wine polyphenols via estrogen receptor, NADPH oxidase and nitric oxide synthase pathways. *Int. J. Biochem. Cell Biol.* **2013**, *45*, 783–791. [CrossRef] [PubMed]

38. Khan, N.; Mukhtar, H. Modulation of signaling pathways in prostate cancer by green tea polyphenols. *Biochem. Pharmacol.* **2013**, *85*, 667–672. [CrossRef] [PubMed]

39. Nunes, C.; Teixeira, N.; Serra, D.; Freitas, V.; Almeida, L.; Laranjinha, J. Red wine polyphenol extract efficiently protects intestinal epithelial cells from inflammation via opposite modulation of JAK/STAT and Nrf2 pathways. *Toxicol. Res.* **2016**, *5*, 53–65. [CrossRef]

40. Scapagnini, G.; Vasto, S.; Abraham, N.G.; Caruso, C.; Zella, D.; Galvano, F. Modulation of Nrf2/ARE pathway by food polyphenols: A nutritional neuroprotective strategy for cognitive and neurodegenerative disorders. *Mol. Neurobiol.* **2011**, *44*, 192–201. [CrossRef] [PubMed]

41. Joven, J.; Micol, V.; Segura-Carretero, A.; Alonso-Villaverde, C.; Menendez, J.A. Bioactive Food Components Platform. Polyphenols and the modulation of gene expression pathways: Can we eat our way out of the danger of chronic disease? *Crit. Rev. Food Sci. Nutr.* **2014**, *54*, 985–1001. [CrossRef] [PubMed]

42. Jayasena, T.; Poljak, A.; Smythe, G.; Braidy, N.; Muench, G.; Sachdev, P. The role of polyphenols in the modulation of sirtuins and other pathways involved in Aalzheimer's disease. *Ageing Res. Rev.* **2013**, *12*, 867–883. [CrossRef] [PubMed]

43. Sloczynska, K.; Powroznik, B.; Pekala, E.; Waszkielewicz, A.M. Antimutagenic compounds and their possible mechanisms of action. *J. Appl. Genet.* **2014**, *55*, 273–285. [CrossRef] [PubMed]

44. Rodeiro, I.; Donato, M.T.; Jimenez, N.; Garrido, G.; Molina-Torres, J.; Menendez, R.; Castell, J.V.; Gomez-Lechon, M.J. Inhibition of human P450 enzymes by natural extracts used in traditional medicine. *Phytother. Res.* **2009**, *23*, 279–282. [CrossRef] [PubMed]

45. Munday, R.; Munday, C.M. Induction of phase II detoxification enzymes in rats by plant-derived isothlocyanates: Comparison of allyl isothiocyanate with sulforaphane and related compounds. *J. Agric. Food Chem.* **2004**, *52*, 1867–1871. [CrossRef] [PubMed]

46. Lu, F.; Zahid, M.; Wang, C.; Saeed, M.; Cavalieri, E.L.; Rogan, E.G. Resveratrol prevents estrogen-DNA adduct formation and neoplastic transformation in MCF-10F cells. *Cancer Prev. Res.* **2008**, *1*, 135–145. [CrossRef] [PubMed]

47. Pérez-Sánchez, A.; Barrajón-Catalán, E.; Herranz-López, M.; Castillo, J.; Micol, V. Lemon balm extract (*Melissa officinalis* L.) promotes melanogenesis and prevents UVB-induced oxidative stress and DNA damage in a skin cell model. *J. Dermatol. Sci.* **2016**, *84*, 169–177. [CrossRef] [PubMed]

48. Grosso, G.; Godos, J.; Lamuela-Raventos, R.; Ray, S.; Micek, A.; Pajak, A.; Sciacca, S.; D'Orazio, N.; Del Rio, D.; Galvano, F. A comprehensive meta-analysis on dietary flavonoid and lignan intake and cancer risk: Level of evidence and limitations. *Mol. Nutr. Food Res.* **2017**, *61*. [CrossRef] [PubMed]

49. Grosso, G.; Micek, A.; Godos, J.; Pajak, A.; Sciacca, S.; Galvano, F.; Giovannucci, E.L. Dietary flavonoid and lignan intake and mortality in prospective cohort studies: Systematic review and dose-response meta-analysis. *Am. J. Epidemiol.* **2017**, *185*, 1304–1316. [CrossRef] [PubMed]

50. Kale, A.; Gawande, S.; Kotwal, S.; Netke, S.; Roomi, M.W.; Ivanov, V.; Niedzwecki, A.; Rath, M. A combination of green tea extract, specific nutrient mixture and quercetin: An effective intervention treatment for the regression of *N*-methyl-*N*-nitrosourea (MNU)-induced mammary tumors in Wistar rats. *Oncol. Lett.* **2010**, *1*, 313–317. [PubMed]

51. Venkatesan, T.; Choi, Y.W.; Mun, S.P.; Kim, Y.K. Pinus radiata bark extract induces caspase-independent apoptosis-like cell death in MCF-7 human breast cancer cells. *Cell Biol. Toxicol.* **2016**, *32*, 451–464. [CrossRef] [PubMed]

52. Chang, C.H.; Ou, T.T.; Yang, M.Y.; Huang, C.C.; Wang, C.J. Nelumbo nucifera gaertn leaves extract inhibits the angiogenesis and metastasis of breast cancer cells by downregulation connective tissue growth factor (CTGF) mediated PI3K/AKT/ERK signaling. *J. Ethnopharmacol.* **2016**, *188*, 111–122. [CrossRef] [PubMed]

53. Studzinska-Sroka, E.; Piotrowska, H.; Kucinska, M.; Murias, M.; Bylka, W. Cytotoxic activity of physodic acid and acetone extract from hypogymnia physodes against breast cancer cell lines. *Pharm. Biol.* **2016**, *54*, 2480–2485. [CrossRef] [PubMed]

54. Sharma, K.; Pachauri, S.D.; Khandelwal, K.; Ahmad, H.; Arya, A.; Biala, P.; Agrawal, S.; Pandey, R.R.; Srivastava, A.; Srivastav, A.; et al. Anticancer effects of extracts from the fruit of morinda citrifolia (noni) in breast cancer cell lines. *Drug Res.* **2016**, *66*, 141–147. [CrossRef] [PubMed]

55. Bouallagui, Z.; Han, J.; Isoda, H.; Sayadi, S. Hydroxytyrosol rich extract from olive leaves modulates cell cycle progression in MCF-7 human breast cancer cells. *Food Chem. Toxicol.* **2011**, *49*, 179–184. [CrossRef] [PubMed]

56. Fogli, S.; Arena, C.; Carpi, S.; Polini, B.; Bertini, S.; Digiacomo, M.; Gado, F.; Saba, A.; Saccomanni, G.; Breschi, M.C.; et al. Cytotoxic activity of oleocanthal isolated from virgin olive oil on human melanoma cells. *Nutr. Cancer Int. J.* **2016**, *68*, 873–877. [CrossRef] [PubMed]

57. Goldsmith, C.D.; Vuong, Q.V.; Sadeqzadeh, E.; Stathopoulos, C.E.; Roach, P.D.; Scarlett, C.J. Phytochemical properties and anti-proliferative activity of *Olea europaea* L. Leaf extracts against pancreatic cancer cells. *Molecules* **2015**, *20*, 12992–13004. [CrossRef] [PubMed]

58. Barrajon-Catalan, E.; Taamalli, A.; Quirantes-Pine, R.; Roldan-Segura, C.; Arraez-Roman, D.; Segura-Carretero, A.; Micol, V.; Zarrouk, M. Differential metabolomic analysis of the potential antiproliferative mechanism of olive leaf extract on the JIMT-1 breast cancer cell line. *J. Pharm. Biomed. Anal.* **2015**, *105*, 156–162. [CrossRef] [PubMed]

59. Dai, X.F.; Li, T.; Bai, Z.H.; Yang, Y.K.; Liu, X.X.; Zhan, J.L.; Shi, B.Z. Breast cancer intrinsic subtype classification, clinical use and future trends. *Am. J. Cancer Res.* **2015**, *5*, 2929–2943. [PubMed]

60. Rakha, E.A.; Reis, J.S.; Ellis, I.O. Combinatorial biomarker expression in breast cancer. *Breast Cancer Res. Treat.* **2010**, *120*, 293–308. [CrossRef] [PubMed]

61. Sotiriou, C.; Pusztai, L. Molecular origins of cancer gene-expression signatures in breast cancer. *N. Engl. J. Med.* **2009**, *360*, 790–800. [CrossRef] [PubMed]

62. Alluri, P.G.; Speers, C.; Chinnaiyan, A.M. Estrogen receptor mutations and their role in breast cancer progression. *Breast Cancer Res.* **2014**, *16*. [CrossRef] [PubMed]

63. Mc Cormack, O.; Harrison, M.; Kerin, M.J.; McCann, A. Role of the progesterone receptor (PR) and the PR isoforms in breast cancer. *Crit. Rev. Oncog.* **2007**, *13*, 283–301. [CrossRef] [PubMed]

64. Mitri, Z.; Constantine, T.; O'Regan, R. The HER2 receptor in breast cancer: Pathophysiology, clinical use, and new advances in therapy. *Chemother. Res. Pract.* **2012**, *2012*, 743193. [CrossRef] [PubMed]

65. Luporsi, E.; André, F.; Spyratos, F.; Martin, P.-M.; Jacquemier, J.; Penault-Llorca, F.; Tubiana-Mathieu, N.; Sigal-Zafrani, B.; Arnould, L.; Gompel, A.; et al. Ki-67: Level of evidence and methodological considerations for its role in the clinical management of breast cancer: Analytical and critical review. *Breast Cancer Res. Treat.* **2012**, *132*, 895–915. [CrossRef] [PubMed]

66. Oakman, C.; Bessi, S.; Zafarana, E.; Galardi, F.; Biganzoli, L.; Di Leo, A. Recent advances in systemic therapy. New diagnostics and biological predictors of outcome in early breast cancer. *Breast Cancer Res.* **2009**, *11*, 205. [CrossRef] [PubMed]

67. Brenton, J.D.; Carey, L.A.; Ahmed, A.A.; Caldas, C. Molecular classification and molecular forecasting of breast cancer: Ready for clinical application? *J. Clin. Oncol.* **2005**, *23*, 7350–7360. [CrossRef] [PubMed]

68. Parker, J.S.; Mullins, M.; Cheang, M.C.U.; Leung, S.; Voduc, D.; Vickery, T.; Davies, S.; Fauron, C.; He, X.P.; Hu, Z.Y.; et al. Supervised risk predictor of breast cancer based on intrinsic subtypes. *J. Clin. Oncol.* **2009**, *27*, 1160–1167. [CrossRef] [PubMed]

69. Lal, S.; McCart Reed, A.E.; de Luca, X.M.; Simpson, P.T. Molecular signatures in breast cancer. *Methods* **2017**. [CrossRef] [PubMed]

70. Fan, C.; Oh, D.S.; Wessels, L.; Weigelt, B.; Nuyten, D.S.A.; Nobel, A.B.; van't Veer, L.J.; Perou, C.M. Concordance among gene-expression-based predictors for breast cancer. *N. Engl. J. Med.* **2006**, *355*, 560–569. [CrossRef] [PubMed]

71. Voduc, K.D.; Cheang, M.C.U.; Tyldesley, S.; Gelmon, K.; Nielsen, T.O.; Kennecke, H. Breast cancer subtypes and the risk of local and regional relapse. *J. Clin. Oncol.* **2010**, *28*, 1684–1691. [CrossRef] [PubMed]

72. Howlader, N.; Altekruse, S.F.; Li, C.I.; Chen, V.W.; Clarke, C.A.; Ries, L.A.G.; Cronin, K.A. US incidence of breast cancer subtypes defined by joint hormone receptor and HER2 status. *JNCI J. Natl. Cancer Inst.* **2014**, *106*. [CrossRef] [PubMed]

73. Arvold, N.D.; Taghian, A.G.; Niemierko, A.; Raad, R.F.A.; Sreedhara, M.; Nguyen, P.L.; Bellon, J.R.; Wong, J.S.; Smith, B.L.; Harris, J.R. Age, breast cancer subtype approximation, and local recurrence after breast-conserving therapy. *J. Clin. Oncol.* **2011**, *29*, 3885–3891. [CrossRef] [PubMed]

74. Haque, R.; Ahmed, S.A.; Inzhakova, G.; Shi, J.X.; Avila, C.; Polikoff, J.; Bernstein, L.; Enger, S.M.; Press, M.F. Impact of breast cancer subtypes and treatment on survival: An analysis spanning two decades. *Cancer Epidemiol. Biomark. Prev.* **2012**, *21*, 1848–1855. [CrossRef] [PubMed]

75. Metzger, O.; Sun, Z.X.; Viale, G.; Price, K.N.; Crivellari, D.; Snyder, R.D.; Gelber, R.D.; Castiglione-Gertsch, M.; Coates, A.S.; Goldhirsch, A.; et al. Patterns of recurrence and outcome according to breast cancer subtypes in lymph node-negative disease: Results from international breast cancer study group trials VIII and IX. *J. Clin. Oncol.* **2013**, *31*, 3083–3090. [CrossRef] [PubMed]

76. Sorlie, T.; Perou, C.M.; Tibshirani, R.; Aas, T.; Geisler, S.; Johnsen, H.; Hastie, T.; Eisen, M.B.; van de Rijn, M.; Jeffrey, S.S.; et al. Gene expression patterns of breast carcinomas distinguish tumor subclasses with clinical implications. *Proc. Natl. Acad. Sci. USA* **2001**, *98*, 10869–10874. [CrossRef] [PubMed]

77. Lund, M.J.; Butler, E.N.; Hair, B.Y.; Ward, K.C.; Andrews, J.H.; Oprea-Ilies, G.; Bayakly, A.R.; O'Regan, R.M.; Vertino, P.M.; Eley, J.W. Age/race differences in HER2 testing and in incidence rates for breast cancer triple subtypes a population-based study and first report. *Cancer* **2010**, *116*, 2549–2559. [PubMed]

78. Koboldt, D.C.; Fulton, R.S.; McLellan, M.D.; Schmidt, H.; Kalicki-Veizer, J.; McMichael, J.F.; Fulton, L.L.; Dooling, D.J.; Ding, L.; Mardis, E.R.; et al. Comprehensive molecular portraits of human breast tumours. *Nature* **2012**, *490*, 61–70. [CrossRef] [PubMed]

79. Zong, Y.; Zhu, L.; Wu, J.; Chen, X.; Huang, O.; Fei, X.; He, J.; Chen, W.; Li, Y.; Shen, K. Progesterone receptor status and Ki-67 index may predict early relapse in luminal B/HER2 negative breast cancer patients: A retrospective study. *PLoS ONE* **2014**, *9*. [CrossRef] [PubMed]

80. Samavat, H.; Kurzer, M.S. Estrogen metabolism and breast cancer. *Cancer Lett.* **2015**, *356*, 231–243. [CrossRef] [PubMed]

81. Tang, Y.; Wang, Y.; Kiani, M.F.; Wang, B. Classification, treatment strategy, and associated drug resistance in breast cancer. *Clin. Breast Cancer* **2016**, *16*, 335–343. [CrossRef] [PubMed]

82. Yang, X.R.; Chang-Claude, J.; Goode, E.L.; Couch, F.J.; Nevanlinna, H.; Milne, R.L.; Gaudet, M.; Schmidt, M.K.; Broeks, A.; Cox, A.; et al. Associations of breast cancer risk factors with tumor subtypes: A pooled analysis from the breast cancer association consortium studies. *J. Natl. Cancer Inst.* **2011**, *103*, 250–263. [CrossRef] [PubMed]

83. Atchley, D.P.; Albarracin, C.T.; Lopez, A.; Valero, V.; Amos, C.I.; Gonzalez-Angulo, A.M.; Hortobagyi, G.N.; Arun, B.K. Clinical and pathologic characteristics of patients with BRCA-positive and BRCA-negative breast cancer. *J. Clin. Oncol.* **2008**, *26*, 4282–4288. [CrossRef] [PubMed]

84. Poljsak, B.; Fink, R. The protective role of antioxidants in the defence against ROS/RNS-mediated environmental pollution. *Oxid. Med. Cell. Longev.* **2014**. [CrossRef] [PubMed]

85. Delfino, R.J.; Staimer, N.; Vaziri, N.D. Air pollution and circulating biomarkers of oxidative stress. *Air Qual. Atmos. Health* **2011**, *4*, 37–52. [CrossRef] [PubMed]

86. Azzam, E.I.; Jay-Gerin, J.-P.; Pain, D. Ionizing radiation-induced metabolic oxidative stress and prolonged cell injury. *Cancer Lett.* **2012**, *327*, 48–60. [CrossRef] [PubMed]

87. Handy, D.E.; Loscalzo, J. Redox regulation of mitochondrial function. *Antioxid. Redox Signal.* **2012**, *16*, 1323–1367. [CrossRef] [PubMed]

88. Aon, M.A.; Cortassa, S.; O'Rourke, B. Redox-optimized ROS balance: A unifying hypothesis. *Biochim. Biophys. Acta Bioenerg.* **2010**, *1797*, 865–877. [CrossRef] [PubMed]

89. Mencalha, A.; Victorino, V.J.; Cecchini, R.; Panis, C. Mapping oxidative changes in breast cancer: Understanding the basic to reach the clinics. *Anticancer Res.* **2014**, *34*, 1127–1140. [PubMed]

90. Jezierska-Drutel, A.; Rosenzweig, S.A.; Neumann, C.A. Role of oxidative stress and the microenvironment in breast cancer development and progression. *Adv. Cancer Res.* **2013**, *119*, 107–125. [PubMed]

91. Gorrini, C.; Harris, I.S.; Mak, T.W. Modulation of oxidative stress as an anticancer strategy. *Nat. Rev. Drug Discov.* **2013**, *12*, 931–947. [CrossRef] [PubMed]

92. Hecht, F.; Pessoa, C.F.; Gentile, L.B.; Rosenthal, D.; Carvalho, D.P.; Fortunato, R.S. The role of oxidative stress on breast cancer development and therapy. *Tumour Biol.* **2016**, *37*, 4281–4291. [CrossRef] [PubMed]

93. Kupsco, A.; Schlenk, D. Oxidative stress, unfolded protein response, and apoptosis in developmental toxicity. *Int. Rev. Cell Mol. Biol.* **2015**, *317*, 1–66. [PubMed]

94. Navarro-Yepes, J.; Burns, M.; Anandhan, A.; Khalimonchuk, O.; del Razo, L.M.; Quintanilla-Vega, B.; Pappa, A.; Panayiotidis, M.I.; Franco, R. Oxidative stress, redox signaling, and autophagy: Cell death versus survival. *Antioxid. Redox Signal.* **2014**, *21*, 66–85. [CrossRef] [PubMed]

95. Samoylenko, A.; Al Hossain, J.; Mennerich, D.; Kellokumpu, S.; Hiltunen, J.K.; Kietzmann, T. Nutritional countermeasures targeting reactive oxygen species in cancer: From mechanisms to biomarkers and clinical evidence. *Antioxid. Redox Signal.* **2013**, *19*, 2157–2196. [CrossRef] [PubMed]

96. Forman, H.J.; Davies, K.J.A.; Ursini, F. How do nutritional antioxidants really work: Nucleophilic tone and para-hormesis versus free radical scavenging in vivo. *Free Radic. Biol. Med.* **2014**, *66*, 24–35. [CrossRef] [PubMed]

97. Khan, N.; Afaq, F.; Mukhtar, H. Cancer chemoprevention through dietary antioxidants: Progress and promise. *Antioxid. Redox Signal.* **2008**, *10*, 475–510. [CrossRef] [PubMed]

98. Stoner, G.D.; Mukhtar, H. Polyphenols as cancer chemopreventive agents. *J. Cell. Biochem. Suppl.* **1995**, *22*, 169–180. [CrossRef] [PubMed]

99. Mileo, A.M.; Miccadei, S. Polyphenols as modulator of oxidative stress in cancer disease: New therapeutic strategies. *Oxid. Med. Cell. Longev.* **2016**. [CrossRef] [PubMed]

100. Vance, T.M.; Su, J.; Fontham, E.T.H.; Koo, S.I.; Chun, O.K. Dietary antioxidants and prostate cancer: A review. *Nutr. Cancer Int. J.* **2013**, *65*, 793–801. [CrossRef] [PubMed]

101. Kim, Y.; Keogh, J.B.; Clifton, P.M. Polyphenols and glycemic control. *Nutrients* **2016**, *8*, 17. [CrossRef] [PubMed]

102. Ge, Q.; Chen, L.; Chen, K. Treatment of diabetes mellitus using ips cells and spice polyphenols. *J. Diabetes Res.* **2017**. [CrossRef] [PubMed]

103. Khan, N.; Khymenets, O.; Urpi-Sarda, M.; Tulipani, S.; Garcia-Aloy, M.; Monagas, M.; Mora-Cubillos, X.; Llorach, R.; Andres-Lacueva, C. Cocoa polyphenols and inflammatory markers of cardiovascular disease. *Nutrients* **2014**, *6*, 844–880. [CrossRef] [PubMed]

104. Hernaez, A.; Remaley, A.T.; Farras, M.; Fernandez-Castillejo, S.; Subirana, I.; Schroeder, H.; Fernandez-Mampel, M.; Munoz-Aguayo, D.; Sampson, M.; Sola, R.; et al. Olive oil polyphenols decrease LDL concentrations and LDL atherogenicity in men in a randomized controlled trial. *J. Nutr.* **2015**, *145*, 1692–1697. [CrossRef] [PubMed]

105. Elbling, L.; Weiss, R.M.; Teufelhofer, O.; Uhl, M.; Knasmueller, S.; Schulte-Hermann, R.; Berger, W.; Mickshe, M. Green tea extract and (−)-epigallocatechin-3-gallate, the major tea catechin, exert oxidant but lack antioxidant activities. *FASEB J.* **2005**, *19*, 807–809. [CrossRef] [PubMed]

106. Hadi, S.M.; Bhat, S.H.; Azmi, A.S.; Hanif, S.; Shamim, U.; Ullah, M.F. Oxidative breakage of cellular DNA by plant polyphenols: A putative mechanism for anticancer properties. *Semin. Cancer Biol.* **2007**, *17*, 370–376. [CrossRef] [PubMed]

107. Decker, E.A. Phenolics: Prooxidants or antioxidants? *Nutr. Rev.* **1997**, *55*, 396–398. [CrossRef] [PubMed]

108. Li, Y.; Trush, M.A. Reactive oxygen-dependent DNA damage resulting from the oxidation of phenolic compounds by a copper-redox cycle mechanism. *Cancer Res.* **1994**, *54*, 1895s–1898s. [PubMed]

109. Denoyer, D.; Masaldan, S.; La Fontaine, S.; Cater, M.A. Targeting copper in cancer therapy: 'Copper that cancer'. *Metallomics* **2015**, *7*, 1459–1476. [CrossRef] [PubMed]

110. Khan, H.Y.; Zubair, H.; Faisal, M.; Ullah, M.F.; Farhan, M.; Sarkar, F.H.; Ahmad, A.; Hadi, S.M. Plant polyphenol induced cell death in human cancer cells involves mobilization of intracellular copper ions and reactive oxygen species generation: A mechanism for cancer chemopreventive action. *Mol. Nutr. Food Res.* **2014**, *58*, 437–446. [CrossRef] [PubMed]

111. Bower, J.J.; Vance, L.D.; Psioda, M.; Smith-Roe, S.L.; Simpson, D.A.; Ibrahim, J.G.; Hoadley, K.A.; Perou, C.M.; Kaufmann, W.K. Patterns of cell cycle checkpoint deregulation associated with intrinsic molecular subtypes of human breast cancer cells. *NPJ Breast Cancer* **2017**, *3*, 9. [CrossRef] [PubMed]

112. Hanahan, D.; Weinberg, R.A. Hallmarks of cancer: The next generation. *Cell* **2011**, *144*, 646–674. [CrossRef] [PubMed]

113. Oikonomou, C.; Cross, F.R. Rising cyclin-CDK levels order cell cycle events. *PLoS ONE* **2011**, *6*, e20788. [CrossRef] [PubMed]

114. Duronio, R.J.; Xiong, Y. Signaling pathways that control cell proliferation. *Cold Spring Harb. Perspect. Biol.* **2013**, *5*. [CrossRef] [PubMed]

115. Shin, S.Y.; Yoon, H.; Ahn, S.; Kim, D.-W.; Bae, D.-H.; Koh, D.; Lee, Y.H.; Lim, Y. Structural properties of polyphenols causing cell cycle arrest at G1 phase in HCT116 human colorectal cancer cell lines. *Int. J. Mol. Sci.* **2013**, *14*, 16970–16985. [CrossRef] [PubMed]

116. Zhao, Q.; Huo, X.-C.; Sun, F.-D.; Dong, R.-Q. Polyphenol-rich extract of salvia chinensis exhibits anticancer activity in different cancer cell lines, and induces cell cycle arrest at the G(0)/G(1)-phase, apoptosis and loss of mitochondrial membrane potential in pancreatic cancer cells. *Mol. Med. Rep.* **2015**, *12*, 4843–4850. [CrossRef] [PubMed]

117. Gonzalez-Sarrias, A.; Ma, H.; Edmonds, M.E.; Seeram, N.P. Maple polyphenols, ginnalins A-C, induce S- and G2/M-cell cycle arrest in colon and breast cancer cells mediated by decreasing cyclins A and D1 levels. *Food Chem.* **2013**, *136*, 636–642. [CrossRef] [PubMed]

118. Liu, S.-M.; Ou, S.-Y.; Huang, H.-H. Green tea polyphenols induce cell death in breast cancer MCF-7 cells through induction of cell cycle arrest and mitochondrial-mediated apoptosis. *J. Zhejiang Univ. Sci. B* **2017**, *18*, 89–98. [CrossRef] [PubMed]

119. Li, T.; Zhu, J.; Guo, L.; Shi, X.; Liu, Y.; Yang, X. Differential effects of polyphenols-enriched extracts from hawthorn fruit peels and fleshes on cell cycle and apoptosis in human MCF-7 breast carcinoma cells. *Food Chem.* **2013**, *141*, 1008–1018. [CrossRef] [PubMed]

120. Chen, H.-S.; Bai, M.-H.; Zhang, T.; Li, G.-D.; Liu, M. Ellagic acid induces cell cycle arrest and apoptosis through TGF-beta/Smad3 signaling pathway in human breast cancer MCF-7 cells. *Int. J. Oncol.* **2015**, *46*, 1730–1738. [CrossRef] [PubMed]

121. Elmore, S. Apoptosis: A review of programmed cell death. *Toxicol. Pathol.* **2007**, *35*, 495–516. [CrossRef] [PubMed]

122. Nagasaka, A.; Kawane, K.; Yoshida, H.; Nagata, S. Apaf-1-independent programmed cell death in mouse development. *Cell Death Differ.* **2010**, *17*, 931–941. [CrossRef] [PubMed]

123. Burgess, D.J. Apoptosis refined and lethal. *Nat. Rev. Cancer* **2013**, *13*, 79. [CrossRef] [PubMed]

124. Miyashita, T.; Krajewski, S.; Krajewska, M.; Wang, H.G.; Lin, H.K.; Liebermann, D.A.; Hoffman, B.; Reed, J.C. Tumor-suppressor p53 is a regulator of bcl-2 and bax gene-expression in vitro in vitro and in vivo in vivo. *Oncogene* **1994**, *9*, 1799–1805. [PubMed]

125. Muller, P.A.J.; Vousden, K.H. Mutant p53 in cancer: New functions and therapeutic opportunities. *Cancer Cell* **2014**, *25*, 304–317. [CrossRef] [PubMed]

126. Jahanafrooz, Z.; Motameh, N.; Bakhshandeh, B. Comparative evaluation of silibinin effects on cell cycling and apoptosis in human breast cancer MCF-7 and T47D cell lines. *Asian Pac. J. Cancer Prev. APJCP* **2016**, *17*, 2661–2665. [PubMed]

127. Song, M.S.; Salmena, L.; Pandolfi, P.P. The functions and regulation of the PTEN tumour suppressor. *Nat. Rev. Mol. Cell Biol.* **2012**, *13*, 283–296. [CrossRef] [PubMed]

128. Gartel, A.L.; Tyner, A.L. The role of the cyclin-dependent kinase inhibitor p21 in apoptosis. *Mol. Cancer Ther.* **2002**, *1*, 639–649. [PubMed]

129. D'Angelo, S.; Martino, E.; Ilisso, C.P.; Bagarolo, M.L.; Porcelli, M.; Cacciapuoti, G. Pro-oxidant and pro-apoptotic activity of polyphenol extract from annurca apple and its underlying mechanisms in human breast cancer cells. *Int. J. Oncol.* **2017**, *51*, 939–948. [CrossRef] [PubMed]

130. Kello, M.; Kulikova, L.; Vaskova, J.; Nagyova, A.; Mojzis, J. Fruit peel polyphenolic extract-induced apoptosis in human breast cancer cells is associated with ros production and modulation of p38MAPK/ERK1/2 and the akt signaling pathway. *Nutr. Cancer* **2017**, *69*, 920–931. [CrossRef] [PubMed]

131. Hassan, Z.K.; Elamin, M.H.; Omer, S.A.; Daghestani, M.H.; Al-Olayan, E.S.; Elobeid, M.A.; Virk, P. Oleuropein induces apoptosis via the p53 pathway in breast cancer cells. *Asian Pac. J. Cancer Prev.* **2013**, *14*, 6739–6742. [CrossRef]

132. Yan, C.-M.; Chai, E.-Q.; Cai, H.-Y.; Miao, G.-Y.; Ma, W. Oleuropein induces apoptosis via activation of caspases and suppression of phosphatidylinositol 3-kinase/protein kinase B pathway in HepG2 human hepatoma cell line. *Mol. Med. Rep.* **2015**, *11*, 4617–4624. [CrossRef] [PubMed]

133. Meeran, S.M.; Patel, S.N.; Chan, T.H.; Tollefsbol, T.O. A novel prodrug of epigallocatechin-3-gallate: Differential epigenetic hTERT repression in human breast cancer cells. *Cancer Prev. Res.* **2011**, *4*, 1243–1254. [CrossRef] [PubMed]

134. Jaiswal, P.K.; Goel, A.; Mittal, R.D. Survivin: A molecular biomarker in cancer. *Indian J. Med. Res.* **2015**, *141*, 389–397. [PubMed]

135. Chen, X.; Li, Y.; Lin, Q.; Wang, Y.; Sun, H.; Wang, J.; Cui, G.; Cai, L.; Dong, X. Tea polyphenols induced apoptosis of breast cancer cells by suppressing the expression of Survivin. *Sci. Rep.* **2014**, *4*. [CrossRef] [PubMed]

136. Thangapazham, R.L.; Passi, N.; Maheshwari, R.K. Green tea polyphenol and epigallocatechin gallate induce apoptosis and inhibit invasion in human breast cancer cells. *Cancer Biol. Ther.* **2007**, *6*, 1938–1943. [CrossRef] [PubMed]

137. Mileo, A.M.; Di Venere, D.; Linsalata, V.; Fraioli, R.; Miccadei, S. Artichoke polyphenols induce apoptosis and decrease the invasive potential of the human breast cancer cell line MDA-MB231. *J. Cell. Physiol.* **2012**, *227*, 3301–3309. [CrossRef] [PubMed]

138. Mileo, A.M.; Di Venere, D.; Abbruzzese, C.; Miccadei, S. Long term exposure to polyphenols of artichoke (*Cynara scolymus* L.) exerts induction of senescence driven growth arrest in the MDA-MB231 human breast cancer cell line. *Oxid. Med. Cell. Longev.* **2015**. [CrossRef] [PubMed]

139. Venkatadri, R.; Muni, T.; Iyer, A.K.V.; Yakisich, J.S.; Azad, N. Role of apoptosis-related miRNAs in resveratrol-induced breast cancer cell death. *Cell Death Dis.* **2016**, *7*. [CrossRef] [PubMed]

140. Srivastava, S.; Somasagara, R.R.; Hegde, M.; Nishana, M.; Tadi, S.K.; Srivastava, M.; Choudhary, B.; Raghavan, S.C. Quercetin, a natural flavonoid interacts with DNA, arrests cell cycle and causes tumor regression by activating mitochondrial pathway of apoptosis. *Sci. Rep.* **2016**, *6*. [CrossRef] [PubMed]

141. Afsar, T.; Trembley, J.H.; Salomon, C.E.; Razak, S.; Khan, M.R.; Ahmed, K. Growth inhibition and apoptosis in cancer cells induced by polyphenolic compounds of *Acacia hydaspica*: Involvement of multiple signal transduction pathways. *Sci. Rep.* **2016**, *6*. [CrossRef] [PubMed]

142. Castillo-Pichardo, L.; Dharmawardhane, S.F. Grape polyphenols inhibit Akt/mammalian target of rapamycin signaling and potentiate the effects of gefitinib in breast cancer. *Nutr. Cancer Int. J.* **2012**, *64*, 1058–1069. [CrossRef] [PubMed]

143. Galluzzi, L.; Pietrocola, F.; Bravo-San Pedro, J.M.; Amaravadi, R.K.; Baehrecke, E.H.; Cecconi, F.; Codogno, P.; Debnath, J.; Gewirtz, D.A.; Karantza, V.; et al. Autophagy in malignant transformation and cancer progression. *EMBO J.* **2015**, *34*, 856–880. [CrossRef] [PubMed]

144. Villar, V.H.; Merhi, F.; Djavaheri-Mergny, M.; Duran, R.V. Glutaminolysis and autophagy in cancer. *Autophagy* **2015**, *11*, 1198–1208. [CrossRef] [PubMed]

145. Mah, L.Y.; Ryan, K.M. Autophagy and cancer. *Cold Spring Harb. Perspect. Biol.* **2012**, *4*. [CrossRef] [PubMed]

146. Yang, Y.; Coleman, M.; Zhang, L.; Zheng, X.; Yue, Z. Autophagy in axonal and dendritic degeneration. *Trends Neurosci.* **2013**, *36*, 418–428. [CrossRef] [PubMed]

147. Plaza-Zabala, A.; Sierra-Torre, V.; Sierra, A. Autophagy and microglia: Novel partners in neurodegeneration and aging. *Int. J. Mol. Sci.* **2017**, *18*, 598. [CrossRef] [PubMed]

148. Xilouri, M.; Stefanis, L. Chaperone mediated autophagy in aging: Starve to prosper. *Ageing Res. Rev.* **2016**, *32*, 13–21. [CrossRef] [PubMed]

149. Chauhan, S.; Mandell, M.A.; Deretic, V. IRGM governs the core autophagy machinery to conduct antimicrobial defense. *Mol. Cell* **2015**, *58*, 507–521. [CrossRef] [PubMed]

150. Chiu, H.-C.; Soni, S.; Kulp, S.K.; Curry, H.; Wang, D.; Gunn, J.S.; Schlesinger, L.S.; Chen, C.-S. Eradication of intracellular francisella tularensis in THP-1 human macrophages with a novel autophagy inducing agent. *J. Biomed. Sci.* **2009**, *16*. [CrossRef] [PubMed]

151. Hasima, N.; Ozpolat, B. Regulation of autophagy by polyphenolic compounds as a potential therapeutic strategy for cancer. *Cell Death Dis.* **2014**, *5*. [CrossRef] [PubMed]

152. Mathew, R.; Karantza-Wadsworth, V.; White, E. Role of autophagy in cancer. *Nat. Rev. Cancer* **2007**, *7*, 961–967. [CrossRef] [PubMed]

153. Huang, H.-C.; Syu, K.-Y.; Lin, J.-K. Chemical composition of solanum nigrum linn extract and induction of autophagy by leaf water extract and its major flavonoids in AU565 breast cancer cells. *J. Agric. Food Chem.* **2010**, *58*, 8699–8708. [CrossRef] [PubMed]

154. Adams, L.S.; Phung, S.; Yee, N.; Seeram, N.P.; Li, L.; Chen, S. Blueberry phytochemicals inhibit growth and metastatic potential of MDA-MB-231 breast cancer cells through modulation of the phosphatidylinositol 3-kinase pathway. *Cancer Res.* **2010**, *70*, 3594–3605. [CrossRef] [PubMed]

155. Sun, T.; Chen, Q.Y.; Wu, L.J.; Yao, X.M.; Sun, X.J. Antitumor and antimetastatic activities of grape skin polyphenols in a murine model of breast cancer. *Food Chem. Toxicol.* **2012**, *50*, 3462–3467. [CrossRef] [PubMed]

156. Nemec, M.J.; Kim, H.; Marciante, A.B.; Barnes, R.C.; Hendrick, E.D.; Bisson, W.H.; Talcott, S.T.; Mertens-Talcott, S.U. Polyphenolics from mango (*Mangifera indica* L.) suppress breast cancer ductal carcinoma in situ proliferation through activation of AMPK pathway and suppression of mTOR in athymic nude mice. *J. Nutr. Biochem.* **2017**, *41*, 12–19. [CrossRef] [PubMed]

157. Tanida, I.; Ueno, T.; Kominami, E. LC3 and autophagy. *Methods Mol. Biol.* **2008**, *445*, 77–88. [PubMed]

158. Scarlatti, F.; Maffei, R.; Beau, I.; Codogno, P.; Ghidoni, R. Role of non-canonical Beclin 1-independent autophagy in cell death induced by resveratrol in human breast cancer cells. *Cell Death Differ.* **2008**, *15*, 1318–1329. [CrossRef] [PubMed]

159. Zhang, L.; Shamaladevi, N.; Jayaprakasha, G.K.; Patil, B.S.; Lokeshwar, B.L. Polyphenol-rich extract of pimenta dioica berries (allspice) kills breast cancer cells by autophagy and delays growth of triple negative breast cancer in athymic mice. *Oncotarget* **2015**, *6*, 16379–16395. [CrossRef] [PubMed]

160. Al Dhaheri, Y.; Attoub, S.; Ramadan, G.; Arafat, K.; Bajbouj, K.; Karuvantevida, N.; AbuQamar, S.; Eid, A.; Iratni, R. Carnosol induces ROS-mediated Beclin1-independent autophagy and apoptosis in triple negative breast cancer. *PLoS ONE* **2014**, *9*. [CrossRef] [PubMed]

161. Fu, Y.; Chang, H.; Peng, X.; Bai, Q.; Yi, L.; Zhou, Y.; Zhu, J.; Mi, M. Resveratrol inhibits breast cancer stem-like cells and induces autophagy via suppressing Wnt/beta-catenin signaling pathway. *PLoS ONE* **2014**, *9*. [CrossRef]

162. Samadi, A.K.; Bilsland, A.; Georgakilas, A.G.; Amedei, A.; Amin, A.; Bishayee, A.; Azmi, A.S.; Lokeshwar, B.L.; Grue, B.; Panis, C.; et al. A multi-targeted approach to suppress tumor-promoting inflammation. *Semin. Cancer Biol.* **2015**, *35*, S151–S184. [CrossRef] [PubMed]

163. Thompson, P.A.; Khatami, M.; Baglole, C.J.; Sun, J.; Harris, S.; Moon, E.-Y.; Al-Mulla, F.; Al-Temaimi, R.; Brown, D.; Colacci, A.; et al. Environmental immune disruptors, inflammation and cancer risk. *Carcinogenesis* **2015**, *36*, S232–S253. [CrossRef] [PubMed]

164. Demaria, S.; Pikarsky, E.; Karin, M.; Coussens, L.M.; Chen, Y.-C.; El-Omar, E.M.; Trinchieri, G.; Dubinett, S.M.; Mao, J.T.; Szabo, E.; et al. Cancer and inflammation: Promise for biologic therapy. *J. Immunother.* **2010**, *33*, 335–351. [CrossRef] [PubMed]

165. Pinho, A.V.; Chantrill, L.; Rooman, I. Chronic pancreatitis: A path to pancreatic cancer. *Cancer Lett.* **2014**, *345*, 203–209. [CrossRef] [PubMed]

166. Higashi, D.; Futami, K.; Kojima, D.; Futatsuki, R.; Ishibashi, Y.; Maekawa, T.; Yano, Y.; Takatsu, N.; Hirai, F.; Matsui, T.; et al. Cancer of the small intestine in patients with crohn's disease. *Anticancer Res.* **2013**, *33*, 2977–2980. [PubMed]

167. Tu, T.; Buhler, S.; Bartenschlager, R. Chronic viral hepatitis and its association with liver cancer. *Biol. Chem.* **2017**, *398*, 817–837. [CrossRef] [PubMed]

168. Wei, X.L.; Luo, H.Y.; Li, C.F.; Jin, Y.; Zeng, Z.L.; Ju, H.Q.; Wu, Q.N.; Wang, Y.; Mao, M.J.; Liu, W.L.; et al. Hepatitis B virus infection is associated with younger median age at diagnosis and death in cancers. *Int. J. Cancer* **2017**, *141*, 152–159. [CrossRef] [PubMed]

169. Kolb, J.M.; Ozbek, U.; Harpaz, N.; Holcombe, R.F.; Ang, C. Effect of helicobacter pylori infection on outcomes in resected gastric and gastroesophageal junction cancer. *J. Gastrointest. Oncol.* **2017**, *8*, 583–588. [CrossRef] [PubMed]

170. Lee, Y.C.; Lin, J.T. Screening and treating Helicobacter pylori infection for gastric cancer prevention on the population level. *J. Gastroenterol. Hepatol.* **2017**, *32*, 1160–1169. [CrossRef] [PubMed]

171. Wang, T.Y.; Cai, H.; Zheng, W.; Michel, A.; Pawlita, M.; Milne, G.; Xiang, Y.B.; Gao, Y.T.; Li, H.L.; Rothman, N.; et al. A prospective study of urinary prostaglandin E2 metabolite, helicobacter pylori antibodies, and gastric cancer risk. *Clin. Infect. Dis.* **2017**, *64*, 1380–1386. [CrossRef] [PubMed]

172. Greten, F.R.; Eckmann, L.; Greten, T.F.; Park, J.M.; Li, Z.W.; Egan, L.J.; Kagnoff, M.F.; Karin, M. IKKbeta links inflammation and tumorigenesis in a mouse model of colitis-associated cancer. *Cell* **2004**, *118*, 285–296. [CrossRef] [PubMed]

173. Coussens, L.M.; Werb, Z. Inflammation and cancer. *Nature* **2002**, *420*, 860–867. [CrossRef] [PubMed]

174. Meira, L.B.; Bugni, J.M.; Green, S.L.; Lee, C.-W.; Pang, B.; Borenshtein, D.; Rickman, B.H.; Rogers, A.B.; Moroski-Erkul, C.A.; McFaline, J.L.; et al. DNA damage induced by chronic inflammation contributes to colon carcinogenesis in mice. *J. Clin. Investig.* **2008**, *118*, 2516–2525. [CrossRef] [PubMed]

175. Benvenuto, M.; Fantini, M.; Masuelli, L.; De Smaele, E.; Zazzeroni, F.; Tresoldi, I.; Calabrese, G.; Galvano, F.; Modesti, A.; Bei, R. Inhibition of ErbB receptors, hedgehog and NF-kappaB signaling by polyphenols in cancer. *Front. Biosci. Landmark* **2013**, *18*, 1291–1311.

176. Ferruelo, A.; de las Heras, M.M.; Redondo, C.; Ramon de Fata, F.; Romero, I.; Angulo, J.C. Wine polyphenols exert antineoplasic effect on androgen resistant PC-3 cell line through the inhibition of the transcriptional activity of COX-2 promoter mediated by NF-k beta. *Actas Urol. Esp.* **2014**, *38*, 429–437. [CrossRef] [PubMed]

177. Willenberg, I.; Meschede, A.K.; Gueler, F.; Jang, M.-S.; Shushakova, N.; Schebb, N.H. Food polyphenols fail to cause a biologically relevant reduction of COX-2 activity. *PLoS ONE* **2015**, *10*. [CrossRef] [PubMed]

178. Lilia Salas, A.; Rosa Alberto, M.; Catiana Zampini, I.; Soledad Cuello, A.; Maldonado, L.; Luis Rios, J.; Schmeda-Hirschmann, G.; Ines Isla, M. Biological activities of polyphenols-enriched propolis from argentina arid regions. *Phytomedicine* **2016**, *23*, 27–31. [CrossRef] [PubMed]

179. Yang, F.; Suo, Y.; Chen, D.; Tong, L. Protection against vascular endothelial dysfunction by polyphenols in sea buckthorn berries in rats with hyperlipidemia. *Biosci. Trends* **2016**, *10*, 188–196. [CrossRef] [PubMed]

180. Surh, Y.-J.; Chun, K.-S.; Cha, H.-H.; Han, S.S.; Keum, Y.-S.; Park, K.-K.; Lee, S.S. Molecular mechanisms underlying chemopreventive activities of anti-inflammatory phytochemicals: Down-regulation of COX-2 and iNOS through suppression of NF-κb activation. *Mutat. Res. Fundam. Mol. Mech. Mutagen.* **2001**, *480–481*, 243–268. [CrossRef]

181. Rodriguez-Ramiro, I.; Ramos, S.; Lopez-Oliva, E.; Agis-Torres, A.; Bravo, L.; Goya, L.; Angeles Martin, M. Cocoa polyphenols prevent inflammation in the colon of azoxymethane-treated rats and in TNF-alpha-stimulated Caco-2 cells. *Br. J. Nutr.* **2013**, *110*, 206–215. [CrossRef] [PubMed]

182. Banerjee, N.; Talcott, S.; Safe, S.; Mertens-Talcott, S.U. Cytotoxicity of pomegranate polyphenolics in breast cancer cells in vitro and vivo: Potential role of miRNA-27a and miRNA-155 in cell survival and inflammation. *Breast Cancer Res. Treat.* **2012**, *136*, 21–34. [CrossRef] [PubMed]

183. Mandal, A.; Bhatia, D.; Bishayee, A. Anti-inflammatory mechanism involved in pomegranate-mediated prevention of breast cancer: The role of NF-kappab and Nrf2 signaling pathways. *Nutrients* **2017**, *9*, 436. [CrossRef] [PubMed]

184. Vuong, T.; Mallet, J.-F.; Ouzounova, M.; Rahbar, S.; Hernandez-Vargas, H.; Herceg, Z.; Matar, C. Role of a polyphenol-enriched preparation on chemoprevention of mammary carcinoma through cancer stem cells and inflammatory pathways modulation. *J. Transl. Med.* **2016**, *14*. [CrossRef] [PubMed]

185. Kang, H.K.; Choi, Y.-H.; Kwon, H.; Lee, S.-B.; Kim, D.-H.; Sung, C.K.; Park, Y.I.; Dong, M.-S. Estrogenic/antiestrogenic activities of a *Epimedium koreanum* extract and its major components: In vitro and in vivo studies. *Food Chem. Toxicol.* **2012**, *50*, 2751–2759. [CrossRef] [PubMed]

186. Camargo, M.S.; Prieto, A.M.; Resende, F.A.; Boldrin, P.K.; Cardoso, C.R.P.; Fernandez, M.F.; Molina-Molina, J.M.; Olea, N.; Vilegas, W.; Cuesta-Rubio, O.; et al. Evaluation of estrogenic, antiestrogenic and genotoxic activity of nemorosone, the major compound found in brown Cuban propolis. *BMC Complement. Altern. Med.* **2013**, *13*. [CrossRef] [PubMed]

187. Collins-Burow, B.M.; Antoon, J.W.; Frigo, D.E.; Elliott, S.; Weldon, C.B.; Boue, S.M.; Beckman, B.S.; Curiel, T.J.; Alam, J.; McLachlan, J.A.; et al. Antiestrogenic activity of flavonoid phytochemicals mediated via the c-jun N-terminal protein kinase pathway. Cell-type specific regulation of estrogen receptor alpha. *J. Steroid Biochem. Mol. Biol.* **2012**, *132*, 186–193. [CrossRef] [PubMed]

188. Korobkova, E.A. Effect of natural polyphenols on CYP metabolism: Implications for diseases. *Chem. Res. Toxicol.* **2015**, *28*, 1359–1390. [CrossRef] [PubMed]

189. Yiannakopoulou, E.C. Interaction of green tea catechins with breast cancer endocrine treatment: A systematic review. *Pharmacology* **2014**, *94*, 245–248. [CrossRef] [PubMed]

190. Way, T.D.; Lee, H.H.; Kao, M.C.; Lin, J.K. Black tea polyphenol theaflavins inhibit aromatase activity and attenuate tamoxifen resistance in HER2/neu-transfected human breast cancer cells through tyrosine kinase suppression. *Eur. J. Cancer* **2004**, *40*, 2165–2174. [CrossRef] [PubMed]

191. Griffiths, K.; Wilson, D.W.; Singh, R.B.; De Meester, F. Effect of dietary phytoestrogens on human growth regulation: Imprinting in health & disease. *Indian J. Med. Res.* **2014**, *140*, 82–90.

192. Patisaul, H.B.; Jefferson, W. The pros and cons of phytoestrogens. *Front. Neuroendocrinol.* **2010**, *31*, 400–419. [CrossRef] [PubMed]

193. Lee, Y.M.; Kim, J.B.; Bae, J.H.; Lee, J.S.; Kim, P.-S.; Jang, H.H.; Kim, H.R. Estrogen-like activity of aqueous extract from agrimonia pilosa ledeb. in MCF-7 cells. *BMC Complement. Altern. Med.* **2012**, *12*. [CrossRef] [PubMed]

194. Paterni, I.; Granchi, C.; Katzenellenbogen, J.A.; Minutolo, F. Estrogen receptors alpha (ERα) and beta (ERβ): Subtype-selective ligands and clinical potential. *Steroids* **2014**, *90*, 13–29. [CrossRef] [PubMed]

195. Chalopin, M.; Tesse, A.; Martinez, M.C.; Rognan, D.; Arnal, J.-F.; Andriantsitohaina, R. Estrogen receptor alpha as a key target of red wine polyphenols action on the endothelium. *PLoS ONE* **2010**, *5*. [CrossRef] [PubMed]

196. Albini, A.; Rosano, C.; Angelini, G.; Amaro, A.; Esposito, A.I.; Maramotti, S.; Noonan, D.M.; Pfeffer, U. Exogenous hormonal regulation in breast cancer cells by phytoestrogens and endocrine disruptors. *Curr. Med. Chem.* **2014**, *21*, 1129–1145. [CrossRef] [PubMed]

197. Lambrinidis, G.; Halabalaki, M.; Katsanou, E.S.; Skaltsounis, A.L.; Alexis, M.N.; Mikros, E. The estrogen receptor and polyphenols: Molecular simulation studies of their interactions, a review. *Environ. Chem. Lett.* **2006**, *4*, 159–174. [CrossRef]

198. De Amicis, F.; Russo, A.; Avena, P.; Santoro, M.; Vivacqua, A.; Bonofiglio, D.; Mauro, L.; Aquila, S.; Tramontano, D.; Fuqua, S.A.W.; et al. In vitro mechanism for downregulation of ER-alpha expression by epigallocatechin gallate in ER+/PR+ human breast cancer cells. *Mol. Nutr. Food Res.* **2013**, *57*, 840–853. [CrossRef] [PubMed]

199. Hallman, K.; Aleck, K.; Quigley, M.; Dwyer, B.; Lloyd, V.; Szmyd, M.; Dinda, S. The regulation of steroid receptors by epigallocatechin-3-gallate in breast cancer cells. *Breast Cancer Targets Ther.* **2017**, *9*, 365–373. [CrossRef] [PubMed]

200. Bak, M.J.; Das Gupta, S.; Wahler, J.; Suh, N. Role of dietary bioactive natural products in estrogen receptor-positive breast cancer. *Semin. Cancer Biol.* **2016**, *40–41*, 170–191. [CrossRef] [PubMed]

201. Wang, Y.; Gho, W.M.; Chan, F.L.; Chen, S.; Leung, L.K. The red clover (*Trifolium pratense*) isoflavone biochanin a inhibits aromatase activity and expression. *Br. J. Nutr.* **2008**, *99*, 303–310. [CrossRef] [PubMed]

202. Wang, Y.; Lee, K.W.; Chan, F.L.; Chen, S.A.; Leung, L.K. The red wine polyphenol resveratrol displays bilevel inhibition on aromatase in breast cancer cells. *Toxicol. Sci.* **2006**, *92*, 71–77. [CrossRef] [PubMed]

203. Lan, Y.; Gho, W.M.; Chan, F.L.; Chen, S.; Leung, L.K. Dietary administration of the licorice flavonoid isoliquiritigenin deters growth of MCF-7 cells overexpressing aromatase. *Int. J. Cancer* **2009**, *124*, 1028–1036. [CrossRef]

204. Moseley, V.R.; Morris, J.; Knackstedt, R.W.; Wargovich, M.J. Green tea polyphenol epigallocatechin 3-gallate, contributes to the degradation of DNMT3A and HDAC3 in HCT 116 human colon cancer cells. *Anticancer Res.* **2013**, *33*, 5325–5333. [PubMed]

205. Kang, I.; Okla, M.; Chung, S. Ellagic acid inhibits adipocyte differentiation through coactivator-associated arginine methyltransferase 1-mediated chromatin modification. *J. Nutr. Biochem.* **2014**, *25*, 946–953. [CrossRef] [PubMed]

206. Fang, M.; Chen, D.; Yang, C.S. Dietary polyphenols may affect DNA methylation. *J. Nutr.* **2007**, *137*, 223S–228S. [PubMed]

207. Pathiraja, T.N.; Stearns, V.; Oesterreich, S. Epigenetic regulation in estrogen receptor positive breast cancer-role in treatment response. *J. Mammary Gland Biol. Neoplasia* **2010**, *15*, 35–47. [CrossRef] [PubMed]

208. Meeran, S.M.; Patel, S.N.; Li, Y.; Shukla, S.; Tollefsbol, T.O. Bioactive dietary supplements reactivate ER expression in ER-negative breast cancer cells by active chromatin modifications. *PLoS ONE* **2012**, *7*. [CrossRef] [PubMed]

209. Carpenter, R.L.; Lo, H.-W. Regulation of apoptosis by HER2 in breast cancer. *J. Carcinog. Mutagen.* **2013**, *2013*. [CrossRef]

210. Zhou, B.H.P.; Liao, Y.; Xia, W.Y.; Zou, Y.Y.; Spohn, B.; Hung, M.C. HER-2/neu induces p53 ubiquitination via Akt-mediated MDM2 phosphorylation. *Nat. Cell Biol.* **2001**, *3*, 973–982. [CrossRef] [PubMed]

211. Siddiqa, A.; Long, L.M.; Li, L.; Marciniak, R.A.; Kazhdan, I. Expression of HER-2 in MCF-7 breast cancer cells modulates anti-apoptotic proteins survivin and Bcl-2 via the extracellular signal-related kinase (ERK) and phosphoinositide-3 kinase (PI3K) signalling pathways. *BMC Cancer* **2008**, *8*. [CrossRef] [PubMed]

212. Boekhout, A.H.; Beijnen, J.H.; Schellens, J.H.M. Trastuzumab. *Oncologist* **2011**, *16*, 800–810. [CrossRef] [PubMed]

213. Menendez, J.A.; Vazquez-Martin, A.; Colomer, R.; Brunet, J.; Carrasco-Pancorbo, A.; Garcia-Villalba, R.; Fernandez-Gutierrez, A.; Segura-Carretero, A. Olive oil's bitter principle reverses acquired autoresistance to trastuzumab (herceptin™) in HER2-overexpressing breast cancer cells. *BMC Cancer* **2007**, *7*. [CrossRef] [PubMed]

214. Menendez, J.A.; Vazquez-Martin, A.; Garcia-Villalba, R.; Carrasco-Pancorbo, A.; Oliveras-Ferraros, C.; Fernandez-Gutierrez, A.; Segura-Carretero, A. Tabanti-HER2 (erbB-2) oncogene effects of phenolic compounds directly isolated from commercial Extra-Virgin Olive Oil (EVOO). *BMC Cancer* **2008**, *8*. [CrossRef] [PubMed]

215. Mahmoodi, N.; Motamed, N.; Paylakhi, S.H.; Mahmoodi, N.O. Comparing the effect of silybin and silybin advanced (tm) on viability and HER2 expression on the human breast cancer SKBR3 cell line by no serum starvation. *Iranian J. Pharm. Res.* **2015**, *14*, 521–530.

216. Way, T.D.; Kao, M.C.; Lin, J.K. Apigenin induces apoptosis through proteasomal degradation of HER2/neu in HER2/neu-overexpressing breast cancer cells via the phosphatidylinositol 3-kinase/Akt-dependent pathway. *J. Biol. Chem.* **2004**, *279*, 4479–4489. [CrossRef] [PubMed]

217. Lai, H.-W.; Chien, S.-Y.; Kuo, S.-J.; Tseng, L.-M.; Lin, H.-Y.; Chi, C.-W.; Chen, D.-R. The potential utility of curcumin in the treatment of HER-2-overexpressed breast cancer: An in vitro and in vivo comparison study with herceptin. *Evid.-Based Complement. Altern. Med.* **2012**. [CrossRef]

218. Sidera, K.; Gaitanou, M.; Stellas, D.; Matsas, R.; Patsavoudi, E. A critical role for HSP90 in cancer cell invasion involves interaction with the extracellular domain of HER-2. *J. Biol. Chem.* **2008**, *283*, 2031–2041. [CrossRef] [PubMed]

219. Citri, A.; Harari, D.; Shohat, G.; Ramakrishnan, P.; Gan, J.; Lavi, S.; Eisenstein, M.; Kimchi, A.; Wallach, D.; Pietrokovski, S.; et al. Hsp90 recognizes a common surface on client kinases. *J. Biol. Chem.* **2006**, *281*, 14361–14369. [CrossRef] [PubMed]

220. Garcia-Carbonero, R.; Carnero, A.; Paz-Ares, L. Inhibition of HSP90 molecular chaperones: Moving into the clinic. *Lancet Oncol.* **2013**, *14*, E358–E369. [CrossRef]

221. Yan, Y.-Y.; Zheng, L.-S.; Zhang, X.; Chen, L.-K.; Singh, S.; Wang, F.; Zhang, J.-Y.; Liang, Y.-J.; Dai, C.-L.; Gu, L.-Q.; et al. Blockade of HER2/neu binding to HSP90 by emodin azide methyl anthraquinone derivative induces proteasomal degradation of HER2/neu. *Mol. Pharm.* **2011**, *8*, 1687–1697. [CrossRef] [PubMed]

222. Chiang, C.-T.; Way, T.-D.; Lin, J.-K. Sensitizing HER2-overexpressing cancer cells to luteolin-induced apoptosis through suppressing p21(WAF1/CLP1) expression with rapamycin. *Mol. Cancer Ther.* **2007**, *6*, 2127–2138. [CrossRef] [PubMed]

223. Vassallo, A.; Vaccaro, M.C.; De Tommasi, N.; Dal Piaz, F.; Leone, A. Identification of the plant compound geraniin as a novel Hsp90 inhibitor. *PLoS ONE* **2013**, *8*. [CrossRef] [PubMed]

224. Giommarelli, C.; Zuco, V.; Favini, E.; Pisano, C.; Dal Piaz, F.; De Tommasi, N.; Zunino, F. The enhancement of antiproliferative and proapoptotic activity of HDAC inhibitors by curcumin is mediated by Hsp90 inhibition. *Cell. Mol. Life Sci.* **2010**, *67*, 995–1004. [CrossRef] [PubMed]

225. Moses, M.A.; Henry, E.C.; Ricke, W.A.; Gasiewicz, T.A. The heat shock protein 90 inhibitor, (−)-epigallocatechin gallate, has anticancer activity in a novel human prostate cancer progression model. *Cancer Prev. Res.* **2015**, *8*, 249–257. [CrossRef] [PubMed]

226. Menendez, J.A.; Lupu, R. Fatty acid synthase and the lipogenic phenotype in cancer pathogenesis. *Nat. Rev. Cancer* **2007**, *7*, 763–777. [CrossRef] [PubMed]

227. Yoon, S.; Lee, M.-Y.; Park, S.W.; Moon, J.-S.; Koh, Y.-K.; Ahn, Y.-H.; Park, B.-W.; Kim, K.-S. Up-regulation of acetyl-CoA carboxylase alpha and fatty acid synthase by human epidermal growth factor receptor 2 at the translational level in breast cancer cells. *J. Biol. Chem.* **2007**, *282*, 26122–26131. [CrossRef] [PubMed]

228. Pizer, E.S.; Thupari, J.; Han, W.F.; Pinn, M.L.; Chrest, F.J.; Frehywot, G.L.; Townsend, C.A.; Kuhajda, F.P. Malonyl-coenzyme-A is a potential mediator of cytotoxicity induced by fatty-acid synthase inhibition in human breast cancer cells and xenograft. *Cancer Res.* **2000**, *60*, 213–218. [PubMed]

229. Zhou, W.; Tu, Y.; Simpson, P.J.; Kuhajda, F.P. Malonyl-CoA decarboxylase inhibition is selectively cytotoxic to human breast cancer cells. *Oncogene* **2009**, *28*, 2979–2987. [CrossRef] [PubMed]

230. Thupari, J.N.; Pinn, M.L.; Kuhajda, F.P. Fatty acid synthase inhibition in human breast cancer cells leads to malonyl-CoA-induced inhibition of fatty acid oxidation and cytotoxicity. *Biochem. Biophys. Res. Commun.* **2001**, *285*, 217–223. [CrossRef] [PubMed]

231. Younesian, O.; Kazerouni, F.; Dehghan-Nayeri, N.; Omrani, D.; Rahimipour, A.; Shanaki, M.; Kalkhoran, M.R.; Cheshmi, F. Effect of curcumin on fatty acid synthase expression and enzyme activity in breast cancer cell line SKBR3. *Iranian J. Cancer Prev.* **2017**, *10*. [CrossRef]

232. Menendez, J.A.; Vazquez-Martin, A.; Oliveras-Ferraros, C.; Garcia-Villalba, R.; Carrasco-Pancorbo, A.; Fernandez-Gutierrez, A.; Segura-Carretero, A. Analyzing effects of extra-virgin olive polyphenols on breast cancer-associated fatty acid synthase protein expression using reverse-phase protein microarrays. *Int. J. Mol. Med.* **2008**, *22*, 433–439. [CrossRef] [PubMed]

233. Oliveras, G.; Blancafort, A.; Urruticoechea, A.; Campuzano, O.; Gómez-Cabello, D.; Brugada, R.; López-Rodríguez, M.L.; Colomer, R.; Puig, T. Novel anti-fatty acid synthase compounds with anti-cancer activity in HER2$^+$ breast cancer. *Ann. N. Y. Acad. Sci.* **2010**, *1210*, 86–92. [CrossRef] [PubMed]

234. Rein, M.J.; Renouf, M.; Cruz-Hernandez, C.; Actis-Goretta, L.; Thakkar, S.K.; da Silva Pinto, M. Bioavailability of bioactive food compounds: A challenging journey to bioefficacy. *Br. J. Clin. Pharmacol.* **2013**, *75*, 588–602. [CrossRef] [PubMed]

235. Gao, S.; Hu, M. Bioavailability challenges associated with development of anti-cancer phenolics. *Mini Rev. Med. Chem.* **2010**, *10*, 550–567. [CrossRef] [PubMed]

236. Watkins, R.; Wu, L.; Zhang, C.; Davis, R.M.; Xu, B. Natural product-based nanomedicine: Recent advances and issues. *Int. J. Nanomed.* **2015**, *10*, 6055–6074.

237. Davidov-Pardo, G.; McClements, D.J. Resveratrol encapsulation: Designing delivery systems to overcome solubility, stability and bioavailability issues. *Trends Food Sci. Technol.* **2014**, *38*, 88–103. [CrossRef]

238. Seguin, J.; Brullé, L.; Boyer, R.; Lu, Y.M.; Ramos Romano, M.; Touil, Y.S.; Scherman, D.; Bessodes, M.; Mignet, N.; Chabot, G.G. Liposomal encapsulation of the natural flavonoid fisetin improves bioavailability and antitumor efficacy. *Int. J. Pharm.* **2013**, *444*, 146–154. [CrossRef] [PubMed]

239. Shaikh, J.; Ankola, D.D.; Beniwal, V.; Singh, D.; Kumar, M.N.V.R. Nanoparticle encapsulation improves oral bioavailability of curcumin by at least 9-fold when compared to curcumin administered with piperine as absorption enhancer. *Eur. J. Pharm. Sci.* **2009**, *37*, 223–230. [CrossRef] [PubMed]

240. Paudel, A.; Worku, Z.A.; Meeus, J.; Guns, S.; Van den Mooter, G. Manufacturing of solid dispersions of poorly water soluble drugs by spray drying: Formulation and process considerations. *Int. J. Pharm.* **2013**, *453*, 253–284. [CrossRef] [PubMed]

241. Aggarwal, B.B.; Kumar, A.; Bharti, A.C. Anticancer potential of curcumin: Preclinical and clinical studies. *Anticancer Res.* **2003**, *23*, 363–398. [PubMed]

242. Bishayee, A. Cancer prevention and treatment with resveratrol: From rodent studies to clinical trials. *Cancer Prev. Res.* **2009**, *2*, 409–418. [CrossRef] [PubMed]

243. Ho, C.K.; Choi, S.W.; Fung, M.S.; Benzie, I.F.F. Tea polyphenols: Absorption, bioavailability and potential toxicity. *CAB Rev. Perspect. Agric. Vet. Sci. Nutr. Nat. Resour.* **2017**, *12*. [CrossRef]

244. Cheng, F.; Li, W.; Zhou, Y.; Shen, J.; Wu, Z.; Liu, G.; Lee, P.W.; Tang, Y. Admetsar: A comprehensive source and free tool for assessment of chemical admet properties. *J. Chem. Inform. Model.* **2012**, *52*, 3099–3105. [CrossRef] [PubMed]

245. Sander, T.; Freyss, J.; Von Korff, M.; Rufener, C. Datawarrior: An open-source program for chemistry aware data visualization and analysis. *J. Chem. Inform. Model.* **2015**, *55*, 460–473. [CrossRef] [PubMed]

246. Spratlin, J.L.; Serkova, N.J.; Eckhardt, S.G. Clinical applications of metabolomics in oncology: A review. *Clin. Cancer Res.* **2009**, *15*, 431–440. [CrossRef] [PubMed]

247. Zhang, X.; Yap, Y.; Wei, D.; Chen, G.; Chen, F. Novel omics technologies in nutrition research. *Biotechnol. Adv.* **2008**, *26*, 169–176. [CrossRef] [PubMed]

248. Del Rio, A.; Da Costa, F.B. Molecular approaches to explore natural and food-compound modulators in cancer epigenetics and metabolism. In *Foodinformatics: Applications of Chemical Information to Food Chemistry*; Springer International Publishing: Gewerbestrasse, Switzerland, 2014; pp. 131–149.

249. Saeed, M.; Kadioglu, O.; Khalid, H.; Sugimoto, Y.; Efferth, T. Activity of the dietary flavonoid, apigenin, against multidrug-resistant tumor cells as determined by pharmacogenomics and molecular docking. *J. Nutr. Biochem.* **2015**, *26*, 44–56. [CrossRef] [PubMed]

250. Paluszczak, J.; Krajka-Kuźniak, V.; Baer-Dubowska, W. The effect of dietary polyphenols on the epigenetic regulation of gene expression in MCF7 breast cancer cells. *Toxicol. Lett.* **2010**, *192*, 119–125. [CrossRef] [PubMed]

251. Vanden Berghe, W. Epigenetic impact of dietary polyphenols in cancer chemoprevention: Lifelong remodeling of our epigenomes. *Pharmacol. Res.* **2012**, *65*, 565–576. [CrossRef] [PubMed]

![antioxidants logo] *antioxidants*

MDPI

Article

Microbial Biotransformation of a Polyphenol-Rich Potato Extract Affects Antioxidant Capacity in a Simulated Gastrointestinal Model

Joelle Khairallah [1], Shima Sadeghi Ekbatan [1], Kebba Sabally [1], Michèle M. Iskandar [1], Raza Hussain [1], Atef Nassar [1,2], Lekha Sleno [3], Laetitia Rodes [4], Satya Prakash [4], Danielle J. Donnelly [2] and Stan Kubow [1,*]

[1] School of Human Nutrition, McGill University, 21111 Lakeshore, Ste. Anne de Bellevue, QC H9X 3V9, Canada; joelle.khairallah@mail.mcgill.ca (J.K.); shima.sadeghi@mail.mcgill.ca (S.S.E.); kebba.sabally@mcgill.ca (K.S.); michele.iskandar@mail.mcgill.ca (M.M.I.); raza.hussain@mail.mcgill.ca (R.H.); atef.nassar@mail.mcgill.ca (A.N.)

[2] Plant Science Department, McGill University, 21111 Lakeshore, Ste. Anne de Bellevue, QC H9X 3V9, Canada; danielle.donnelly@mcgill.ca

[3] Chemistry Department, University of Quebec at Montreal, 2101 rue Jeanne-Mance, Montreal, QC H2X 2J6, Canada; sleno.lekha@uqam.ca

[4] Department of Biomedical Engineering, Duff Medical Building, McGill University, 3775 Rue University, Montreal, QC H3A 2B4, Canada; laetitia.rodes@mail.mcgill.ca (L.R.); satya.prakash@mcgill.ca (S.P.)

* Correspondence: stan.kubow@mcgill.ca; Tel.: +1-514-398-7754; Fax: +1-514-398-7739

Received: 26 February 2018; Accepted: 15 March 2018; Published: 20 March 2018

Abstract: A multistage human gastrointestinal model was used to digest a polyphenol-rich potato extract containing chlorogenic acid, caffeic acid, ferulic acid, and rutin as the primary polyphenols, to assess for their microbial biotransformation and to measure changes in antioxidant capacity in up to 24 h of digestion. The biotransformation of polyphenols was assessed by liquid chromatography–mass spectrometry. Antioxidant capacity was measured by the ferric reducing antioxidant power (FRAP) assay. Among the colonic reactors, parent (poly)phenols were detected in the ascending (AC), but not the transverse (TC) or descending (DC) colons. The most abundant microbial phenolic metabolites in all colonic reactors included derivatives of propionic acid, acetic acid, and benzoic acid. As compared to the baseline, an earlier increase in antioxidant capacity ($T = 8$ h) was seen in the stomach and small intestine vessels as compared to the AC ($T = 16$ h) and TC and DC ($T = 24$ h). The increase in antioxidant capacity observed in the DC and TC can be linked to the accumulation of microbial smaller-molecular-weight phenolic catabolites, as the parent polyphenolics had completely degraded in those vessels. The colonic microbial digestion of potato-based polyphenols could lead to improved colonic health, as this generates phenolic metabolites with significant antioxidant potential.

Keywords: phenolic metabolites; *Solanum tuberosum*; biotransformation; polyphenols; potato; antioxidant; digestion; gastrointestinal model

1. Introduction

Dietary (poly)phenolic compounds are secondary metabolites widespread in fruits, vegetables, seeds, and plant-derived beverages such as tea and coffee [1]. Among the wide variety of (poly)phenolics found in foods, phenolic acids and flavonoids have received the greatest research attention for their antioxidant, antiobesity, anti-inflammatory, and antidiabetic properties [2–4]. Potatoes contain significant amounts of (poly)phenolics including phenolic acids as well as flavonols such as rutin [5,6], in addition to other bioactive phytochemicals including glycoalkaloids [7,8]. Due to their high consumption as a food staple, potatoes (*Solanum tuberosum* L.) are an important dietary source of (poly)phenolics in the French, American, and Greek diets [9,10]. A (poly)phenolic-rich potato extract (PRPE), containing chlorogenic acid, caffeic acid, ferulic acid, and rutin as the major (poly)phenolic constituents, has been demonstrated to protect against glucose intolerance and obesity in male and female mice fed a high-fat diet [11].

Since (poly)phenolics are present in foods as esters, polymers, or glycosides, they are generally poorly absorbed in the upper gastrointestinal (GI) tract [12–14]. For example, only one-third of ingested chlorogenic acid and 2–15% of flavonoids were estimated to be absorbed in the stomach and small intestine (SI) [15]. Due to their poor bioavailability, substantial amounts of (poly)phenolics reach the large intestine, where they undergo degradation by gut microflora into a range of smaller-molecular-weight compounds [16]. The gut microbial-mediated biotransformation of (poly)phenolics could contribute to their systemic health benefits, as their smaller-molecular-weight by-products are more bioavailable [17]. This phenomenon is evidenced by the appearance of microbial metabolites in the plasma of healthy participants with an intact colon approximately 6–12 h following the ingestion of supplements of chlorogenic acid, caffeic acid, or rutin [18–20]. Health-promoting effects have been related to the antioxidant capacity of plant foods, which has been partly attributed to (poly)phenolics [21]. However, the relationship of antioxidant activities of (poly)phenolics to their microbial by-products has not been well studied. Consequently, investigations regarding the identification and quantification of microbial colonic metabolites and their impact on antioxidant potential is highly relevant to the possible health benefits of (poly)phenolics.

An experimental approach to investigate the biotransformation of (poly)phenolics has involved simulated in vitro GI digestion models. Most GI model studies evaluating (poly)phenolic metabolism have focused on the enzymatic and chemical digestive conditions of the upper GI tract [22,23]. Alternatively, some studies have used a basic fermenter involving human or rat fecal incubations to assess microbial action on (poly)phenolics [24–27]. A more comprehensive approach that stimulates upper and lower GI digestion involves multistage in vitro models consisting of reactors pertaining to the stomach, SI, and the three colonic compartments of the ascending colon (AC), transverse colon (TC), and descending colon (DC) [28]. There have been limited studies of (poly)phenolic biotransformation involving a multistage GI stimulator. A recent study demonstrated that the feeding of pure reference (poly)phenolic compounds into a multireactor GI system led to microbial phenolic metabolite profiles and antioxidant activities that varied among the three colonic reactors [29]. Similarly, the multireactor-GI-stimulated digestion of red wine showed differing microbial phenolic metabolite profiles among the three colonic vessels [30].

In the present study, PRPE was subjected to digestion via a computer-controlled multistage GI model to examine phenolic catabolites in the colonic reactors using liquid chromatography-mass spectrometry (LC-MS). The antioxidant potential of the digests of PRPE was assessed at different stages of digestion using an antioxidant capacity assay.

2. Materials and Methods

2.1. Phenolic Extraction

PRPE was generated by POS Bio-Sciences (POS Bio-Sciences, Saskatoon, SK, Canada) as previously described [11]. In brief, 20 kg of "Onaway" potatoes were diced, freeze-dried, and extracted by agitation with 200 L of a 90% (v/v) aqueous ethanol solution for 1 h at room temperature. The ratio of powder to aqueous ethanol was 1:10 (w/v). After extraction, the extract was separated from the solids by centrifugation at $1076 \times g$ for 10 min. The extract was then concentrated under vacuum at 40–50 °C, until the volume was reduced to approximately 15 L and the ethanol percentage, measured using a hydrometer, was less than 10%. Water was added back during evaporation to attain a low ethanol content. Afterwards, the concentrate was freeze-dried to generate powdered PRPE with a phenolic content (mg/g) of chlorogenic acid (8.9), caffeic acid (0.6), ferulic acid (0.2), and rutin (1.2). PRPE was stored at −80 °C until use for the simulated GI model digestion studies.

2.2. In Vitro Digestion of PRPE

In vitro digestion of PRPE in the computer-controlled gastrointestinal model followed a previously published method [28]. The GI model system involved five consecutive double-jacketed reactors, which represented the stomach, SI, AC, TC, and DC. The unit was fully computer-controlled using LabVIEW® software (National Instruments, Austin, TX, USA), which controls the addition of food to the stomach, buffers for pH adjustment to the vessels, and pancreatic juice to the SI. The passage of food in the stomach was simulated by the addition of gastric solution (0.1 M HCl and pepsin; Sigma-Aldrich, Oakville, ON, Canada). The SI vessel was stimulated by addition of a pancreatic solution containing (12 g/L) NaHCO$_3$ (S7277, Sigma-Aldrich, Oakville, ON, Canada), (6 g/L) Oxgall (DF0128-17-8, Fisher Scientific, Nepean, ON, Canada), and (0.9 g/L) pancreatin (P1750-100 g, Sigma-Aldrich). The flow of intestinal content between all reactors was automatically computer-controlled. The total transit time was 8 h, i.e., 2 h in the stomach and SI vessels, followed by 4 h in the colonic vessels. The total stabilization period was 2 weeks, whereby freshly collected fecal slurry samples were inoculated into the three colonic reactors. The fecal samples were obtained from five healthy volunteers having no history of GI disease or any use of antibiotics within the previous 6 months. During the stabilization period, the GI system was continuously fed three times per day with the GI nutrient solution. The GI nutrient solution was composed of (g/L): arabinogalactan (1), pectin (2), xylan (1), starch (3), glucose (0.4), yeast extract (3), peptone (1), mucin (4), and cysteine powder (0.5); as previously established to stabilize the microbial community in the colonic vessels [28]. Conditions in all the vessels of the simulated gut were kept strictly anaerobic by flushing nitrogen for 20 min into the air space daily and at the time of each manipulation. The nutritive media was stored at 4 °C until the time of injection. Temperature-controlled water flowed between the double-glass-jacketed reactors to keep the temperature at 37 °C. The pH in the vessels was continuously measured by pH probes and automatically adjusted using 0.2 M HCl (AC423795000, Fisher Scientific, Ottawa, ON, Canada) or 0.5 M NaOH solutions (415413, Sigma-Aldrich, Oakville, ON, Canada) to keep a pH of 2.0 in the stomach and 6.5 in the SI. The pH ranges in the reactors of the AC, TC, and DC were maintained at 5.60–6.40, 6.20–6.40, and 6.60–6.80, respectively.

After a 2-week stabilization period, 130 g of PRPE, containing 1.4 g of total (poly)phenolics, was administered into the gut model system. This amount is slightly above the average daily total (poly)phenolic intake of 1 g that is generally consumed daily through food or supplement intake [31]. The extract was dissolved using 25 mL of methanol and solubilized into the GI food as described in other studies involving the feeding of (poly)phenolic solutions to microflora [15,32]. The fluid in the vessels was continuously stirred using a magnetic stirrer and the vessels were wrapped with aluminum foil to prevent polyphenol photodecomposition. Digestion lasted 24 h and samples were taken from all the vessels before PRPE addition ($T = 0$ h) and every 8 h during the 24 h digestion. To ensure the stability of the polyphenols and inactivate digestive enzymes after the removal of digests, 5 mL of 0.5 M HCl solution was added to decrease the pH (pH ≤ 2) of the withdrawn samples. The aliquots were centrifuged at 1000× *g* for 20 min and stored in 15 mL Falcon tubes at −80 °C for later analysis.

2.3. LC-MS Analysis of Phenolic Metabolites

Samples were thawed, vortexed, and filtered with 25 mm syringe filters (0.45 μm, MCE, sterile; Fisher Scientific Ottawa, ON, Canada) into 1 mL glass vials before analysis. Phenolic metabolites were separated and analysed using LC-MS analysis as previously described [29]. In brief, a Gemini-NX (5 μm, 100 mm × 4.6 mm) column (Phenomenex, Torrance, CA, USA) and a 4.6 mm × 2.0 mm guard column (Phenomenex, Torrance, CA, USA) were used with two mobile phases: solvent A (10 mM formic acid, pH 3.5) and solvent B (5 mM ammonium formate solution in 100% methanol). The following gradient elution was used: 0 min 5% B, 2 min 5% B, 5 min 30% B, 7 min 70% B, 9 min

100% B, and 12 min 100% B, with a flow rate of 1.0 mL/min, and 20 μL of sample was injected. LC-MS analysis used a 6210 LC-MS Time of Flight system (Agilent Technologies, Santa Clara, CA, USA) in negative electrospray ionization mode with internal calibration using calibrant ions at *m/z* 119.0363 and 966.0007 at a scanning speed of 1 spectrum/s. Source conditions were as follows: capillary voltage of −4000 V, gas temperature of 350 °C, drying gas flow of 12 L/min, nebulizer gas at 50 psi, and fragmentor and skimmer voltages at 100 V and 60 V, respectively. The relative abundance of metabolized compounds was measured relative to the quantification of 3-hydroxyphenylacetic acid as the reference peak, as 3-hydroxyphenylacetic acid was noted to be present at the same concentration in all but the ascending colonic vessel at *T* = 0.

2.4. Ferric Reducing Antioxidant Power (FRAP) Assay

The FRAP assay was used to determine the total antioxidant capacity of the sample through the reduction of the ferric tripyridyl triazine complex to a ferrous complex [33]. The FRAP reagents were prepared as previously described [34]. The reaction was carried out using a 96-well plate upon the addition of 30 μL H$_2$O, 10 μL standards or samples, and 200 μL FRAP solution. Samples were left to react at room temperature for 20 min and absorbance was read at 593 nm in a microplate reader (Infinite PRO 200 series, Tecan Group, San Jose, CA, USA). Ferrous sulfate solution was used as an external standard with a calibration curve range of 0.1 to 10 mM. The results were expressed as ferrous sulfate equivalents.

2.5. Statistical Analysis

Data were tested for normality and significant least square means (LSMeans) were compared using Tukey's honest significant difference (HSD) posthoc test. Statistical significance was set at $p < 0.05$ and all statistical analyses were performed using SAS 9.2 (SAS Institute Inc., Cary, NC, USA). Data are expressed as means ± SEM.

3. Results and Discussion

3.1. Biotransformation of Polyphenols

Table 1 illustrates the profiles of the parent (poly)phenolics and their microbial metabolites in the three colonic vessels of the GI model at baseline (*T* = 0 h) and after 24 h digestion of PRPE. A total of 11 microbial metabolites were detected in the colonic vessels.

Table 1. Polyphenols and their metabolites after human-stimulated intestinal digestion at baseline (T0) and after 24 h (T24) postdigestion [1].

Theoretical Mass (m/z) [2]	Measured Mass	Mass Accuracy (ppm)	Retention Time (min)	Common Name	Systematic Name	AC		TC		DC	
						T0	T24	T0	T24	T0	T24
609.1461	609.1422	6.4	8.7	Rutin	Quercetin-3-O-rutinoside	-	9.62	-	-	-	-
353.0878	353.0863	4.3	7.5	Chlorogenic acid	(1S,3R,4R,5R)-3-[[(2E)-3(3,4-Dihydroxyphenyl)prop-2enoyl]oxy]1,4,5trihydroxycyclohexanecarboxylic acid	-	37.96	-	-	-	-
301.0354	301.0395	13.7	8	Quercetin	2-(3,4-Dihydroxyphenyl)-3,5,7-trihydroxy-4H-chromen-4-one	-	1.05	-	-	-	0.15
193.0506	193.0499	3.8	8.5	Ferulic acid	3-(4-Hydroxy-3-methoxy-phenyl)prop-2-enoic acid	-	2.03	-	0.16	-	-
191.0561	191.0544	2.3	1.7	Quinic acid	(1S,3R,4S,5R)-1,3,4,5-Tetrahydroxycyclohexanecarboxylic acid	-	642.68	-	28.74	-	1.39
181.0506	181.0505	0.7	7.7	Dihydrocaffeic acid	3-(3',4'-Dihydroxyphenyl) propionic acid	-	136.83	36.16	2.92	13.70	37.05
179.0350	179.0339	6.1	8	Caffeic acid	3,4-Dihydroxycinnamic acid	-	327.36	-	-	-	0.50
167.035	167.0346	2.2	6.6	Vanillic acid	4-Hydroxy-3-methoxybenzoic acid	-	0.10	-	0.03	-	0.02
165.0557	165.0557	0.1	8.4	3-Hydroxyphenyl propionic acid	3-(3'-Hydroxyphenyl)propionic acid	-	36.69	-	28.59	-	7.56
163.0401	163.0401	0.2	8.4	Coumaric acid	The isomer is not specified from our data	-	12.34	-	1.05	-	-
153.0193	153.0204	6.9	7.2	Protocatechuic acid	3,4-Dihydroxybenzoic acid	-	3.65	-	3.26	-	2.64
151.0401	151.0409	5.5	7.7	3-Hydroxyphenyl acetic acid	3-Hydroxyphenylacetic acid [3]	-	1.00	1.00	1.00	1.00	1.00
149.0608	149.0604	2.7	9.4	3-Phenylpropionic acid	3-Phenylpropionic acid	-	0.00	5.01	1.55	1.53	1.04
147.0452	147.0453	1	8.5	Cinnamic acid	3-Phenylprop-2-enoic acid	-	0.74	0.46	-	-	-
137.0244	137.0245	0.6	7.2	3-Hydroxybenzoic acid	3-Hydroxybenzoic acid	-	9.02	-	8.88	-	3.03

AC = ascending colon; TC = transverse colon; DC = descending colon; (–) absent; masses are shown as [M − H]. [1] Determined by liquid chromatography-mass spectrometry (LC-MS) analysis. [2] Identification based on previous literature [19,25,32]. [3] The quantities of the polyphenols and their metabolites are calculated relative to the concentration of 3-hydroxyphenyl acetic acid.

Figure 1 shows the extracted ion chromatogram from the 24 h simulated digestion of PRPE in the AC compartment, which is a representative vessel that contained the native (poly)phenolics and their microbial metabolites.

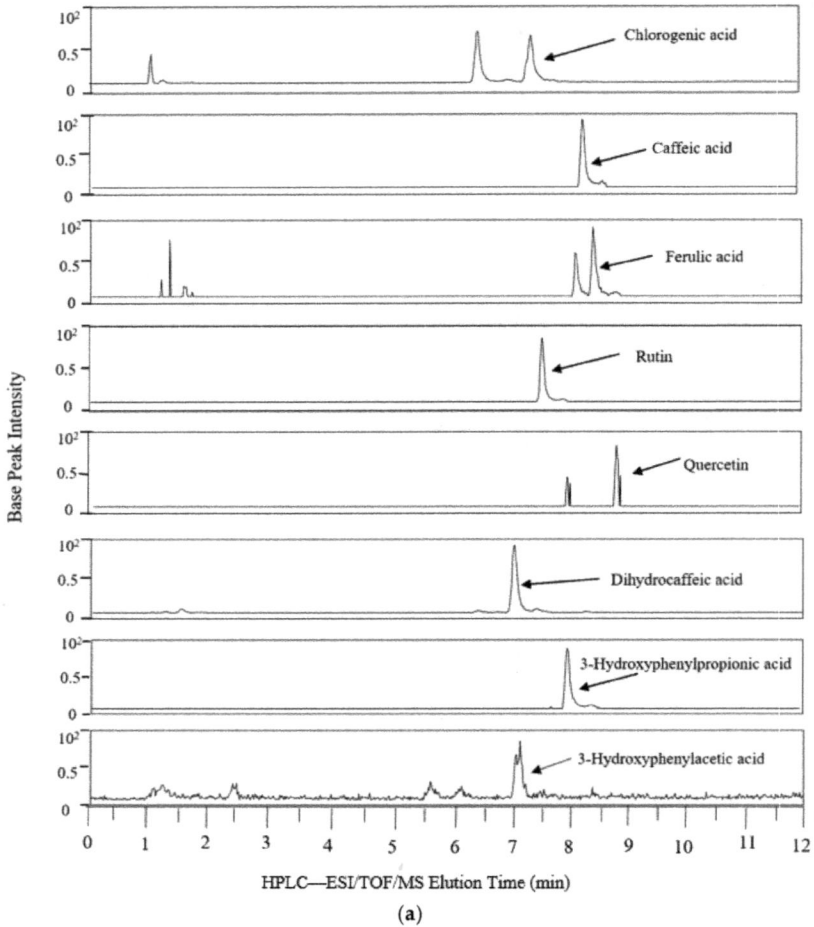

Figure 1. *Cont.*

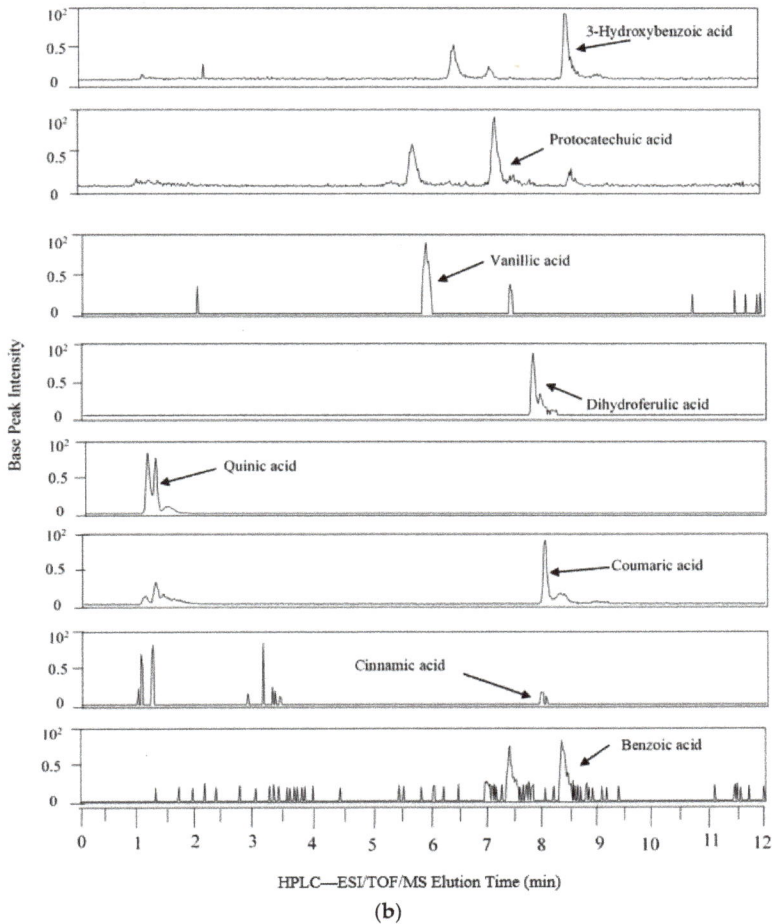

Figure 1. Representative extracted ion chromatograms from colonic vessels after 24 h of PRPE fermentation. HPLC-ESI/TOF/MS: High Performance Liquid Chromatography-Electrospray Ionization/Time of Flight/Mass Spectrometry. (**a**) AC; (**b**) AC continued.

The phenolic profile differences among the colonic vessels could be attributed partly to different microbial communities, as the pattern of (poly)phenolic degradation has been previously noted to depend upon varying gut microbial profiles caused by the different pH range in each of the reactors [35]. Because the AC was the first reactor exposed to the parent (poly)phenolics, it contained relatively high levels of chlorogenic acid, caffeic acid, ferulic acid, and rutin in comparison with the other colonic vessels. Likewise, as parent (poly)phenolics begin to undergo microbial catabolism in the AC, this reactor accumulated the largest number of phenolic metabolites with greater relative abundance (except for 3-phenylpropionic acid) compared to the other colonic vessels. The major microbial catabolites produced from the fermentation of PRPE-derived (poly)phenolics included caffeic acid, dihydrocaffeic acid, phenylpropionic acid, 3-hydroxyphenylpropionic acid, 3-hydroxyphenylacetic acid, 3-hydroxybenzoic acid, and coumaric acid, which all have been detected in human fecal water and in fermentation studies using human fecal slurry [28,36]. The concurrent presence of native (poly)phenolics and their microbial metabolites in the AC following the 24 h digestion contrasts with batch fermentation studies with human fecal slurry showing the complete degradation of chlorogenic acid, caffeic acid, and rutin that can occur as early as 30 min to 5 h following incubation [24,25,37,38]. The high levels of quinic acid in the AC can be generated via bacterial cinnamoyl esterase-mediated cleavage of the ester bond between caffeic acid and quinic acid in chlorogenic acid [35,39,40]. The catabolism of chlorogenic acid would explain the high amounts of caffeic acid in the AC, in addition to the caffeic acid provided from PRPE itself. Caffeic acid undergoes microbial reduction of the double bond to generate dihydrocaffeic acid [32], which was also detectable at high levels in the AC. Dihydrocaffeic acid undergoes bacterial dehydroxylation to generate 3-hydroxylphenylpropionic acid, which was present at substantially lower levels in the AC; while 3-phenylpropionic acid, the microbial degradation product of 3-hydroxylphenylpropionic acid, was not detected. The above findings are in concert with Rechner et al. [32], who reported that larger doses of chlorogenic acid (500 to 1500 mg/L), similar to that provided in the present study, strongly shifted the composition of chlorogenic end-products to metabolites generated at the first catabolic steps. Therefore, high concentrations of caffeic and dihydrocaffeic acid generated from a large input of chlorogenic acid appear to inhibit their subsequent degradation to 3-hydroxylphenylpropionic acid and 3-phenylpropionic acid. Dihydrocaffeic acid can also be transformed via microbial α-oxidation or β-oxidation reactions to produce protocatechuic acid [32,38,41,42], detected in the AC. The PRPE supplement provided rutin, observed in the AC along with quercetin, a microbial degradation product of rutin [43], which was noted in low amounts. The presence of 3-hydroxylphenylpropionic acid and protocatechuic acid in the AC can be partly attributed to the bacterial degradation of quercetin [44]. The presence of ferulic acid in the AC can be obtained from the PRPE supplement as well as bacterial methylation of caffeic acid which can generate ferulic acid [45]. Ferulic acid can undergo further microbial metabolism to form vanillic acid [19], which was detected at low amounts in all the colonic reactors, indicating that this was a product of a relatively minor catabolic event. The parent (poly)phenolics provided by PRPE were absent in the TC and DC, apart from ferulic acid, which was seen at barely detectable levels (Table 1). Prior to PRPE addition, dihydrocaffeic acid, 3-hydroxyphenylacetic acid, 3-phenylpropionic acid, cinnamic acid, and benzoic acid were measured in the colonic reactors. The presence of such catabolites has also been described in previous fecal fermentation studies prior to (poly)phenolic addition [29,30,35,38], and may be the result of protein and carbohydrate fermentation of the nutritive medium [30], as the two-week stabilization would likely have eliminated residual phenolics from the original fecal matter. The diminution of most of the microbial metabolites seen in the DC and TC was probably secondary to microbial β-oxidation reactions, which lead to the degradation of phenylacetic, phenylpropionic, and benzoic acids [46]. A major increase was seen in the levels of dihydrocaffeic acid in the DC compartment as compared to the relatively low levels in the TC, which suggests that further microbial phenolic biotransformation occurs in the DC. Overall, the detection of significant amounts of phenolic acids in the TC and DC coincides with previous multistage in vitro digestion models involving extracts of (poly)phenolic-rich red beverages such as wine, grape juice, and black tea [47,48].

3.2. Antioxidant Capacity

Figure 2 shows the change in FRAP antioxidant capacity with time in the five compartments of the GI model after PRPE addition.

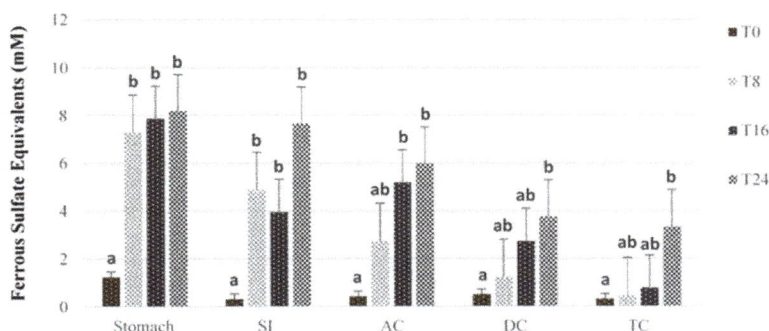

Figure 2. FRAP antioxidant capacity after human stimulated intestinal digestion at baseline (T0) and after 8 h (T8), 16 h (T16) and 24 h (T24) of digestion. SI = small intestine; AC = ascending colon; TC = transverse colon; DC = descending colon. Data are represented as means ± SE. Bars that do not share the same letter within the same vessel are significantly ($p < 0.05$) different from each other, based on Tukey's honest significant difference (HSD) posthoc test.

In general, an increase in antioxidant activity was observed in all GI vessels upon the addition of PRPE, although the timing of the increase differed among the reactors. After 8 h of PRPE digestion, the stomach and SI reactors had significantly higher antioxidant capacity compared to the baseline ($p < 0.05$). In contrast, antioxidant capacity in the colonic vessels increased more slowly; this was greater after 16 h for the AC and 24 h for the DC and TC. As chlorogenic acid, caffeic acid, and rutin have antioxidant properties [43,49], the more rapid increase in antioxidant capacity in the stomach and SI reactors can be attributed to the parent (poly)phenolics, which showed no apparent degradation in those vessels (see Supplementary Material). Similarly, the relatively shorter time for increased antioxidant capacity following PRPE addition in the AC compartment versus the DC and TC can be attributed to the concurrent presence of both the parent (poly)phenolics and their microbial metabolites in the AC. Hence, microbial catabolism of the parent (poly)phenolics appeared to initially reduce antioxidant activity, which agrees with previous findings involving the fecal microbial catabolism of (poly)phenolic compounds [13,29]. On the other hand, the increase in antioxidant capacity observed at 24 h in the DC and TC can be linked to the accumulation of microbial smaller-molecular-weight phenolic catabolites, as the parent (poly)phenolics had completely degraded in those vessels. This finding coincides with that of an earlier study showing increased antioxidant capacity with the generation of microbial phenolic metabolites after the degradation of pure reference (poly)phenolics using an identical in vitro digestion model [29]. Several studies have reported effective antioxidant properties in relation to the microbial phenolic metabolites detected in the present study, including coumaric acid, 3-hydroxyphenylpropionic acid, vanillic acid, dihydrocaffeic acid, and protocatechuic acid [29,50].

4. Conclusions

In conclusion, the association of antioxidant activities with the simpler microbial phenolic acid metabolites generated from PRPE digestion highlights the utility of simulated digestion and fermentation for the assessment of bioactive properties of polyphenol-rich plant food extracts. Further in vivo investigations and mechanistic studies are needed to determine the physiological relevance of the free-radical-scavenging activities associated with the microbial metabolites of potato-derived polyphenols.

Supplementary Materials: The following are available online at http://www.mdpi.com/2076-3921/7/3/43/s1. The supplementary material file includes data acquired via HPLC depicting the catabolism of parent (poly)phenolics and appearance of metabolites throughout the five vessels of the GI model after 24 h in vitro digestion, in two figures (Figures S1 and S2). The methodology for acquisition of HPLC data is also described. Figure S1: HPLC Chromatograms of (poly)phenolics in colonic vessels at baseline and after 24 h of fermentation. Figure S2: HPLC Chromatograms of polyphenols in (A) digested potato extract, (B) stomach vessel (V1), and (C) small intestine vessel (V2) after 24 h of digestion.

Acknowledgments: The authors wish to thank Leanne Ohlund for technical assistance with LC-MS analysis. This study was supported by the Discovery Grant Program from the Natural Sciences and Engineering Council of Canada to Stan Kubow.

Author Contributions: Joelle Khairallah and Shima Sadeghi Ekbatan performed the in vitro digestion experiments, FRAP assays, and data anlaysis; Kebba Sabally and Lekha Sleno performed the HPLC and LC-MS asssays; Joelle Khairallah, Michèle M. Iskandar and Raza Hussain wrote sections of the manuscript; Atef Nassar interpreted the LC-MS data; Laetitia Rodes and Satya Prakash provided the equipment, expert guidance, and troubleshooting assistance for the in vitro digestions; Stan Kubow and Danielle J. Donnelly conceived and designed the experiments; Stan Kubow wrote the paper.

Conflicts of Interest: Stan Kubow and Danielle J. Donnelly hold a U.S. patent (Methods for treating type-2 diabetes and obesity; U.S. Patent 9,446,063 B2).

References

1. Tsao, R. Chemistry and biochemistry of dietary polyphenols. *Nutrients* **2010**, *2*, 1231–1246. [CrossRef] [PubMed]
2. Solayman, M.; Ali, Y.; Alam, F.; Islam, M.A.; Alam, N.; Khalil, M.I.; Gan, S.H. Polyphenols: Potential future arsenals in the treatment of diabetes. *Curr. Pharm. Des.* **2016**, *22*, 549–565. [CrossRef] [PubMed]
3. Wang, S.; Moustaid-Moussa, N.; Chen, L.; Mo, H.; Shastri, A.; Su, R.; Bapat, P.; Kwun, I.; Shen, C.-L. Novel insights of dietary polyphenols and obesity. *J. Nutr. Biochem.* **2014**, *25*, 1–18. [CrossRef] [PubMed]
4. Zarrelli, A.; Sgambato, A.; Petito, V.; De Napoli, L.; Previtera, L.; Di Fabio, G. New C-23 modified of silybin and 2,3-dehydrosilybin: Synthesis and preliminary evaluation of antioxidant properties. *Bioorg. Med. Chem. Lett.* **2011**, *21*, 4389–4392. [CrossRef] [PubMed]
5. Reddivari, L.; Hale, A.L.; Miller, J.C. Determination of phenolic content, composition and their contribution to antioxidant activity in specialty potato selections. *Am. J. Potato Res.* **2007**, *84*, 275–282. [CrossRef]
6. Camire, M.E.; Kubow, S.; Donnelly, D.J. Potatoes and human health. *Crit. Rev. Food Sci. Nutr.* **2009**, *49*, 823–840. [CrossRef] [PubMed]
7. Romanucci, V.; Di Fabio, G.; Di Marino, C.; Davinelli, S.; Scapagnini, G.; Zarrelli, A. Evaluation of new strategies to reduce the total content of α-solanine and α-chaconine in potatoes. *Phytochem. Lett.* **2018**, *23*, 116–119. [CrossRef]
8. Romanucci, V.; Pisanti, A.; Di Fabio, G.; Davinelli, S.; Scapagnini, G.; Guaragna, A.; Zarrelli, A. Toxin levels in different variety of potatoes: Alarming contents of α-chaconine. *Phytochem. Lett.* **2016**, *16*, 103–107. [CrossRef]
9. Chun, O.K.; Kim, D.O.; Smith, N.; Schroeder, D.; Han, J.T.; Lee, C.Y. Daily consumption of phenolics and total antioxidant capacity from fruit and vegetables in the American diet. *J. Sci. Food Agric.* **2005**, *85*, 1715–1724. [CrossRef]
10. Brat, P.; George, S.; Bellamy, S.; Du Chaffaut, L.; Scalbert, A.; Mennen, L.; Arnault, N.; Amiot, M.J. Daily polyphenol intake in France from fruit and vegetables. *J. Nutr.* **2006**, *136*, 2368–2373. [CrossRef] [PubMed]
11. Kubow, S.; Hobson, L.; Iskandar, M.M.; Sabally, K.; Donnelly, D.J.; Agellon, L.B. Extract of Irish potatoes (*Solanum tuberosum* L.) decreases body weight gain and adiposity and improves glucose control in the mouse model of diet-induced obesity. *Mol. Nutr. Food Res.* **2014**, *58*, 2235–2238. [CrossRef] [PubMed]
12. Hollman, P.C.; de Vries, J.H.M.; van Leeuwen, S.D.; Mengelers, M.J.; Katan, M.B. Absorption of dietary quercetin glycosides and quercetin in healthy ileostomy volunteers. *Am. J. Clin. Nutr.* **1995**, *62*, 1276–1282. [CrossRef] [PubMed]
13. Olthof, M.R.; Hollman, P.C.; Buijsman, M.N.; van Amelsvoort, J.M.; Katan, M.B. Chlorogenic acid, quercetin-3-rutinoside and black tea phenols are extensively metabolized in humans. *J. Nutr.* **2003**, *133*, 1806–1814. [CrossRef] [PubMed]

14. Stalmach, A.; Steiling, H.; Williamson, G.; Crozier, A. Bioavailability of chlorogenic acids following acute ingestion of coffee by humans with an ileostomy. *Arch. Biochem. Biophys.* **2010**, *510*, 98–105. [CrossRef] [PubMed]

15. Gao, K.; Xu, A.; Krul, C.; Venema, K.; Liu, Y.; Niu, Y.; Lu, J.; Bensoussan, L.; Seeram, N.P.; Heber, D.; et al. Of the major phenolic acids formed during human microbial fermentation of tea, citrus, and soy flavonoid supplements, only 3,4-dihydroxyphenylacetic acid has antiproliferative activity. *J. Nutr.* **2006**, *136*, 52–57. [CrossRef] [PubMed]

16. Possemiers, S.; Bolca, S.; Verstraete, W.; Heyerick, A. The intestinal microbiome: A separate organ inside the body with the metabolic potential to influence the bioactivity of botanicals. *Fitoterapia* **2011**, *82*, 53–66. [CrossRef] [PubMed]

17. Kemperman, R.A.; Bolca, S.; Roger, L.C.; Vaughan, E.E. Novel approaches for analysing gut microbes and dietary polyphenols: Challenges and opportunities. *Microbiology* **2010**, *156*, 3224–3231. [CrossRef] [PubMed]

18. Graefe, E.U.; Wittig, J.; Mueller, S.; Riethling, A.K.; Uehleke, B.; Drewelow, B.; Pforte, H.; Jacobasch, G.; Derendorf, H.; Veit, M. Pharmacokinetics and bioavailability of quercetin glycosides in humans. *J. Clin. Pharmacol.* **2001**, *41*, 492–499. [CrossRef] [PubMed]

19. Farah, A.; Monteiro, M.; Donangelo, C.M.; Lafay, S. Chlorogenic acids from green coffee extract are highly bioavailable in humans. *J. Nutr.* **2008**, *138*, 2309–2315. [CrossRef] [PubMed]

20. Renouf, M.; Guy, P.A.; Marmet, C.; Fraering, A.L.; Longet, K.; Moulin, J.; Enslen, M.; Barron, D.; Dionisi, F.; Cavin, C.; et al. Measurement of caffeic and ferulic acid equivalents in plasma after coffee consumption: Small intestine and colon are key sites for coffee metabolism. *Mol. Nutr. Food Res.* **2010**, *54*, 760–766. [CrossRef] [PubMed]

21. Richelle, M.; Tavazzi, I.; Offord, E. Comparison of the antioxidant activity of commonly consumed polyphenolic beverages (coffee, cocoa, and tea) prepared per cup serving. *J. Agric. Food Chem.* **2001**, *49*, 3438–3442. [CrossRef] [PubMed]

22. Bermúdez-Soto, M.J.; Tomás-Barberán, F.A.; García-Conesa, M.T. Stability of polyphenols in chokeberry (*Aronia melanocarpa*) subjected to in vitro gastric and pancreatic digestion. *Food Chem.* **2007**, *102*, 865–874. [CrossRef]

23. Tagliazucchi, D.; Verzelloni, E.; Bertolini, D.; Conte, A. In vitro bio-accessibility and antioxidant activity of grape polyphenols. *Food Chem.* **2010**, *120*, 599–606. [CrossRef]

24. Kroon, P.A.; Faulds, C.B.; Ryden, P.; Robertson, J.A.; Williamson, G. Release of covalently bound ferulic acid from fiber in the human colon. *J. Agric. Food Chem.* **1997**, *45*, 661–667. [CrossRef]

25. Gonthier, M.P.; Remesy, C.; Scalbert, A.; Cheynier, V.M.; Souquet, J.; Poutanen, K.; Aura, A.M. Microbial metabolism of caffeic acid and its esters chlorogenic and caftaric acids by human faecal microbiota in vitro. *Biomed. Pharmacother.* **2006**, *60*, 536–540. [CrossRef] [PubMed]

26. Saura-Calixto, F.; Serrano, J.; Goñi, I. Intake and bioaccessibility of total polyphenols in a whole diet. *Food Chem.* **2007**, *101*, 492–501. [CrossRef]

27. Gumienna, M.; Lasik, M.; Czarnecki, Z. Bioconversion of grape and chokeberry wine polyphenols during simulated gastrointestinal in vitro digestion. *Int. J. Food Sci. Nutr.* **2011**, *62*, 226–233. [CrossRef] [PubMed]

28. Molly, K.; Woestyne, M.V.; Verstraete, W. Development of a 5-step multichamber reactor as a simulation of the human intestinal microbial ecosystem. *Appl. Microbiol. Biotechnol.* **1993**, *39*, 254–258. [CrossRef] [PubMed]

29. Sadeghi Ekbatan, S.; Sleno, L.; Sabally, K.; Khairallah, J.; Azadi, B.; Rodes, L.; Prakash, S.; Donnelly, D.J.; Kubow, S. Biotransformation of polyphenols in a dynamic multistage gastrointestinal model. *Food Chem.* **2016**, *204*, 453–462. [CrossRef] [PubMed]

30. Cueva, C.; Jiménez-Girón, A.; Muñoz-González, I.; Esteban-Fernández, A.; Gil-Sánchez, I.; Dueñas, M.; Martín-Alvarez, P.J.; Pozo-Bayón, M.A.; Bartolomé, B.; Moreno-Arribas, M.V. Application of a new dynamic gastrointestinal simulator (SIMGI) to study the impact of red wine in colonic metabolism. *Food Res. Int.* **2015**, *72*, 149–159. [CrossRef]

31. Scalbert, A.; Williamson, G. Dietary intake and bioavailability of polyphenols. *J. Nutr.* **2000**, *130*, 2073–2085. [CrossRef]

32. Rechner, A.R.; Smith, M.A.; Kuhnle, G.; Gibson, G.R.; Debnam, E.S.; Srai, S.K.S.; Moore, K.P.; Rice-Evans, C.A. Colonic metabolism of dietary polyphenols: Influence of structure on microbial fermentation products. *Free Radic. Biol. Med.* **2004**, *36*, 212–215. [CrossRef] [PubMed]

33. Benzie, I.F.F.; Strain, J.J. The ferric reducing ability of plasma (FRAP) as a measure of "antioxidant power": The FRAP assay. *Anal. Biochem.* **1996**, *239*, 70–76. [CrossRef] [PubMed]

34. Kubow, S.; Iskandar, M.M.; Melgar-Bermudez, E.; Sleno, L.; Sabally, K.; Azadi, B.; How, E.; Prakash, S.; Burgos, G.; Felde, T.Z. Effects of simulated human gastrointestinal digestion of two purple-fleshed potato cultivars on anthocyanin composition and cytotoxicity in colonic cancer and non-tumorigenic cells. *Nutrients* **2017**, *9*, 953. [CrossRef] [PubMed]

35. Bolca, S.; Van de Wiele, T.; Possemiers, S. Gut metabotypes govern health effects of dietary polyphenols. *Curr. Opin. Biotechnol.* **2013**, *24*, 220–225. [CrossRef] [PubMed]

36. Aura, A.M. Microbial metabolism of dietary phenolic compounds in the colon. *Phytochem. Rev.* **2008**, *7*, 407–429. [CrossRef]

37. Juániz, I.; Ludwig, I.A.; Bresciani, L.; Dall'Asta, M.; Mena, P.; Del Rio, D.; Cid, C.; de Peña, M.-P. Bioaccessibility of (poly)phenolic compounds of raw and cooked cardoon (*Cynara cardunculus* L.) after simulated gastrointestinal digestion and fermentation by human colonic microbiota. *J. Funct. Food* **2017**, *32*, 195–207. [CrossRef]

38. Ludwig, I.A.; de Peña, M.-P.; Cid, C.; Crozier, A. Catabolism of coffee chlorogenic acids by human colonic microbiota. *Biofactors* **2013**, *39*, 623–632. [CrossRef] [PubMed]

39. Andreasen, M.F.; Kroon, P.A.; Williamson, G.; García-Conesa, M.T. Esterase activity to hydrolyze dietary antioxidant hydroxycinnamates is distributed along the intestine of mammals. *J. Agric. Food Chem.* **2001**, *49*, 5679–5684. [CrossRef] [PubMed]

40. Couteau, D.; McCartney, A.L.; Gibson, G.R.; Williamson, G.; Faulds, C.B. Isolation and characterization of human colonic bacteria able to hydrolyse chlorogenic acid. *J. Appl. Microbiol.* **2001**, *90*, 873–881. [CrossRef] [PubMed]

41. Konishi, Y.; Hitomi, Y.; Yoshioka, E. Intestinal absorption of *p*-coumaric and gallic acids in rats after oral administration. *J. Agric. Food Chem.* **2004**, *52*, 2527–2532. [CrossRef] [PubMed]

42. Peppercorn, M.A.; Goldman, P. Caffeic acid metabolism by gnotobiotic rats and their intestinal bacteria. *PNAS* **1972**, *69*, 1413–1415. [CrossRef] [PubMed]

43. Jaganath, I.B.; Mullen, W.; Lean, M.E.J.; Edwards, C.A.; Crozier, A. In vitro catabolism of rutin by human fecal bacteria and the antioxidant capacity of its catabolites. *Free Radic. Biol. Med.* **2009**, *47*, 1180–1189. [CrossRef] [PubMed]

44. Parkar, S.G.; Trower, T.M.; Stevenson, D.E. Fecal microbial metabolism of polyphenols and its effects on human gut microbiota. *Anaerobe* **2013**, *23*, 12–19. [CrossRef] [PubMed]

45. Rechner, A.R.; Spencer, J.P.E.; Kuhnle, G.; Harn, U.; Rice-Evans, C.A. Novel biomarkers of the metabolism of caffeic acid derivatives in vivo. *Free Radic. Biol. Med.* **2001**, *30*, 1213–1222. [CrossRef]

46. Gross, G.; Jacobs, D.M.; Peters, S.; Possemiers, S.; Van Duynhoven, J.; Vaughan, E.E.; van de Wiele, T. In vitro bioconversion of polyphenols from black tea and red wine/grape juice by human intestinal microbiota displays strong interindividual variability. *J. Agric. Food Chem.* **2010**, *58*, 10236–10246. [CrossRef] [PubMed]

47. Barroso, E.; Cueva, C.; Peláez, C.; Martínez-Cuesta, M.C.; Requena, T. Development of human colonic microbiota in the computer-controlled dynamic simulator of the gastrointestinal tract SIMGI. *LWT Food Sci. Technol.* **2015**, *61*, 283–289. [CrossRef]

48. Van Dorsten, F.A.; Peters, S.; Gross, G.; Gomez-Roldan, V.; Klinkenberg, M.; de Vos, R.C.; Vaughan, E.E.; van Duynhoven, J.P.; Possemiers, S.; van de Wiele, T.; et al. Gut microbial metabolism of polyphenols from black tea and red wine/grape juice is source-specific and colon-region dependent. *J. Agric. Food Chem.* **2012**, *60*, 11331–11342. [CrossRef] [PubMed]

49. Kikuzaki, H.; Hisamoto, M.; Hirose, K.; Akiyama, K.; Taniguchi, H. Antioxidant properties of ferulic acid and its related compounds. *J. Agric. Food Chem.* **2002**, *50*, 2161–2168. [CrossRef] [PubMed]

50. Gómez-Ruiz, J.A.; Leake, D.S.; Ames, J.M. In vitro antioxidant activity of coffee compounds and their metabolites. *J. Agric. Food Chem.* **2007**, *55*, 6962–6969. [CrossRef] [PubMed]

antioxidants

MDPI

Review

Phytochemical Constituents, Health Benefits, And Industrial Applications of Grape Seeds: A Mini-Review

Zheng Feei Ma [1],* and Hongxia Zhang [2],*

[1] Department of Public Health, Xi'an Jiaotong-Liverpool University, Suzhou 215123, China
[2] Department of Food Science, University of Otago, Dunedin 9054, New Zealand
*   Correspondence: Zhengfeei.Ma@xjtlu.edu.cn (Z.F.M.); Zhanghongxia326@hotmail.com (H.Z.);
    Tel.: +86-512-8188-4938 (Z.F.M.); +64-3-470-9198 (H.Z.)

Received: 10 August 2017; Accepted: 12 September 2017; Published: 15 September 2017

Abstract: Grapes are one of the most widely grown fruits and have been used for winemaking since the ancient Greek and Roman civilizations. Grape seeds are rich in proanthocyanidins which have been shown to possess potent free radical scavenging activity. Grape seeds are a complex matrix containing 40% fiber, 16% oil, 11% proteins, and 7% complex phenols such as tannins. Grape seeds are rich sources of flavonoids and contain monomers, dimers, trimers, oligomers, and polymers. The monomeric compounds includes (+)-catechins, (−)-epicatechin, and (−)-epicatechin-3-O-gallate. Studies have reported that grape seeds exhibit a broad spectrum of pharmacological properties against oxidative stress. Their potential health benefits include protection against oxidative damage, and anti-diabetic, anti-cholesterol, and anti-platelet functions. Recognition of such health benefits of proanthocyanidins has led to the use of grape seeds as a dietary supplement by the consumers. This paper summarizes the studies of the phytochemical compounds, pharmacological properties, and industrial applications of grape seeds.

Keywords: grape seeds; grapes; proanthocyanidins; flavonoids; catechins

1. Introduction

Grapes are one of the most widely grown fruits and the total production of grapes worldwide is approximately 60 million tons [1]. The major producers of grapes are the USA, China, Italy, and France [1]. Grapes can be categorized into grapes with edible seeds, seedless, wine grapes, table grapes, and raisin grapes [2]. North American grapes (*Vitis labrusca* and *Vitis rotundifolia*), European grapes (*Vitis vinifera*), and French hybrids are the main species of grapes [2].

Grapes have been used for winemaking since the ancient Greek and Roman civilizations [3]. In the winemaking process, over 0.3 kg of solid byproducts are produced per kg of grapes crushed [4]. Grape marc is the main byproduct which accounts for about two third of the solids [4]. Grape marc consists of grape skins (50%), seeds (25%), and stalks (25%) [4]. Therefore, grape seeds are the industrial byproduct of the winemaking process. Although grape seeds are a relatively inexpensive source of antioxidant compounds [5], they amount to 38–52% on a dry matter basis [6]. Grape seeds are treated as waste if extracts are not made and it is estimated that about 10–12 kg of grape seeds in 100 kg of wet residues are produced by the industry [1].

In recent years, grape seed extract has become increasingly popular on the market as a nutritional supplement especially in the Australia, Korea, Japan and the United States [7]. This is because grape seeds are rich in phenolic compounds [8] and have potentially beneficial effects for human health such as protection against peptic ulcers [8,9]. Grape seeds have been reported to exhibit scavenge superoxide radicals [10]. Grape seeds are rich in flavan-3-ol, including proanthocyanidins and catechins [8]. They contain high concentration of polyphenol proanthocyanidins, which are the oligomers of flavan-3-ol units including catechin and epicatechin [11]. Proanthocyanidins, which belong to condensed tannins, are present as procyanidins and prodelphinidins in grape seeds and skins [12]. Procyanidins and prodelphinidins are extracted from the seeds and skins during fermentation into wine [12]. Anthocyanins and proanthocyanidins play a significant role in the stability, taste, and color of the red wines [13]. The simplest proanthocyanidins are dimeric proanthocyanidins which have 4→8 linked monomers [14].

Therefore, the aim of this work was to summarize the phytochemical constituents, pharmacological activities, and industrial applications of grape seeds. We hope that this work would be a valuable reference resource to provide beneficial inspiration for the future studies.

Search Strategy

An electronic literature search was conducted using PubMed, Medline (OvidSP), and Google Scholar until May 2017. Additional articles were identified from references in the retrieved articles. Search terms included combinations of the following: "grape", "grape seed", "phytochemical", "antioxidant", "anthocyanins", "proanthocyanidins", and "pharmacological". The search was restricted to articles in English that addressed the phytochemical constituents and pharmacological properties of grape seeds. For the purpose of this review, studies involving human subjects will be given priority.

2. Phytochemical Compounds in Grape Seeds

Grape seeds contain protein (11%), fiber (35%), minerals (3%), and water (7%) [15]. In addition, the lipid content of grape seeds ranges from 7 to 20% [15]. The oil content of grape seeds is traditionally extraction using mechanical methods or organic solvents [4]. In mechanical extraction, although the quality of product is superior, the extraction gives a lower yield. While organic solvent extraction gives a higher yield, it requires solvent recovery through distillation and the final product contains traces of residual solvent [4]. While the supercritical method is regarded as a promising method which can produce similar quantity and better quality of oil yield than mechanical and organic solvent extractions [4]. Cold-pressing is used to extract the oil from grape seeds without chemical treatment or heat [16]. Although the cold pressing usually gives a lower yield than other conventional solvent extraction, it may retain more bioactive components and be safer because there are no solvent residues in the grape seed oil [15].

Several studies have been conducted on grape seeds, in order to determine their bioactive compounds [8,17–21]. Grape seed extracts contain a heterogeneous mixture of monomers (5–30%), oligomers (17–63%), and polymers (11–39%) composed of proanthocyanidins [22]. Proanthocyanidins are the major compound in grape seed extracts [23]. The red color and astringent taste of grape seed extracts can be attributed to proanthocyanidins. However, higher concentrations of proanthocyanidins may affect the sensory and color properties of the product [24].

Negro et al. [25] reported that the total phenols (gallic acid equivalent (GAE)), total flavonoids (catechin equivalent (CE)), catechin equivalent (CE), and proanthocyanidins (cyanidin equivalent (CyE)) in red grape seed extracts were 8.58 g/100 g dry matter (DM), 8.36 g/100 g DM, 6.41 g/100 g, DM and 5.95 g/100 g DM, respectively. Another study by Ghouila et al. [26] documented that the total phenolic content and total flavonoid content of Ahmeur Bouamer grape seeds were 265.15 mg GAE/g of dry mass and 14.08 mg catechin equivalent (CE)/g of dry mass, respectively.

In a study analyzing 26 samples of six white grape varieties and 44 of four red grape varieties from different areas of Castilla-La Mancha, Spain, Rodríguez Montealegre et al. [8] reported that the grape seeds were shown to contain catechin, epicatechin, epicatechin gallate, protocatechic acid, procyanidin B1, procyanidin B2, procyanidin B3, and procyanidin B4. Small quantities of galic acid and protocatechic acid were also present in grape seeds [8].

Similarly, another study by Escribano-Bailon et al. [17] reported that the major constituents in *V. vinifera* (Tintal del pais) grape seeds were (+)-catechin (11%) followed by (−)-epicatechin

(10%), (−)-epicatechin-3-*O*-gallate (9%), epicatechin 3-*O*-gallate-(4β→8)-catechin (B1-3-*O*-gallate) (7%), and epicatechin-(4β→8)-epicatechin (dimer B2) (6%). These results were also similar with previous studies conducted on the polyphenols of grape seeds from other varieties of grape species in Greek islands [21,27,28]. Anastasiadi et al. [21] reported that the total phenolic content in grape seeds ranged from 325 to 812 mg/g gallic acid equivalents.

There are 14 dimeric, 11 trimeric procyanidins, and 1 tetrameric procyanidin that have been identified from grape seeds [18]. In a study that investigated the quantities of flavan-3-ols in the seeds of 17 grape cultivars grown in Ontario, Canada using high-performance liquid chromatography (HPLC), Fuleki and Ricardo da Silva [18] documented the presence of monomers of (+)-catechin, (−)-epicatechin-3-*O*-gallate, and (−)-epicatechin. Considerable quantities of highly polymerized procyanidins were also found in grape seeds [18]. Similarly, Prieur et al. [29] also reported the presence of catechin, epicatechin-3-*O*-gallate, and epicatechin were determined by HPLC following the thiolysis degradation, providing evidence that procyanidins are the grape seed tannins.

Gabetta et al. [19] documented the presence of monomers, dimers, trimers, tetramers, pentamers, hexamers, heptamer, and their gallates in grape seeds. A study by Freitas et al. [30] obtained several fractions of oligomeric and polymeric procyanidins from grape seeds using gel chromatography. The authors [30] also identified several grape seed flavan-3-ol derivatives and phenol acids including protocatechic, caffeic acids, trimer epi-(4β-8 or 6)-epi-(4β-6 or 8)-epi, trimer cat-(4α-8)-epi-(4β-8)-epi, and trimer cat-(4α-8)-epi-(4β-6)-cat.

Some factors such as the variety of grapevine, viticultural, and environmental factors can influence the concentration of phenolic compounds in grapes [8]. Epicatechin, (+)-catechin, and gallic acid were the major phenolic compounds detected in the muscadine grape seeds [20]. Pastrana-Bonilla et al. [20] reported that the concentration of epicatechin, (+)-catechin, gallic acid, total anthocyanin, and total phenolics in muscadine grape seeds were 1299 mg/100 g of fresh weight (FW), 558 mg/100 g of FW, 6.9 mg/100 g of FW, 4.3 mg/100 g of FW, and 2179 mg/g GAE, respectively.

3. Pharmacological Properties

Proanthocyanidins are present in substantial amounts in grape seeds and have attracted attention of consumers because of their potential health effects [31]. In vitro, proanthocyanidins have been shown to exhibit strong antioxidant activity and scavenge reactive oxygen and nitrogen species, modulate immune function and platelet activation, and produce vasorelaxation by inducing nitric oxide (NO) release from endothelium [31]. In addition, proanthocyanidins also inhibits the progression of atherosclerosis and prevent the increase of low-density lipoprotein (LDL) cholesterol concentration [32].

3.1. Anti-Diabetic

Montagut et al. [33] reported Wister female rats treated with 25 mg grape seed procyanidin extract/kg body weight per day for 30 days had an improved homeostatis model assessment-insulin resistance index accompanied by downregulation of primers Glut4, Irs1, and Pparg2 in mesenteric white adipose tissue (WAT), suggesting that grape seed procyanidin has a positive long term-effect on glucose homeostasis.

Another study by Montagut et al. [34] demonstrated that the oligomeric structures of grape seed procyanidin extracts activated the insulin receptor by interacting and inducing the autophosphorylation of the insulin receptor in order to stimulate the glucose uptake.

3.2. Antioxidant

Different methods including the 1,1-diphenyl-2-picryhidrazyl (DPPH) method and oxygen radical absorbance capacity (ORAC) have been employed to evaluate the antioxidant capacity of phenolic compounds from grape seeds. Poudel et al. [35] reported that the antioxidant capacity of grape seed extract using DPPH ranged between 17 and 92 mmol Trolox® antioxidant equivalent (TE)/g. While the ORAC method was 42 mmol TE/g [35]. Pastrana-Bonilla et al. [20] demonstrated that the antioxidant capacity was the highest in grape seeds (281 μM Trolox equivalent antioxidant capacity (TEAC)/g of FW) followed by leaves (236 μM TEAC/g of FW), skins (13 μM TEAC/g of FW), and pulps (2.4 μM TEAC/g of FW), suggesting that the grape seeds extracts are promising antioxidant for dietary supplement. Similarly, two studies [36,37] also confirmed the antioxidant activity of grape seed extract by using β-carotene linoleate, linoleic acid peroxidation, DPPH, and phosphomolybdenum complex methods.

Puiggròs et al. [38] reported that grape seed procyanidin extracts modulated the expression of antioxidant systems, suggesting that grape seed procyanidin extracts might improve the cellular redox status via glutathione synthesis pathways. Another study by Guo et al. [39] reported that grape seed procyanidins provided protective effect against the ethanol induced toxicity in mouse brain cells. Vinson et al. [40] reported that the supplementation of 50 mg/kg and 100 mg/kg grape seed proanthocyanidin extracts in hamsters reduced the formation of foam cells by 50% and 63%, respectively. The development of foam cells is the indicator of early stages of atherosclerosis [40]. In a randomized, double-blind, placebo-controlled study, Preuss et al. [41] reported that the supplementation of grape seed proanthocyanidin extracts and chromium polynicotinate in hypercholesterolemic participants decreased the concentrations of total cholesterol, oxidized LDL, and LDL significantly after two months.

3.3. Anti-Platelet

Grape seed extracts were shown to exhibit anti-platelet action [42]. Olas et al. [42] reported that grape seed extracts (5–50 μg/mL) showed a reduction in platelet aggregation, adhesion, and generation of superoxide anion. In addition, the authors [42] also found that grape seed extract was more effective than the pure resveratrol solution in the reduction of platelet processes.

3.4. Anti-Cholesterol

Cetin et al. [43] reported that there were no significant changes in levels of superoxide dismutase and catalase of rats supplemented with grape seed extracts when methotrexate (MTX) was administered. The authors [43] suggested that the supplementation of grape seed extract may be protective against oxidative damage induced by MTX which is used as cytotoxic chemotherapeutic agent [44].

In a 12-week of supplementation containing 0, 200, or 400 mg grape seed extract to 61 healthy subjects with LDL cholesterol level of 100 to 180 mg/dL, Sano et al. [45] reported that the 200 mg and 400 mg group reported a significant decrease in malondialdehyde-modified LDL (MDA-LDL) when compared to the baseline level after 12 weeks ($p = 0.008$ and 0.009, respectively). MDA-LDL is involved in the pathogenesis of arteriosclerosis [46].

In a study that investigated the effect of supplementing a single high-fat meal with 300 mg of proanthocyanidin-rich grape seed extracts in eight male adults, Natella et al. [47] reported that grape seed extracts reduce postprandial oxidative stress by increasing the plasma antioxidant concentration and preventing the increase of lipid hydroperoxides. Consequently, this improves the resistance to oxidative modification of LDL cholesterol [47]. It is suggested that the polyphenols in grape seed extracts activate serum paraoxonase (PON) which prevents the postprandial increase in lipid peroxides [47]. Serum paraoxonase is an enzyme associated with high-density lipoprotein (HDL) that hydrolyzes lipoprotein peroxides and inhibits LDL oxidation [48].

3.5. Anti-Inflammation

In a study that evaluated the effect of procyanidin from grape seeds on the inflammatory mediators in rat fed with a high fat diet, Terra et al. [49] reported that rats fed with a high fat diet supplemented with procyanidins from grape seeds (345 mg/kg feed) for 19 weeks had a lower plasma C-reactive protein (CRP) level than rat fed with high-fat diet, suggesting that the decrease in plasma CRP is related to a downregulation of CRP mRNA expression in the liver and mesenteric white adipose tissue (WAT). In addition, the authors [49] also reported a decrease in the expression of the proinflammatory cytokine tumour necrosis factor alpha (TNF-α) and interleukin 6 (IL-6) in the mesenteric WAT of rats fed with a high fat diet supplemented with procyanidins from grape seeds. IL-6 is a stress-induced inflammatory cytokine, which is directly implicated in atherogenesis [49].

Studies have revealed that the concentrations of proinflammatory mediators are elevated in the conditions such as insulin resistant states of diabetes and obesity [50]. In a study that investigated whether procyanidins play a role in modulating inflammation, Chacon et al. [51] reported that macrophage-like (THP-1) cell lines and human adipocytes (SGBS) pre-treated with grape-seed procyanidin extracts had a reduction in IL-6 and monocyte chemoattracting protein (MCP-1) expression after inflammatory stimulus. In addition, the authors [51] also reported that the translocation of NF-κB to the nucleus was partially inhibited in macrophage-like (THP-1) cell lines and human adipocytes (SGBS) pre-treated with grape-seed procyanidin extracts.

Terra et al. [52] reported that rats fed with grape-seed procyanidin extracts had a reduced body weight and plasmatic systemic markers of TNF-α and CRP. In addition, grape-seed procyanidin extracts also increased adiponectin expression and decreased IL-6 [52]. Grape-seed procyanidin extracts are also linked to a reduced expression of epidermal growth factor module-containing mucin-like receptor 1 (EMR1) (specific marker of macrophage F4/80), suggesting a reduced macrophage infiltration of WAT [52]. Therefore, the regular consumption of food containing procyanidins might help to prevent low-grade inflammatory-related diseases in obesity characterized by macrophage accumulation in WAT and abnormal cytokine production [52].

In a study that investigated the effect of grape seed procyanidins on glucose metabolism in streptozotocin-induced diabetic rats, Pinent et al. [53] reported that grape seed procyanidins stimulated the glucose uptake in L6e9 myotubes and 3T3-L1 adipocytes in a dose-dependent manner. In addition, grape seed procyanidins also stimulated glucose transported-4 translocation to the plasma membrane [53]. The authors [53] suggested that procyanidins possesses insulin-like effects in insulin sensitive cells. Therefore, all these findings suggested that procyanidins in grape seeds inhibited mRNA levels and decreased the risk of diseases associated to obesity and high fat diets.

3.6. Anti-Aging

In a study investigating the effect of grape seed extracts on the aged-associated accumulation of oxidative DNA damage in male albino rats of Wister strain, Balu et al. [54] reported that grape seed extracts inhibited the accumulation of age-related oxidative DNA damage products such as 8-hydroxy-2'-deoxyguanosine (8-OHdG) and DNA protein cross-links in spinal cord and in various brain regions including the striatum, cerebral cortex, and hippocampus.

Another study by Balu et al. [55] reported that lipid peroxidation and antioxidant defenses in male albino rats supplemented with grape seed extract were normalized, suggesting that grape seed extracts could be used to improve the antioxidant status and decreased the incidence of free radical-induced lipid peroxidation in the central nervous system of aged rats.

3.7. Anti-Microbial

Since grape seeds are rich sources of polyphenols, grape seeds have also shown promise as novel microbial agents. Jayaprakasha et al. [37] reported that the defatted grape seed extracts were shown to exhibit anti-bacterial effect against *Bacillus cereus* (*B. cereus*), *Bacillus subtilis* (*B. subtilis*), *Staphylococcus aureus* (*S. aureus*), *Bacillus coagulans* (*B. coagulans*), *Escherichia coli* (*E. coli*), and *Pseudomonas aeruginosa* (*P. aeruginosa*). In addition, the authors demonstrated that these extracts completely inhibited Gram-positive bacteria and Gram-negative bacteria at 850–1000 ppm and 1250–1500 ppm, respectively. A study by Ghouila et al. [26] demonstrated that grape seed extracts exhibited antimicrobial activity to *Microccocus luteus* ATCC®9341, *S. aureus* ATCC®29213, *E. coli* ATCC®25992, *P. aeruginosa* ATCC®27853, *Aspergillus niger*, and *Fusarium oxysporum*, with a diameter of the inhibition growth zone ranged from 15–20 mm.

Baydar et al. [56] reported that grape seed extracts acted against *S. aureus* after 48 h and *Aeromonas hydrophila* after 1 h. A study by Anastasiadi et al. [21] reported that the minimum inhibition concentration of grape seed extract against *Listeria monocytogenes* (*L. monocytogenes*) was 0.26, suggesting that grape seeds could be used as an inexpensive source of natural antilisterial mixtures. Future research should be conducted on screening the grape seed extracts for natural anti-microbial compounds.

Another study by Brown et al. [57] reported that muscadine grape seed extracts were effective in inhibiting *H. pylori* in vitro. The authors [57] demonstrated that although muscadine grape seed extracts had the highest total phenolic content (646 mg GAE/g dry weight (dw)) than skin extracts (135 mg GAE/g dw), the skin extracts had a better efficacy than seed extracts against *H. pylori* [57]. Therefore, it is suggested that the anti-*H. pylori* activity might be influenced by the type and content of phenolics [57]. *H. pylori* infection is associated with several gastroduodenal diseases such as gastric cancer [58].

3.8. Anti-Tumour

Grape seed extracts have also exhibited anti-tumor properties. Promising results have been shown by several studies performed in human colorectal carcinoma [59], head and neck squamous cell carcinoma [60], and prostate cancer cells [61]. Therefore, it is suggested that the supplementation of grape seed extracts might be an effective anti-tumor agent in clinical settings.

3.9. Other Pharmacological Properties

A study by Feng et al. [62] reported that grape seed extracts reduced brain weight loss from 20% in vehicle pups to 3% in treated pups ($p < 0.01$). The loss of brain weight was used to measure the brain damage in pups with neontal hypoxic ischemic brain injury. In addition, grape seed extract also improved the histopathologic brain score in cortex, hippocampus, and thalamus. The authors [62] suggested that the treatment with grape seed extracts could be used to inhibit lipid peroxidation and reduce hypoxic ischemic brain injury.

3.10. Toxicity

Grape seed extract has been approved as generally recognized as safe (GRAS) by Food and Drug Administration (FDA) and is sold commercially as a dietary supplement listed on the Everything Added to Food in the United States (EAFUS) database [63]. Bentivegna and Whitney [64] reported that no-observed-adverse effect level (NOAEL) of grape seed extracts determined in the rats was 1.78 g/kg body weight/day and the normal amount of grape seed extract used in food applications was 0.01–1%.

Sano et al. [45] also reported that there were no abnormal changes in the physiological and clinical laboratory tests in participants after taking tablets containing 200 mg and 400 mg grape seed extracts. In addition, no problematic results were reported in the urinary sedimentation of these participants, suggesting that the tablets containing 200 and 400 mg grape seed extracts are safe to consume [45]. Similarly, a subchronic oral toxicity study by Yamakoshi et al. [7] documented that there was no noticeable sign of toxicity in rat fed with grape seed extract as a dietary admixture at 0.02% (low dose group), 0.2% (middle dose group), and 2% (high dose group) (w/w) for 90 days. The authors [7] reported that the NOAEL of grape seed extract used in the subchronic oral toxicity study was 2% in the diet, which was equivalent to 1410 mg/kg body weight in males and 1501 mg/kg body weight/day in females. Therefore, these studies indicate a lack of toxicity and more systematic toxicological studies on grape seed should be conducted.

4. Industrial Applications

Lipid peroxidation is one of the major reasons for the deterioration of food products during processing and storage. Therefore, antioxidants have been added to food products to extend their shelf life. Synthetic antioxidants such as tertiary butylhydroquinone (TBHQ), butylated hydroxyanisole (BHA), and butylated hydroxytoluene (BHT) have restricted use in food products because of their carcinogenic and toxic effects [65]. Therefore, the search for natural antioxidants, particularly of plant origin, has greatly increased in recent years.

Grape seeds can also be exploitable for the preservation of food products. For example, grape seed extracts have been used as a food additive in Japan [7]. A study by Jayaprakasha et al. [36] reported that the grape seed extracts showed good reducing power at a 500 µg/L concentration when using potassium ferricyanide reduction methods. In addition, grape seed extracts also showed 65–90% antioxidant activity at 100 ppm.

Grape seed extracts have been evaluated for their antioxidative effect on meat products [66,67]. Ahn et al. [66] reported that the addition of grape seed extracts had an improved oxidative stability and reduced the hexanal concentration by 97% in cooked ground beef after three days of refrigerated storage when compared to the control without added antioxidant. Similarly, Lau and King [67] also reported similar improve oxidative stability in turkey patties. In addition, the authors [67] also reported that when the grape seed extracts were added at 1.0% and 2.0% to turkey patties, the thiobarbituric acid reactive substances (TBARS) values were decreased nearly 10-fold when compared to the control. The grape seed extracts added to turkey patties were also reported to exhibit wine odor and slightly bitter aftertaste [67].

In a study that investigated the effect of four different concentrations of grape seed extracts (i.e., 0.0, 0.4, 0.8, and 1.6 g/kg) on cooked turkey breast meat, Mielnik et al. [68] reported that the addition of grape seed extracts before cooking significantly improved the lipid stability of turkey breast meat during cooking process and chill storage. The authors [68] reported that the efficiency of grape seed extract to prevent oxidation increased when the concentration of antioxidant was from 0.4 g/kg to 1.6 g/kg. Therefore, it is suggested that grape seed extracts can be a very effective antioxidant in inhibiting lipid oxidation of cooked turkey meat during chill-storage. Similarly, Brannan [69] also demonstrated a reduction in the biomarkers of lipid oxidation and sensory scores for key attributes in ground chicken when the grape seed extracts were added during refrigerated storage.

Brannan and Mah [70] documented that 0.1% grape seed extract was effective in inhibiting the formation of primary (e.g., hexanal and lipid hydroperoxides) and secondary oxidation products (e.g., TBARS) in both raw and cooked meat systems during refrigerated and frozen storage. It is suggested that in vivo mechanism of antioxidant activity for grape seed extracts includes the inhibition of nitrositive stress, stimulation of enzymatic production of nitric oxide [71], and oxygen radical scavenging [72].

The antioxidant activity of grape seed extracts can be influenced by heating conditions such as temperature and time. In a study that evaluated the effect of heating conditions on the antioxidant activity of grape seed extracts, Kim et al. [73] reported that heat treatment significantly increased the concentrations of caffeine and gallocatechin gallate in grape seed extracts, suggesting that heating could be used as a method to improve the antioxidant activity of grape seed extracts. Since polyphenols in grape seeds possess antibacterial activity, grape seeds could be incorporated into the foods in order to prevent the growth of foodborne pathogens such as *L. monocytogenes* [21].

Grape seed extracts are also included in the formulations of cosmetic products for anti-aging reasons. This is because grape seed extracts are rich in proanthocyanidins which have strong free radical scavenging properties [74].

5. Conclusions and Future Research

Grape seeds have high antioxidant potential. Their potential health benefits include protection against oxidative damage, and anti-diabetic, anti-cholesterol, and anti-platelet properties. The health benefits of grape seed consumption are thought to arise mainly from bioactivities of their polyphenols. Potential health benefits of dietary polyphenols on major chronic non-communicable diseases have been shown in several meta-analyses [75–78]. Therefore, the screening of individual polyphenol constituents that exhibit health-promoting properties in grape seed requires further investigation. This is because a cause-effect relationship between the intake of grape seed and its health effects can only be established when the composition of grape seeds is properly characterized and standardized. Furthermore, extensive investigation is needed to examine the effects of adding these beneficial

Antioxidants **2017**, *6*, 71

polyphenol constituents from grape seeds using advanced technologies into food systems. Further research is needed to evaluate the effectiveness of grape seed extract in the food ecosystem and to establish their role as an antimicrobial agent in food safety.

Acknowledgments: The authors received no specific funding for this work.

Author Contributions: Zheng Feei Ma wrote the first draft of the paper. Zheng Feei Ma and Hongxia Zhang revised the paper.

Conflicts of Interest: The authors declare no conflict of interest.

References

1. Matthäus, B. Virgin grape seed oil: Is it really a nutritional highlight? *Eur. J. Lipid Sci. Technol.* **2008**, *110*, 645–650. [CrossRef]
2. Girard, B.; Mazza, G. Functional grape and citrus products. In *Functional Foods—Biochemical and Processing Aspects*; Mazza, G., Ed.; Technomic Publishing Co., Inc.: Lancaster, PA, USA, 1998; Volume 1, pp. 139–191. ISBN 9781566764872.
3. Estreicher, S.K. Wine. In *The Encyclopedia of Ancient History*; Bagnall, R.S., Brodersen, K., Champion, C.B., Erskine, A., Huebner, S.R., Eds.; John Wiley & Sons, Inc.: Oxford, UK, 2013; pp. 1–5. ISBN 9781444338386.
4. Duba, K.S.; Fiori, L. Supercritical CO_2 extraction of grape seed oil: Effect of process parameters on the extraction kinetics. *J. Supercrit. Fluids* **2015**, *98*, 33–43. [CrossRef]
5. Cádiz-Gurrea, M.; Borrás-Linares, I.; Lozano-Sánchez, J.; Joven, J.; Fernández-Arroyo, S.; Segura-Carretero, A. Cocoa and grape seed byproducts as a source of antioxidant and anti-inflammatory proanthocyanidins. *Int. J. Mol. Sci.* **2017**, *18*, 376. [CrossRef] [PubMed]
6. Maier, T.; Schieber, A.; Kammerer, D.R.; Carle, R. Residues of grape (*Vitis vinifera* L.) seed oil production as a valuable source of phenolic antioxidants. *Food Chem.* **2009**, *112*, 551–559. [CrossRef]
7. Yamakoshi, J.; Saito, M.; Kataoka, S.; Kikuchi, M. Safety evaluation of proanthocyanidin-rich extract from grape seeds. *Food Chem. Toxicol.* **2002**, *40*, 599–607. [CrossRef]
8. Rodríguez Montealegre, R.; Romero Peces, R.; Chacón Vozmediano, J.L.; Martínez Gascueña, J.; García Romero, E. Phenolic compounds in skins and seeds of ten grape *Vitis vinifera* varieties grown in a warm climate. *J. Food Comp. Anal.* **2006**, *19*, 687–693. [CrossRef]
9. Kim, T.H.; Jeon, E.J.; Cheung, D.Y.; Kim, C.W.; Kim, S.S.; Park, S.H.; Han, S.W.; Kim, M.J.; Lee, Y.S.; Cho, M.L. Gastroprotective effects of grape seed proanthocyanidin extracts against nonsteroid anti-inflammatory drug-induced gastric injury in rats. *Gut Liver* **2013**, *7*, 282–289. [CrossRef] [PubMed]
10. El-Beshbishy, H.A.; Mohamadin, A.M.; Abdel-Naim, A.B. In vitro evaluation of the antioxidant activities of grape seed (*Vitis vinifera*) extract, blackseed (*Nigella sativa*) extract and curcumin. *J. Taibah Univ. Med. Sci.* **2009**, *4*, 23–35.
11. Weseler, A.R.; Bast, A. Masquelier's grape seed extract: From basic flavonoid research to a well-characterized food supplement with health benefits. *Nutr. J.* **2017**, *16*, 5. [CrossRef] [PubMed]
12. Sun, B.; Spranger, M.I. Review: Quantitative extraction and analysis of grape and wine proanthocyanidins and stilbenes. *Ciência e Técnica Vitivinícola* **2005**, *20*, 59–89.
13. He, F.; Liang, N.; Mu, L.; Pan, Q.; Wang, J.; Reeves, M.J.; Duan, C. Anthocyanins and their variation in red wines I. Monomeric anthocyanins and their color expression. *Molecules* **2012**, *17*, 1571–1601. [CrossRef] [PubMed]
14. Yilmaz, Y.; Toledo, R.T. Health aspects of functional grape seed constituents. *Trends Food Sci. Technol.* **2004**, *15*, 422–433. [CrossRef]
15. Shinagawa, F.B.; Santana, F.C.D.; Torres, L.R.O.; Mancini-Filho, J. Grape seed oil: A potential functional food? *Food Sci. Technol.* **2015**, *35*, 399–406. [CrossRef]
16. Parry, J.; Hao, Z.; Luther, M.; Su, L.; Zhou, K.; Yu, L. Characterization of cold-pressed onion, parsley, cardamom, mullein, roasted pumpkin, and milk thistle seed oils. *J. Am. Oil Chem. Soc.* **2006**, *83*, 847–854. [CrossRef]
17. Escribano-Bailon, T.; Gutierrez-Fernandez, Y.; Rivas-Gonzalo, J.C.; Santos-Buelga, C. Characterization of procyanidins of *Vitis vinifera* variety Tinta del Pais grape seeds. *J. Agric. Food Chem.* **1992**, *40*, 1794–1799. [CrossRef]
18. Fuleki, T.; Ricardo da Silva, J.M. Catechin and procyanidin composition of seeds from grape cultivars grown in Ontario. *J. Agric. Food Chem.* **1997**, *45*, 1156–1160. [CrossRef]

19. Gabetta, B.; Fuzzati, N.; Griffini, A.; Lolla, E.; Pace, R.; Ruffilli, T.; Peterlongo, F. Characterization of proanthocyanidins from grape seeds. *Fitoterapia* **2000**, *71*, 162–175. [CrossRef]

20. Pastrana-Bonilla, E.; Akoh, C.C.; Sellappan, S.; Krewer, G. Phenolic content and antioxidant capacity of muscadine grapes. *J. Agric. Food Chem.* **2003**, *51*, 5497–5503. [CrossRef] [PubMed]

21. Anastasiadi, M.; Chorianopoulos, N.G.; Nychas, G.J.E.; Haroutounian, S.A. Antilisterial activities of polyphenol-rich extracts of grapes and vinification byproducts. *J. Agric. Food Chem.* **2009**, *57*, 457–463. [CrossRef] [PubMed]

22. Waterhouse, A.L.; Ignelzi, S.; Shirley, J.R. A comparison of methods for quantifying oligomeric proanthocyanidins from grape seed extracts. *Am. J. Enol. Vitic.* **2000**, *51*, 383–389.

23. Hernández-Jiménez, A.; Gómez-Plaza, E.; Martínez-Cutillas, A.; Kennedy, J.A. Grape skin and seed proanthocyanidins from Monastrell × Syrah grapes. *J. Agric. Food Chem.* **2009**, *57*, 10798–10803. [CrossRef] [PubMed]

24. Monteleone, E.; Condelli, N.; Dinnella, C.; Bertuccioli, M. Prediction of perceived astringency induced by phenolic compounds. *Food Qual. Prefer.* **2004**, *15*, 761–769. [CrossRef]

25. Negro, C.; Tommasi, L.; Miceli, A. Phenolic compounds and antioxidant activity from red grape marc extracts. *Bioresour. Technol.* **2003**, *87*, 41–44. [CrossRef]

26. Ghouila, Z.; Laurent, S.; Boutry, S.; Vander Elst, L.; Nateche, F.; Muller, R.; Baaliouamer, A. Antioxidant, antibacterial and cell toxicity effects of polyphenols from Ahmeur Bouamer grape seed extracts. *J. Fund. Appl. Sci.* **2017**, *9*, 392–410. [CrossRef]

27. Guendez, R.; Kallithraka, S.; Makris, D.P.; Kefalas, P. An analytical survey of the polyphenols of seeds of varieties of grape (*Vitis vinifera*) cultivated in Greece: Implications for exploitation as a source of value-added phytochemicals. *Phytochem. Anal.* **2005**, *16*, 17–23. [CrossRef] [PubMed]

28. Guendez, R.; Kallithraka, S.; Makris, D.P.; Kefalas, P. Determination of low molecular weight polyphenolic constituents in grape (*Vitis vinifera* sp.) seed extracts: Correlation with antiradical activity. *Food Chem.* **2005**, *89*, 1–9. [CrossRef]

29. Prieur, C.; Rigaud, J.; Cheynier, V.; Moutounet, M. Oligomeric and polymeric procyanidins from grape seeds. *Phytochemistry* **1994**, *36*, 781–784. [CrossRef]

30. De Freitas, V.A.P.; Glories, Y.; Bourgeois, G.; Vitry, C. Characterisation of oligomeric and polymeric procyanidins from grape seeds by liquid secondary ion mass spectrometry. *Phytochemistry* **1998**, *49*, 1435–1441. [CrossRef]

31. Varzakas, T.; Zakynthinos, G.; Verpoort, F. Plant food residues as a source of nutraceuticals and functional foods. *Foods* **2016**, *5*, 88. [CrossRef] [PubMed]

32. Quesada, H.; del Bas, J.M.; Pajuelo, D.; Diaz, S.; Fernandez-Larrea, J.; Pinent, M.; Arola, L.; Salvado, M.J.; Blade, C. Grape seed proanthocyanidins correct dyslipidemia associated with a high-fat diet in rats and repress genes controlling lipogenesis and VLDL assembling in liver. *Int. J. Obes.* **2009**, *33*, 1007–1012. [CrossRef] [PubMed]

33. Montagut, G.; Bladé, C.; Blay, M.; Fernández-Larrea, J.; Pujadas, G.; Salvadó, M.J.; Arola, L.; Pinent, M.; Ardévol, A. Effects of a grapeseed procyanidin extract (GSPE) on insulin resistance. *J. Nutr. Biochem.* **2010**, *21*, 961–967. [CrossRef] [PubMed]

34. Montagut, G.; Onnockx, S.; Vaqué, M.; Bladé, C.; Blay, M.; Fernández-Larrea, J.; Pujadas, G.; Salvadó, M.J.; Arola, L.; Pirson, I.; et al. Oligomers of grape-seed procyanidin extract activate the insulin receptor and key targets of the insulin signaling pathway differently from insulin. *J. Nutr. Biochem.* **2010**, *21*, 476–481. [CrossRef] [PubMed]

35. Poudel, P.R.; Tamura, H.; Kataoka, I.; Mochioka, R. Phenolic compounds and antioxidant activities of skins and seeds of five wild grapes and two hybrids native to Japan. *J. Food Comp. Anal.* **2008**, *21*, 622–625. [CrossRef]

36. Jayaprakasha, G.K.; Singh, R.P.; Sakariah, K.K. Antioxidant activity of grape seed (*Vitis vinifera*) extracts on peroxidation models in vitro. *Food Chem.* **2001**, *73*, 285–290. [CrossRef]

37. Jayaprakasha, G.K.; Selvi, T.; Sakariah, K.K. Antibacterial and antioxidant activities of grape (*Vitis vinifera*) seed extracts. *Food Res. Int.* **2003**, *36*, 117–122. [CrossRef]

38. Puiggròs, F.; Llópiz, N.; Ardévol, A.; Bladé, C.; Arola, L.; Salvadó, M.J. Grape seed procyanidins prevent oxidative injury by modulating the expression of antioxidant enzyme systems. *J. Agric. Food Chem.* **2005**, *53*, 6080–6086. [CrossRef] [PubMed]

39. Guo, L.; Wang, L.H.; Sun, B.; Yang, J.Y.; Zhao, Y.Q.; Dong, Y.X.; Spranger, M.I.; Wu, C.F. Direct in vivo evidence of protective effects of grape seed procyanidin fractions and other antioxidants against ethanol-induced oxidative DNA damage in mouse brain cells. *J. Agric. Food Chem.* **2007**, *55*, 5881–5891. [CrossRef] [PubMed]

40. Vinson, J.A.; Mandarano, M.A.; Shuta, D.L.; Bagchi, M.; Bagchi, D. Beneficial effects of a novel IH636 grape seed proanthocyanidin extract and a niacin-bound chromium in a hamster atherosclerosis model. *Mol. Cell. Biochem.* **2002**, *240*, 99–103. [CrossRef] [PubMed]

41. Preuss, H.G.; Wallerstedt, D.; Talpur, N.; Tutuncuoglu, S.O.; Echard, B.; Myers, A.; Bui, M.; Bagchi, D. Effects of niacin-bound chromium and grape seed proanthocyanidin extract on the lipid profile of hypercholesterolemic subjects: A pilot study. *J. Med.* **2000**, *31*, 227–246. [PubMed]

42. Olas, B.; Wachowicz, B.; Tomczak, A.; Erler, J.; Stochmal, A.; Oleszek, W. Comparative anti-platelet and antioxidant properties of polyphenol-rich extracts from: Berries of *Aronia melanocarpa*, seeds of grape and bark of *Yucca schidigera* in vitro. *Platelets* **2008**, *19*, 70–77. [CrossRef] [PubMed]

43. Cetin, A.; Kaynar, L.; Kocyigit, I.; Hacioglu, S.K.; Saraymen, R.; Ozturk, A.; Sari, I.; Sagdic, O. Role of grape seed extract on methotrexate induced oxidative stress in rat liver. *Am. J. Chin. Med.* **2008**, *36*, 861–872. [CrossRef] [PubMed]

44. Seigers, R.; Loos, M.; Van Tellingen, O.; Boogerd, W.; Smit, A.B.; Schagen, S.B. Neurobiological changes by cytotoxic agents in mice. *Behav. Brain Res.* **2016**, *299*, 19–26. [CrossRef] [PubMed]

45. Sano, A.; Uchida, R.; Saito, M.; Shioya, N.; Komori, Y.; Tho, Y.; Hashizume, N. Beneficial effects of grape seed extract on malondialdehyde-modified LDL. *J. Nutr. Sci. Vitaminol.* **2007**, *53*, 174–182. [CrossRef] [PubMed]

46. Amaki, T.; Suzuki, T.; Nakamura, F.; Hayashi, D.; Imai, Y.; Morita, H.; Fukino, K.; Nojiri, T.; Kitano, S.; Hibi, N.; et al. Circulating malondialdehyde modified LDL is a biochemical risk marker for coronary artery disease. *Heart* **2004**, *90*, 1211–1213. [CrossRef] [PubMed]

47. Natella, F.; Belelli, F.; Gentili, V.; Ursini, F.; Scaccini, C. Grape seed proanthocyanidins prevent plasma postprandial oxidative stress in humans. *J. Agric. Food Chem.* **2002**, *50*, 7720–7725. [CrossRef] [PubMed]

48. Aviram, M.; Rosenblat, M.; Bisgaier, C.L.; Newton, R.S.; Primo-Parmo, S.L.; La Du, B.N. Paraoxonase inhibits high-density lipoprotein oxidation and preserves its functions. A possible peroxidative role for paraoxonase. *J. Clin. Investig.* **1998**, *101*, 1581–1590. [CrossRef] [PubMed]

49. Terra, X.; Montagut, G.; Bustos, M.; Llopiz, N.; Ardevol, A.; Blade, C.; Fernandez-Larrea, J.; Pujadas, G.; Salvado, J.; Arola, L.; et al. Grape-seed procyanidins prevent low-grade inflammation by modulating cytokine expression in rats fed a high-fat diet. *J. Nutr. Biochem.* **2009**, *20*, 210–218. [CrossRef] [PubMed]

50. Dandona, P.; Aljada, A.; Bandyopadhyay, A. Inflammation: The link between insulin resistance, obesity and diabetes. *Trends Immunol.* **2004**, *25*, 4–7. [CrossRef] [PubMed]

51. Chacon, M.R.; Ceperuelo-Mallafre, V.; Maymo-Masip, E.; Mateo-Sanz, J.M.; Arola, L.; Guitierrez, C.; Fernandez-Real, J.M.; Ardevol, A.; Simon, I.; Vendrell, J. Grape-seed procyanidins modulate inflammation on human differentiated adipocytes in vitro. *Cytokine* **2009**, *47*, 137–142. [CrossRef] [PubMed]

52. Terra, X.; Pallarés, V.; Ardèvol, A.; Bladé, C.; Fernández-Larrea, J.; Pujadas, G.; Salvadó, J.; Arola, L.; Blay, M. Modulatory effect of grape-seed procyanidins on local and systemic inflammation in diet-induced obesity rats. *J. Nutr. Biochem.* **2011**, *22*, 380–387. [CrossRef] [PubMed]

53. Pinent, M.; Blay, M.; Bladé, M.C.; Salvadó, M.J.; Arola, L.; Ardévol, A. Grape seed-derived procyanidins have an antihyperglycemic effect in streptozotocin-induced diabetic rats and insulinomimetic activity in insulin-sensitive cell lines. *Endocrinology* **2004**, *145*, 4985–4990. [CrossRef] [PubMed]

54. Balu, M.; Sangeetha, P.; Murali, G.; Panneerselvam, C. Modulatory role of grape seed extract on age-related oxidative DNA damage in central nervous system of rats. *Brain Res. Bull.* **2006**, *68*, 469–473. [CrossRef] [PubMed]

55. Balu, M.; Sangeetha, P.; Haripriya, D.; Panneerselvam, C. Rejuvenation of antioxidant system in central nervous system of aged rats by grape seed extract. *Neurosci. Lett.* **2005**, *383*, 295–300. [CrossRef] [PubMed]

56. Baydar, N.G.; Sagdic, O.; Ozkan, G.; Cetin, S. Determination of antibacterial effects and total phenolic contents of grape (*Vitis vinifera* L.) seed extracts. *Int. J. Food Sci. Technol.* **2006**, *41*, 799–804. [CrossRef]

57. Brown, J.C.; Huang, G.; Haley-Zitlin, V.; Jiang, X. Antibacterial effects of grape extracts on Helicobacter pylori. *Appl. Environ. Microbiol.* **2009**, *75*, 848–852. [CrossRef] [PubMed]

58. Ma, Z.F.; Majid, N.A.; Yamaoka, Y.; Lee, Y.Y. Food allergy and Helicobacter pylori infection: A systematic review. *Front. Microbial.* **2016**, *7*, 368. [CrossRef] [PubMed]

59. Kaur, M.; Singh, R.P.; Gu, M.; Agarwal, R.; Agarwal, C. Grape seed extract inhibits in vitro and in vivo growth of human colorectal carcinoma cells. *Clin. Cancer Res.* **2006**, *20*, 6194–6202. [CrossRef] [PubMed]

60. Sun, Q.; Prasad, R.; Rosenthal, E.; Kativar, S.K. Grape seed proanthocyanidins inhibit the invasive potential of head and neck cutaneous squamous cell carcinoma cells by targeting EGFR expression and epithelial-to-mesenchymal transition. *BMC Complement. Altern. Med.* **2011**, *11*, 134. [CrossRef] [PubMed]

61. Park, S.Y.; Lee, Y.H.; Choi, K.C.; Seong, A.R.; Choi, H.K.; Lee, O.H.; Hwang, H.J.; Yoon, H.G. Grape seed extract regulates androgen receptor-mediated transcription in prostate cancer cells through potent anti-histone acetyltransferase activity. *J. Med. Food* **2011**, *14*, 9–16. [CrossRef] [PubMed]

62. Feng, Y.; Liu, Y.M.; Fratkins, J.D.; LeBlanc, M.H. Grape seed extract suppresses lipid peroxidation and reduces hypoxic ischemic brain injury in neonatal rats. *Brain Res. Bull.* **2005**, *66*, 120–127. [CrossRef] [PubMed]

63. Perumalla, A.V.S.; Hettiarachchy, N.S. Green tea and grape seed extracts—Potential applications in food safety and quality. *Food Res. Int.* **2011**, *44*, 827–839. [CrossRef]

64. Bentivegna, S.S.; Whitney, K.M. Subchronic 3-month oral toxicity study of grape seed and grape skin extracts. *Food Chem. Toxicol.* **2002**, *40*, 1731–1743. [CrossRef]

65. Taghvaei, M.; Jafari, S.M. Application and stability of natural antioxidants in edible oils in order to substitute synthetic additives. *J. Food Sci. Technol.* **2015**, *52*, 1272–1282. [CrossRef] [PubMed]

66. Ahn, J.; Grün, I.U.; Fernando, L.N. Antioxidant properties of natural plant extracts containing polyphenolic compounds in cooked ground beef. *J. Food Sci.* **2002**, *67*, 1364–1369. [CrossRef]

67. Lau, D.W.; King, A.J. Pre- and post-mortem use of grape seed extract in dark poultry meat to inhibit development of thiobarbituric acid reactive substances. *J. Agric. Food Chem.* **2003**, *51*, 1602–1607. [CrossRef] [PubMed]

68. Mielnik, M.B.; Olsen, E.; Vogt, G.; Adeline, D.; Skrede, G. Grape seed extract as antioxidant in cooked, cold stored turkey meat. *Food Sci. Technol.* **2006**, *39*, 191–198. [CrossRef]

69. Brannan, R.G. Effect of grape seed extract on descriptive sensory analysis of ground chicken during refrigerated storage. *Meat Sci.* **2009**, *81*, 589–595. [CrossRef] [PubMed]

70. Brannan, R.G.; Mah, E. Grape seed extract inhibits lipid oxidation in muscle from different species during refrigerated and frozen storage and oxidation catalyzed by peroxynitrite and iron/ascorbate in a pyrogallol red model system. *Meat Sci.* **2007**, *77*, 540–546. [CrossRef] [PubMed]

71. Roychowdhury, S.; Wolf, G.; Keilhoff, G.; Bagchi, D.; Horn, T. Protection of primary glial cells by grape seed proanthocyanidin extract against nitrosative/oxidative stress. *Nitric Oxide* **2001**, *5*, 137–149. [CrossRef] [PubMed]

72. Bagchi, D.; Bagchi, M.; Stohs, S.J.; Das, D.K.; Ray, S.D.; Kuszynski, C.A.; Joshi, S.S.; Pruess, H.G. Free radicals and grape seed proanthocyanidin extract: Importance in human health and disease prevention. *Toxicology* **2000**, *148*, 187–197. [CrossRef]

73. Kim, S.Y.; Jeong, S.M.; Park, W.P.; Nam, K.C.; Ahn, D.U.; Lee, S.C. Effect of heating conditions of grape seeds on the antioxidant activity of grape seed extracts. *Food Chem.* **2006**, *97*, 472–479. [CrossRef]

74. Allemann, I.B.; Baumann, L. Antioxidants used in skin care formulations. *Skin Therapy Lett.* **2008**, *13*, 5–9.

75. Grosso, G.; Micek, A.; Godos, J.; Pajak, A.; Sciacca, S.; Galvano, F.; Giovannucci, E.L. Dietary flavonoid and lignan intake and mortality in prospective cohort studies: Systematic review and dose-response meta-analysis. *Am. J. Epidemiol.* **2017**. [CrossRef] [PubMed]

76. Menezes, R.; Rodriguez-Mateos, A.; Kaltsatou, A.; González-Sarrías, A.; Greyling, A.; Giannaki, C.; Andres-Lacueva, C.; Milenkovic, D.; Gibney, E.R.; Dumont, J.; et al. Impact of flavonols on cardiometabolic biomarkers: A meta-analysis of randomized controlled human trials to explore the role of inter-individual variability. *Nutrients* **2017**, *9*, 117. [CrossRef] [PubMed]

77. Wang, X.; Ouyang, Y.Y.; Liu, J.; Zhao, G. Flavonoid intake and risk of CVD: A systematic review and meta-analysis of prospective cohort studies. *Br. J. Nutr.* **2014**, *111*, 1–11. [CrossRef] [PubMed]

78. Liu, Y.J.; Zhan, J.; Liu, X.L.; Wang, Y.; Ji, J.; He, Q.Q. Dietary flavonoids intake and risk of type 2 diabetes: A meta-analysis of prospective cohort studies. *Clin. Nutr.* **2014**, *33*, 59–63. [CrossRef] [PubMed]

MDPI

St. Alban-Anlage 66

4052 Basel

Switzerland

Tel. +41 61 683 77 34

Fax +41 61 302 89 18

www.mdpi.com

Antioxidants Editorial Office

E-mail: antioxidants@mdpi.com

www.mdpi.com/journal/antioxidants

www.ingramcontent.com/pod-product-compliance
Lightning Source LLC
Chambersburg PA
CBHW051844210326
41597CB00033B/5774